Black Liberation

BLACK LIBERATION

■ ■ ■ ■ ■ ■ ■

A Comparative History of Black Ideologies in the United States and South Africa

George M. Fredrickson

New York Oxford
OXFORD UNIVERSITY PRESS
1995

Oxford University Press

Oxford New York
Athens Auckland Bangkok Bombay
Calcutta Cape Town Dar es Salaam Delhi
Florence Hong Kong Istanbul Karachi
Kuala Lumpur Madras Madrid Melbourne
Mexico City Nairobi Paris Singapore
Taipei Tokyo Toronto

and associated companies in
Berlin Ibadan

Published by Oxford University Press, Inc.
200 Madison Avenue, New York, New York 10016

Oxford is a registered trademark of Oxford University Press

Library of Congress Cataloging-in-Publication Data
Fredrickson, George M., 1934–
Black liberation : a comparative history of Black ideologies
in the United States and South Africa / George M. Fredrickson.
p. cm. Includes bibliographical references (p.) and index.
ISBN 0-19-505749-X
1. Civil rights movements—United States—History. 2. Civil
rights movements—South Africa—History. 3. Afro-Americans—
Politics and government. 4. United States—Race relations.
5. Blacks—South Africa—Politics and government. 6. South
Africa—Race relations. 7. Pan-Africanism—History. 8. Black
nationalism—United States—History. 9. Black nationalism—South
Africa—History. I. Title.
E185.61.F836 a1995 973'.0496073—dc20 94-37504

2 4 6 8 9 7 5 3 1

Printed in the United States of America
on acid-free paper

For Lynne, Peter, and Sean

Acknowledgments

The research, writing, and editing that made this book possible could not have been accomplished without the generous help of many people and institutions. The principal research was done in the following repositories: the Stanford University Library, especially the library of the Hoover Institution; the Rhodes and Bodleian libraries of the University of Oxford; in South Africa, the libraries of the Universities of the Witwatersrand, Cape Town, Durban-Westville, and South Africa (UNISA); and the Harvard University Library. I hope that the many librarians and curators who rendered me special service will forgive me for not mentioning each of them. In my zeal to get at the documents they provided, I often—and inexcusably—failed to note their names. I can, however, name two scholars who assisted my research in a direct and exceptional way: Louis Harlan and Robert Edgar provided me with primary materials from their own files that turned out to be of enormous value. I also received help in locating sources from Thomas F. Jackson and Ian Solomon.

Intensive work on this project began in 1988–89 when I was Harmsworth Professor of American History at Oxford. Dr. Stanley Trapido offered me the hospitality of his seminar in South African History and Politics, and I had the rare opportunity of presenting my preliminary findings to four successive meetings of the seminar. The feedback that I received from Stan Trapido and the other members of this distinguished group of South Africanists was invaluable. I would also like to thank Professor Shula Marks of the School of Oriental and African Studies at the University of London for inviting me to present a paper describing my project to her seminar and for the useful criticism that I received on that occasion. During my research trip to South Africa in the spring of 1989, I benefited from the hospitality and advice of several South African scholars, especially Bruce Murray, Charles van Onselen, Tim Couzens, and Philip Bonner of the University of the Witwatersrand; Christopher Saunders, Hermann Giliomee, David Welsh, Francis Wilson, and Helen Bradford of the University of Cape Town; and Colin Bundy of the University of the Western Cape. While in Cape Town, I learned much about the links between American and South African freedom struggles during an extended interview with Allan Boesak. On a subsequent visit to South Africa in the summer of 1992, I received input from Tom Lodge, Fatima Meer, Mewa Ramgobin, Greg Cuthbertson, and William Freund, among others.

The manuscript was completed and revised while I was a fellow of the W.E.B. Du Bois Institute for Afro-American Research at Harvard University in 1993. I wish to thank the Institute and its director, Henry Louis Gates, Jr., for the financial support that enabled me to bring the work to fruition. The staff of the Du Bois Institute, especially its associate director, Randall Burkett, rendered me every possible day-to-day assistance, and I also learned a great deal from the weekly seminars that provided me and other fellows with a chance to present our work to a discerning and critical audience.

Portions of the manuscript dealing with black religious nationalism were read by James Campbell, Randall Burkett, and Richard Newman, all of whom provided extensive and perceptive comments. A complete draft of the book benefited from the close scrutiny of Anthony Marx, Clayborne Carson, and Sterling Stuckey. I often, but not always, followed the advice of these highly valued friends and colleagues. Consequently, they deserve much of the credit for the strengths of the book but bear no responsibility for its shortcomings.

Sheldon Meyer of Oxford University Press played his customary role as peerless editor, providing a full measure of the encouragement and friendly guidance that gives an author confidence and sense of direction. I would also like to thank Stephanie Sakson for a highly professional job of copy editing, and Andrew Albanese and Joellyn Ausanka for guiding the book through the press. Last but not least, my wife Hélène deserves much credit for accepting gracefully the disruptions in our life that pursuit of this project entailed and for proofreading the text in its various incarnations, employing her acute sense of linguistic propriety to catch many errors.

G.M.F.

Contents

Introduction 3

1. "Palladium of the People's Liberties": The Suffrage Question and the Origins of Black Protest 14

   Black Voting in the Nineteenth Century; The Ballot in African-American Protest Thought, 1840–1905; The Cape Suffrage and the Origins of Black Protest Politics in South Africa; Suffrage Struggles: Connections and Comparisons

2. "Ethiopia Shall Stretch Forth Her Hands": Black Christianity and the Politics of Liberation 57

   The Problem of Religion and Resistance; Ethiopianist Thought in the Nineteenth Century; Popular Ethiopianism and African-American Missions to Africa; Ethiopianism in South Africa; What Happened to American Ethiopianism?

3. Protest of "The Talented Tenth": Black Elites and the Rise of Segregation 94

   The Making of Segregation; African-Americans Mobilize Against Segregation; The National Congress in Comparative Perspective; Resisting the High Tide of Segregation, 1913–1919

4. "Africa for the Africans": Pan-Africanism and Black Populism, 1918–1930 137

   Working-Class Protest and Middle-Class Organizations, 1918–1921; Elite Pan-Africanism; Populist Pan-Africanism: The Garvey Movement; Black Populism in South Africa: Garveyism and the ICU; Two Black Populisms: Comparing the UNIA and the ICU

5. "Self-Determination for Negroes": Communists and Black Freedom Struggles, 1928–1948 179

Reds and Blacks: Introduction and Overview; Rise of the Black Self-Determination Policy; Blacks and the United Front, 1934–1939; The Second World War and the Parting of the Ways

6. "We Shall Not Be Moved": Nonviolent Resistance to White Supremacy, 1940–1965 225

The Gandhian Tradition; Nonviolence in South Africa; Martin Luther King, Jr., and Nonviolence in the American South; Comparing Nonviolent Struggles

7. "Black Man You Are on Your Own": Black Power and Black Consciousness 277

Pan-Africanism in South Africa, 1944–1960; The Rise of Black Power in the United States; Black Consciousness in South Africa; Comparing Black Power and Black Consciousness

Epilogue 319

Notes 325

Index 367

Black Liberation

Introduction

This book might be considered a sequel to my earlier study *White Supremacy: A Comparative Study in American and South African History*, but it is not one that I planned or expected to write when that work was published in 1981. *White Supremacy* compared the attitudes, ideologies, and policies associated with white or European domination over blacks and other people of color in the history of both societies. In the introduction, I acknowledged the "obvious limitations" of this approach: "Comparative studies of nonwhite responses and resistance movements would be enormously valuable and should be done. But a useful prelude to such a work is awareness of what nonwhites were up against. . . ."[1]

At the time, nothing was further from my mind than writing my own study of the other side of the black-white confrontation. I thought of *White Supremacy* as a one-shot excursion into comparative history involving South Africa, after which I would revert to my previous vocation as simply an historian of the United States with a special interest in black-white relations. But two things happened to change my mind. The first was my growing sense of the unlikelihood that anyone else would undertake a study such as the one I had proposed. It eventually dawned on me that the broad understanding of South African history that I had acquired in writing *White Supremacy* had prepared me exceptionally well to carry on such work and that if I did not do it no one else was likely to in the foreseeable future. The second new consideration was the remarkable course of events in South Africa in the 1980s. I watched with fascination as a massive resistance movement challenged white supremacy as it had never been challenged before. Like most other observers writing in the previous decade, I had not expected such a development; my studies of the ideas and institutions associated with white domination had made it seem that the apartheid regime was backed by sufficient white power and determination to make it, if not invulnerable, at least in control of the situation for several years to come. As events unfolded in South Africa, I became intensely curious about the history of the black opposition—how it had developed, what its governing ideas were, and how it compared with the African-American freedom struggle that had been central to my public consciousness since the 1950s and that I had long wanted to study and write about.

The resulting book is not in fact a full comparative history of black resistance to white supremacy. It does not attempt to deal with slave resistance or with the struggles of independent Africans against the original establishment of European imperial hegemony, and it limits most of its attention to the past hundred years or so. It focuses on the protest of black people within the structures of domination established by whites, once these were in place, and presumes that blacks had some right or opportunity to make their grievances known. Protest, as distinguished from violent or military resistance, can occur only when a group is given some access to the central institutions of a society and some standing in its legal and political system, however limited and pro forma such "rights" may be. When slaves are freed, their new status requires some legal recognition and formal protection, and this minimal entitlement can become the basis for demanding a more substantive kind of freedom. When indigenous peoples are conquered by intruders, they are not as a rule literally enslaved or denied all legal-political standing. Although equal rights with their conquerors may be flagrantly denied, they at least have some chance to agitate for improvements in their status and conditions. Of course, ex-slaves or conquered indigenes may conclude that reformist protest is unavailing and that resort or reversion to violent resistance is warranted. But, as this study will demonstrate, violent resistance, although sometimes discussed and justified in principle, played little actual role in the African-American freedom struggle and, before the 1960s, was not a central feature of the black South African battle against segregation and apartheid. Justifiably, therefore, this book is mainly a study of protest on behalf of racial equality rather than of revolutionary action to "put the bottom rail on top." That this should be true is somewhat surprising in the case of the black South African majority. One of the aims of this study is to make comprehensible the remarkable congruence between the struggle of a racial minority for equal rights in the United States and of a majority for its rightful share of power in South Africa.

In using the term "black" for the protagonists in this study, I am aware of a certain ambiguity in applying the term to South Africa. As a synonym for African-American, the term presents no problem, but the South African meaning of the word has broadened since the 1970s. Previously referring only to the descendants of African tribesmen who spoke Bantu languages, it now may include Asian South Africans, who are mostly of Indian descent, and the mixed-race group known as "Coloreds" (a hybrid population of European, Asian, and southern African aboriginal derivation that has historically been concentrated in the Western Cape and considered distinct from both Europeans and Africans)—in other words all those previously categorized as "nonwhite." Most of the time, I use "black" in the limited traditional sense as a synonym for "African" and am mainly concerned with persons or organizations who would have identified themselves in both ways. But I also consider some of the efforts of "non-Europeans" to establish a united front against white supremacy and treat "Colored" and Indian movements in relation to such efforts, sometimes using the term black in the contemporary sense to describe the coalitions that were advocated or achieved. In any case, the context should make clear whether "black" is being used in a restrictive or an inclusive sense. Like most other historians, I have been unable to find a satisfactory substitute for the problematic term "Colored" when referring specifically to the mixed-race group that emerged in the Western Cape. (Although I will not put quotes around the word when it subsequently appears, they may be presumed.)

The fact that this book finds so much similarity between the two struggles may be surprising to readers of *White Supremacy*, as it was to me once I became aware of it. In the earlier study, I stressed the structural and demographic differences between twentieth-century black-white relations in the two societies and concluded that the contrast was so great that there was little basis for detailed comparison, unless one focused, as I proceeded to do, on the treatment of two racial minorities with similar origins —African-Americans and South African Coloreds. But when I began to read more deeply in the literature of African and African-American protest I became aware that the two discourses addressed common questions, often in remarkably similar ways. The ideologies of black advancement or liberation that emerged from these discourses were much closer to each other than the external circumstances had led to me to expect.

The reasons for this discursive congruence will become fully apparent in the pages that follow. Generally speaking, however, there were four main sources of ideological parallelism. The first, which might have been anticipated from *White Supremacy*, is the broad similarity in the arguments and rationalizations of the white supremacists. Racists in both societies claimed that black people were intellectually and morally inferior, incapable of self-government, and therefore unfit to vote, hold office, and associate with whites on a basis of equality. To counter such arguments effectively and to neutralize their potentially demoralizing effect on blacks, the case had to be made for the natural equality of all human beings, and specifically black or African achievements and capabilities had to be demonstrated. If African-Americans often pointed with pride to precolonial African civilizations, black South Africans often cited the progress of blacks in the United States to show what black people were capable of accomplishing.

This Pan-African frame of reference was the second main source of congruence. Politically aware African Americans and black South Africans did not conceive of themselves as simply engaged in isolated combat with their own particular sets of white oppressors. From the early to mid-nineteenth century onward, they were keenly aware of the larger struggle of Africans and people of African descent throughout the world against the efforts of Europeans or people of European ancestry to enslave, colonize, disfranchise, and segregate people designated as black. Hence African-Americans and black South Africans shared in a larger Pan-African discourse and responded to changing international currents of black thought and opinion. As a result of this black internationalism, they influenced each other and also responded creatively to the same ideologies and movements, some specifically Pan-African and others anti–imperialist in a broader sense, that may have originated in neither the United States nor South Africa (such as Marxist-Leninism and Gandhian nonviolence).

Some post-modernists to the contrary notwithstanding, however, discourse does not exist independently of external realities or circumstances. I am enough of a materialist to assume that the degree of similarity that I found in the two discourses would not have existed if the two contexts had not had something in common beyond producing standard white supremacist libels against black people that had to be refuted. Here I am forced to retreat somewhat from the claim in *White Supremacy* that demographic differences come close to obviating useful comparisons of black-white relations in twentieth century America and South Africa. The third and perhaps most fundamental source of the ideological parallelism that I discovered was a comparable sense of *minority* status within white-dominated, multi-racial societies. Although blacks were of course an overwhelming numerical majority in South Africa, they were until 1994 even more of a minority in terms of their access to political, social, and economic power than the African-American fraction, usually about one tenth, of the total population of the United States. Numbers are not the whole story, or even the most important part of it. What matters is power or leverage, and it was the unholy achievement of South African white domination to have made blacks even more powerless relative to whites than were African-Americans after the abolition of slavery. The elaborate mechanisms by which this end was achieved are described in *White Supremacy*, but the point that needs additional stress is that territorial and political segregation served to make a numerical majority into a functional minority that had little or no power within the South African state or polity. Black Americans, of course, have been both a minority in the demographic sense and an oppressed group in the socio-political sense, which meant they were denied a proportionate or fair share of status, wealth, and power. This assumption of a similar, if not identical, form of unjust and inequitable minority status permitted blacks in both societies to make persuasive analogies without statistical caveats.

Of course, black thinking about the shape of future liberation could scarcely avoid being influenced by the differing population ratios—and the effects of this contrast will be shown—but in most black projections of a liberated South Africa whites remained an important, if no longer dominant, segment of the population. With the exception of a radical-separatist minority, African-Americans have also assumed that the future of their hopes would continue to involve significant interactions with whites. It is this predominant assumption of biracialism or multi-racialism that makes me suppose that comparisons between black ideologies in the United States and South Africa can be more fruitful, at least for some purposes, than comparisons of either with anti-colonialist movements in typical black African countries.

The fourth basic similarity is the comparable social and cultural position within the black communities of those who normally articulated the ideologies and led the movements. The emergence in the United States and South Africa during the nineteenth and early twentieth centuries of a black educated elite or middle class that had absorbed many elements of European or Euro-American culture was an essential prerequisite for the kind of modern political thought and action that is the primary concern of this study. These elites were faced with similar challenges. They often had the task of mediating between the modernist beliefs and values inculcated by their education and the traditional or folk culture of the mass of the people whose grievances they sought to articulate. If they became too enamored of their own superiority over the less-educated masses, they were likely to be challenged by leaders and movements of a populist type. Ideological and tactical tensions, resulting from social and cultural differences between leaders and followers, or between elites and masses, were a common feature of the history of black ideologies and movements in the two societies that provided me with opportunities for comparison.

This concern with elites and leadership reveals the essential character and inescapable limitations of this study. I do not pretend to have written a comprehensive comparative *social* history of African-Americans and black South Africans, and the work will not satisfy those who believe that all history should be written "from the bottom up." It focuses on the thought and actions of exceptional rather than ordinary people. But it is not exactly "history from the top down," and might be described more accurately, if a bit cumbersomely, as "history from the top of the bottom up *and* down." It would be possible to defend this historiographic stance in general terms, but it may suffice to suggest that the best opportunities for controlled comparison of American and South African freedom struggles comes from such a vantage point. A comparative history of popular *mentalities* would be likely to confront particularities that are incommensurable. A comparative sociological history of black population groups stressing the effects of industrialization, urbanization, and segregation on previously rural or peasant folk would be possible to write, but it would be mainly the story of people being buffeted by impersonal and apparently irresistible forces. To do full justice to black agency and to the creativity of thought and action called forth by the struggle against white supremacy, I thought it preferable to attempt what might be described as an intellectual/political study.

As the subtitle suggests, the focus of this study is on ideology and ideologists. But the term ideology makes sense only in a political context and clearly implies an element of social determination, although not necessarily a thoroughgoing social determinism. In the classic Marxian formulation, ideology is a rationale or rationalization for ruling class domination that may be internalized as "false consciousness" by the dominated and exploited class or classes. I am not using it in that fashion but rather as a non-pejorative term for the political consciousness and programmatic thinking of any social group. I view ideology as a rationalized and action-oriented form of consciousness that is not be confused with culture in the broader sense as the total pattern of beliefs, symbols, myths, and behaviors that characterize a nation, class, community, or status group. Ideology is a part of culture but not the whole of it.[2] Consequently, this work is closer in method to intellectual history than to social history or cultural history as currently practiced. Its main protagonists might even be considered intellectuals, but most of them would be, in Antonio Gramsci's terminology, "organic" rather than "traditional" intellectuals, which means that their thought was consciously and explicitly rooted in the struggles of the social group to which they belonged rather than being presented as the Olympian reflections of putatively independent minds.[3] My examination of their rhetoric, principles, and programs is not, therefore, the kind of "history of ideas" that makes the "life of the mind" a world unto itself that transcends the economic, social, and political formations that impinge upon consciousness and limit the options of historical actors. Using the term ideology rather than "thought" or "mind" is meant to convey both social constraints and political purposes. In short, the ideas or ideologies of black assertion and liberation discussed in the book are treated contextually—in terms of the circumstances surrounding their emergence—and pragmatically—in relation to the purposes they served and how well they served them.

I hope, therefore, that the primary focus on ideological processes will not obscure the relevance of the book to an understanding of the larger social and political histories of black freedom struggles in the two societies. Variations on common ideological themes or the differing practical consequences of similarly conceived movements can be explained only with reference to specific contexts—social, economic, political, and cultural. Drawing heavily on the specialized work that many other scholars have done on relevant aspects of American and South African history, I have tried to provide contextual explanations and interpretations of the public expressions of ideology that were my principal primary sources.

The book is organized around a sequence of ideological and political responses to white supremacy that applies to both societies. Chapter One focuses on the African and African-American versions of Victorian political liberalism and more specifically on the struggle that originated in the nineteenth century to secure the ballot on the same basis as whites. Chapter Two, which overlaps somewhat with One chronologically, describes the close connections between the black religious nationalism or Ethiopianism that emerged in the United States and South Africa during the nineteenth and early twentieth centuries and attempts to deal with the complex and problematic relationship between separatist black Christianity and political struggles for freedom and equality in the twentieth century. Chapter Three focuses on the response of black educated elites to the rise of legalized segregation in the two societies between the turn of the century and the end of World War I. It juxtaposes and compares the founding and early policies of the National Association for the Advancement of Colored People and the South African Native National Congress. Chapter Four considers Pan-Africanist and black populist movements during the 1920s, with particular attention to the ideas and actions of Marcus Garvey and his Universal Negro Improvement Organization in both countries and those of a comparable mass movement of black South Africans, the Industrial and Commercial Workers Union of Africa (ICU). The fifth chapter examines Marxist or Communist interpretations of, and contributions to, black protest movements between the 1920s and the 1940s. Chapter Six treats the ideology and practice of nonviolence between the 1940s and the 1960s and compares the American nonviolent civil rights movement with the African National Congress's Defiance Campaign of the 1950s. The last chapter probes the relationship between the American Black Power movement of the late 1960s and early '70s and the Black Consciousness ideology that came to distinguish the domestic opposition to apartheid in the 1970s.

Since this is an historical work deriving its authority from the hindsight that comes only after the dust has settled to some extent —and from the existence of a reliable body of historical scholarship to be mined and synthesized—I have not felt obliged or competent to treat the 1980s and the early '90s in as much detail as earlier decades. But Chapters Six and Seven both look ahead to take account of more recent events, and a brief epilogue offers some tentative comparative perspectives on the current situation of blacks in the two societies.

A secondary aim of this study is to trace some of the actual interactions or connections between black political thought and ideology in the two countries. The fact that there was a degree of mutual awareness and some borrowing of ideas and rhetoric from the other side of the Atlantic and the Equator—especially by the South Africans—adds substance and credibility to the comparison. At times, I have been in the fortunate position of being able to build on the comparisons actually being made by the historical actors rather than having them depend entirely on my own ingenuity. But I have not attempted to write the full history of the cultural and intellectual transactions between black America and black South Africa. Since my concern is with operational political ideas and ideologies, I have dealt with cultural exchanges only to the extent that I could discern a political impact. I thought it was important to deal with black religion in its relation to political thought and action but have not felt obliged, for example, to treat the fascinating subject of musical and literary influences. Other scholars have produced, and are producing, work of high quality on a broad range of specific encounters and interactions between African-Americans and black South Africans, and I have made use of such work for my purposes in addition to probing some previously unexplored connections uncovered by my own research in primary sources.[4] But a full treatment of the cross-cultural exchanges would require a great deal of additional research in rare and scattered materials. I can imagine a splendid book on the subject that might use this one as a point of departure. But my concern with the detailing of links and influences has been subordinated to my interest in comparison. Exactly how an African-American idea got to South Africa, or vice versa, matters less to me than what happened to it when it got there. I am interested in how broadly similar conceptualizations of black liberation were modified or reinterpreted to suit local circumstances and in what occurred when similar ideologies were acted upon under conditions that were in some ways very different.

The analytical and interpretive aspects of the study are consistent with a general theory of race relations that will be familiar to readers of my earlier book, *The Arrogance of Race*.[5] But since the focus there was mainly on white racism, I need to say something about how these assumptions could be applied to my treatment of *black* racial consciousness. I do not use the theory in an explicit and systematic way in the main body of the work and most of what I say there does not stand or fall on the reader's judgment of its validity. But it does represent one way of making sense, in a very abstract and general way, of my principal findings. From Max Weber, I have learned to think of "status" as a basis for social grouping and stratification that differs from "class." Status consciousness is the sense shared by members of a social group of how much esteem, honor, and prestige they possess relative to other groups in the same society; class consciousness is the awareness of a shared economic interest possessed by groups that have a distinctive relationship to the marketplace or the means of production. These two sources of identity and inequality often coincide and become mutually reinforcing, but they can vary independently. In multi-ethnic or multi-racial societies a principal form of status consciousness is based on ethnicity in the broad sense of real or presumed differences in group ancestry. Societies like the United States and South Africa have traditionally had the character of ethnic status hierarchies, as well as being divided into social classes. An African-American physician in the early twentieth century, to take one example, would have a complex social position. He would be economically middle class, the member of a lower ethnic status group, and of high professional or educational status within the group. Which social identity loomed largest in his consciousness and motivated his actions at a particular moment would depend on the situation in which he found himself. In his relations with whites, he would be made to feel his lower ethnic status; in seeking payment from poor patients his class identity might be uppermost in his mind; and in the affairs of the black community his in-group status would give him a claim to prestige and leadership. My understanding of white racism is that it is fundamentally a commitment to maintaining, through discriminatory action or inaction, a higher ethnic status than blacks and other people of color. The efforts of blacks to enhance their ethnic status requires a conscious repudiation of the sense of social inferiority that the hierarchical structure of racial and ethnic relations is meant to inculcate.

Logically speaking, the oppositional status consciousness of a subordinated group could take one of three political forms: the assertion of a claim to equal status in a common, nonracial society (a universalist or cosmopolitan response), a reversal of the ethnic hierarchy that would place formerly subordinated groups in a position of dominance over their erstwhile superiors (an ethnic supremacist response), or the achievement of group self-esteem by ceasing to belong to the same society as the former ethnic superordinates (a separatist response). The latter consummation might be achieved by exterminating, expelling, or deporting the other group, by seceding or emigrating from its domain, or by partitioning what had been a common territory.

This Weberian thought experiment conveys a sense of the theoretical possibilities of a rebellious black consciousness in the United States and South Africa, but it is devoid of the power to predict which of the alternatives will be pursued most strenuously and consistently. As Weber would have been the first to acknowledge, only the kind of history that acknowledges a complex multiplicity of causes and interactions can provide the answers. Much of this study can be seen as an account of how the alternative approaches to black liberation were discussed, debated, and put into action with varying consequences. But it will also show how they could lose their abstract purity and fuse or amalgamate in various ways. For example, the separatist impulse could play a role, perhaps an indispensable one, in movements that aimed ultimately at equal status in a common society. It seems evident that the cosmopolitan or universalist response to white supremacy has been historically predominant in the formal ideologies of both struggles, which is somewhat more surprising in the case of South Africa than of the United States. Blacks in South Africa were never emancipated by force of arms and promised equal citizenship in a colorblind republic. (The reneging on this commitment in the United States after Reconstruction could not destroy the dream that it engendered.) A major concern will be how to explain the failure of black supremacist or separatist perspectives to become ascendant among a brutally oppressed majority, as well as the continued viability and vitality of some forms of black separatism among a minority that was promised equal civil and political rights in the 1860s and actually achieved them a century later.

It is also clear that full-blown black supremacism—do to the whites as they have done to us—was rarely, if ever, seriously advocated by credible black leaders and intellectuals in either society. In the United States, such a turning of the tables was of course a physical impossibility—except perhaps in the Deep South during Reconstruction, when whites complained of a "black domination" that never actually occurred. But in South Africa it was to a fear of black supremacy that the architects of segregation and apartheid appealed as justification for their actions. But no evidence survives that black resisters ever proposed their own upside-down version of apartheid. The most extreme and draconian measure that was ever publicly advocated—and this was a rarity—was that when blacks took power the whites would be "pushed into the sea." The current slogan of a tiny minority of black extremists—"one settler, one bullet"—probably expresses an intense commitment to racial separatism rather than a fixed intent to commit genocide.

The usual poles of the debate on the goals of black liberation in both the United States and South Africa were equitable incorporation into a common society with whites, on the one hand, and the creation of a self-determining black nation on the other. In the case of South Africa, the alternatives came to be designated as "nonracialism" and "Africanism"; in the United States as "integrationism" and "black nationalism." A way to describe the dichotomy in language that would apply to both societies would be to call it a tension between color-blind universalism and racially defined nationalism. But it would be misleading to conceive of a repetitive controversy between fixed and clear-cut alternatives. Although there were some consistent advocates of both positions, many of those who tried to define black liberation were actually seeking to combine the essential insights provided by both orientations. White supremacy was, after all, both a violation of universalist conceptions of human rights *and* the oppression of one historically constructed "racial" group by another. A main concern of this study is to understand and compare the efforts of black liberationists in both societies to resolve the issue of the proper balance in the struggle of universalist and ethnocentric perspectives, as well as to probe the political consequences of the solutions that they embraced.

The issue often arose in the context of discussions about what role whites should play in the black struggle. Roughly parallel debates in both countries developed on the issue of whether black movements should welcome or exclude whites who professed to be friendly to the cause of liberation. Often, but not always, the decision made reflected whether liberation itself was conceived as the triumph of an inclusive universalism or primarily in terms of ethnic or racial nationalism.

Besides trying to conceive the ultimate ends of the freedom struggle, movement leaders and intellectuals also confronted comparable questions of the means to be used in the struggle. The abstract alternatives in both cases were threefold: the pursuit of gradualist reform using the channels available to blacks within the legal-constitutional system established by whites, nonviolent resistance to challenge that system in the hope of forcing more rapid and fundamental change, and revolutionary violence to overthrow the system. All three of these methods of protest or resistance could be associated with either a universalist or a nationalist goal. Black Marxists and Communists have usually advocated a universalist revolution, at least in the long run, and some advocates of Black Power or Black Consciousness have in fact called for an ethnocentric reformism. In the confrontations that led to the Sharpeville massacre in 1960, the Pan-Africanist Congress was confronting white power on behalf of an exclusionary African nationalism, while the African National Congress engaged in similar actions under the banner of a nonracial universalism, resembling the one being projected at the same time by the nonviolent civil rights movement in the American South.

How this common set of ideological and tactical issues was debated and resolved, and how the theoretical and practical results of these deliberations affected the historical trajectory of the respective black struggles, is the principal concern of this book. A further question for Americans and South Africans in the 1990s is whether or not the ideologies of liberation that have come to predominate in our own time have proved, or are proving, adequate to the tasks they now face. It strains the competence of the historian to answer such a question, but I will reflect upon it briefly in the Epilogue, making a tentative effort to apply the lessons of the foregoing comparative history to the current situation of blacks in the United States and South Africa.

· 1 ·

"Palladium of the People's Liberties": The Suffrage Question and the Origins of Black Protest

Black Voting in the Nineteenth Century

Since power and equality in representative political systems depend on the right to vote, it should come as no surprise that black freedom struggles in the United States and South Africa have made access to the ballot box a main objective in the battle against white domination. Once slavery had been abolished in the United States and the conquest of indigenous societies had been completed in South Africa, agitation for the vote on the same basis as whites became central to the cause of "equal rights." For more than a century blacks protested against the arrogance and unfairness of political systems that paid lip service to democracy but actually permitted whites to monopolize power. The contradiction between liberal-democratic political theory and racist practice gave to black orators and writers the chance to seize high moral ground by exposing the injustice and hypocrisy of their oppressors. In 1965 blacks in the United States gained their greatest triumph when Congress passed legislation that made their right to vote enforceable in all the states of the Union. In South Africa, the African National Congress finally achieved its goal of "one person, one vote" in 1994.

The origins of the contemporary struggles for the suffrage can be found in the mid- to late nineteenth century. Some of the righteous intensity of subsequent campaigns for voting rights derived from the fact that color-blind political equality was not unprecedented in either society but was in fact a legacy from mid-Victorian political liberalism. Ex-slaves in the United States had been enfranchised shortly after their emancipation, and the Constitution was changed in 1870 to prohibit denial of the right to vote on grounds of "race, color, or previous condition of servitude." In the Cape Colony of South Africa—the center of British power in the region and the cradle of white South African society—ex-slaves and conquered indigenes were enfranchised after 1854 if they could meet a relatively low property and educational qualification that also applied to white settlers. These mid-nineteenth century manifestations of racial liberalism raised hopes among blacks that they could gain substantive equality and a measure of power to shape their own destinies through engaging in electoral politics.[1]

Toward the end of the century, an intensifying racism fed white fears of black political empowerment and led to self-conscious and systematic efforts to restrict or nullify black voting rights. In protesting these changes as the betrayal of earlier pledges or guarantees, black leaders and reformers could argue for political inclusion on grounds that might persuade liberal or moderate whites that nothing revolutionary or contrary to generally accepted political norms was being proposed. But the immediate failure to protect or extend the suffrage also led to a reconsideration of the tactics and strategy of black advancement. If blacks in the southern states and in most of South Africa could not, for the time being, expect to exercise full rights of citizenship, what might they do to strengthen themselves for a future bid for political equality? Central to the debates over strategies to replace or supplement the thwarted efforts to achieve equality through political agitation and participation were the kind of prescriptions for group progress through self-help, education, and economic efficiency that are usually associated with the name of Booker T. Washington.

Liberal conceptions of polity and citizenship were originally based on the tacit assumption that "the people" being provided with "equal rights" were citizens of a culturally and racially homogeneous national community. However inclusive the language of equal rights in the Declaration of Independence might seem to be, the new American republic tolerated black slavery and limited citizenship through naturalization to Caucasians. In the minds of most white Americans of the pre-Civil War era an invisible color qualification appeared between "all" and "men" in the Declaration's epoch-making statement about who was created equal and endowed with unalienable rights. In nineteenth-century Britain, the color question did not arise as a domestic issue, and the power of class perceptions was such that a black American abolitionist who dressed and acted like a gentleman was normally treated like one; occurrences of this kind fostered an illusion that Britain, in stark contrast to the United States, was a society without color prejudice. But Britons operating in colonial contexts took for granted the right of Europeans to rule over "lesser breeds"; and John Stuart Mill, the foremost spokesman for mid-century liberalism, maintained with little fear of contradiction that representative institutions could function successfully only in societies that had attained "a sufficiently advanced state" and were "not divided by strong antipathies of race, language, or nationality."[2]

Without the insistence of black protesters that the rights recognized by liberal-democratic theory were truly universal in their application, it is doubtful that significant numbers of American republicans or British liberals would ever have gone beyond an ethnically qualified conception of equality. Black liberation struggles also helped to enlarge the general meaning of liberty and equality; the number of rights to which human beings were normatively entitled would increase partly as a result of their efforts. But the struggle for a conception of human rights that would outlaw all forms of racism would be a long and frustrating one. In the nineteenth century, liberal theory was still close to its roots as an ideology linked to the early development of capitalism and the rise of the bourgeoisie. Above all, liberal equality meant an equal right to acquire and protect private property; it most definitely did not mean economic equality or interference with the "natural" distribution of wealth that was thought to result from the unfettered operation of the market. People had to be "equal under the law" so that they could freely contract with one another; for the laboring classes, this meant the sometimes dubious freedom to work for the best wages available, which required the right to move about in search of work, unrestrained by serfdom, slavery, or any other form of legal restraint. This liberal conception of economic freedom was essential to the success of movements for the abolition of slavery in Britain and the United States; it gave a practical, materialist justification for a crusade that was begun in the name of religious humanitarianism but could scarcely have triumphed over powerful vested economic interests if its appeal had been limited to philanthropists, Christian perfectionists, and romantic reformers. But the abolition of archaic, precapitalist labor systems did not eliminate racial discrimination or provide those emancipated with the resources and education that would have enabled them to exercise their freedom in a positive, self-fulfilling fashion.[3]

Nineteenth-century liberals did not normally view the right to vote as an intrinsic or essential right of citizenship. Women, for example, were thought to possess some of the basic rights of citizens without, of course, being granted the suffrage. Unlike such fundamental or "natural" rights as equality before the law, freedom of contract, and protection of person and property, voting was generally viewed as a privilege, to be granted or withheld as the best interests of the community—as determined by its already enfranchised members—dictated. In principle, nineteenth-century liberal theory had no objection to educational, property, or gender qualifications for the suffrage; it is in this respect that one can differentiate between liberal ideology and the kind of radical democratic theory that defended universal manhood suffrage (and later universal adult suffrage) as a matter of principle.[4]

A comparison of the actual suffrage in mid-nineteenth-century Britain and the United States might suggest a difference in basic political beliefs, since the franchise was still highly restrictive in Britain while universal white manhood suffrage was prevalent in the Yankee democracy on the other side of the Atlantic. But such a contrast would be misleading.

Beginning with the Parliamentary Reform Act of 1832 Britain was gradually extending the suffrage down the economic and social scale, a process that did not result in universal suffrage until well into the twentieth century. This gradualism was in perfect accord with the liberal conception of the vote as a privilege to be granted as a matter of policy rather than a fundamental right.

Practice in the United States developed quite differently but not in such a way as to signify the unequivocal triumph of universal adult or even male suffrage as a democratic principle. The mass enfranchisement of white males that took place in "the Jacksonian era" was based on a conception of the "people" that was not only limited by race and gender but was also predicated on the assumption that most white males were actual or potential property owners—as indeed they were at a time when landowning farmers outnumbered wage earners and agricultural tenants in the white male population. Hence universal manhood suffrage was more a product of American conditions than an application of radical democratic principles. In the version of American "republicanism" that was still dominant at the time, Thomas Jefferson's view that the "independence" that came from private property was essential to a responsible and virtuous electorate remained influential. If some men without property now gained the suffrage it was on the assumption that most of them would soon attain the independence necessary to good citizenship by hard work or migration to "vacant lands" in the West. What was radical about America was not its willingness to enfranchise the working classes but rather its expansive belief that virtually all white males could rise into the propertied and entrepreneurial classes. When the industrial revolution and mass immigration of the mid- to late nineteenth century brought genuine proletarianization and destroyed dreams of a democracy of yeoman farmers and small entrepreneurs, there were serious efforts to abandon universal manhood suffrage and introduce property and educational qualifications. They failed to achieve their aims, except to the extent that they were able to do so by such indirect means as tougher registration requirements and secret ballots that the voter had to read for himself, not so much because of the strength of radical democratic theory as because uneducated and relatively impoverished voters and the (mostly Democratic) politicians whom they had elected to office refused to consent to their own disfranchisement or loss of power.[5]

The enfranchisement of southern blacks that took place in the South under the Reconstruction Acts of 1867 and '68 was an implicit assertion of a more radical conception of democracy than nineteenth-century liberals or republicans had previously been willing to contemplate. It appeared to build on the American precedent of universal manhood suffrage, but its context was very different from the one that had enfranchised all white males earlier in the century. For the first time, an uneducated, propertyless local majority was given the vote and the potential power to rule over an educated and propertied minority. When the time came to amend the

Constitution to guarantee black voting rights, however, the call of a handful of radicals to embody the principle of universal manhood suffrage in the fundamental law of the land was rejected. The Fifteenth Amendment mandated only a racially impartial suffrage, and the way was open for disfranchisement of blacks through a variety of tests that were seemingly nonracial but actually designed to have a discriminatory impact. This decision was not so much a reflection of racist intent as of a refusal to move from a liberal to a radical-democratic conception of the suffrage.[6]

If it did not require universal suffrage, nineteenth century political liberalism nevertheless found it increasingly difficult to justify voting restrictions based purely on race. The ideal of civil equality and the growing commitment to equality of opportunity implied that ascriptive disabilities should have no part in determining political participation. Limitations on the political rights of women could perhaps be justified on the grounds that they were represented in civic affairs by their husbands, but no such rationale was available in the case of blacks. Consequently there was a tension, and at times an open conflict, between liberal political theory and racist ideologies that would base voting rights on membership in a superior, "self-governing" race. Liberalism, carried to its nonracial conclusion, was an ideology to which blacks and their white allies could appeal in the struggles for equal citizenship in late nineteenth-century America and South Africa.

In his speech at the annual meeting of the Massachusetts Antislavery Society in 1865, Frederick Douglass made the case for black manhood suffrage. Blacks want the vote, Douglass argued, "because ours is a peculiar government based on a peculiar idea, and that idea is universal suffrage. If I were in a monarchical government, or an autocratic or aristocratic government, where the few bore rule and the many were subject, there would be no special stigma resting upon me because I did not exercise the elective suffrage . . . , but here where universal suffrage is the rule, where that is the fundamental idea of government, to rule us out is to make us an exception, to brand us with the stigma of inferiority. . . ."[7]

Here was the heart of the case for black suffrage as put forth by black and white Radicals during the Reconstruction era. Although mainstream liberal theory, even in the United States, did not require universal manhood suffrage, it offered no clear basis for enfranchising all males of one race and not of another. The only way to restrict black voting rights would be to transform general American practice and establish educational or property tests that applied impartially to black and white alike. The Reconstruction Acts of 1867 and '68 seemed to follow Douglass's logic, but in fact they deviated from it by making loyalty to the Union a qualification for the suffrage. Ex-slaves who were putatively loyal were enfranchised, but ex-Confederate leaders who were considered of doubtful loyalty were denied the right to vote and hold office during the Reconstruction process. Most congressional Republicans who voted for black suffrage in the South were not radical democrats or firm believers in racial equality. They were conventional nineteenth-century liberals and moderate white supremacists who were persuaded that the extraordinary challenge of reconstructing the Union on a permanent basis and insuring the future success of the Union-saving Republican party required measures that in other circumstances they would have deemed unwise or unjustified. It was no wonder then that they held back from making universal male suffrage a part of the Constitution when they framed the Fifteenth Amendment in 1869. The half-hearted and conditional nature of the northern white support for black suffrage in the South also revealed itself in the retreat from voting-rights enforcement efforts in the 1870s. In the absence of a consistent and determined federal force to police elections, the struggle of white supremacists to gain control of southern state governments by terrorizing and intimidating black voters could not be resisted. Once in power, the "Redeemers" began to restrict black suffrage through a variety of devices that made no direct reference to race, a process that culminated in the massive constitutional disfranchisements of the period 1890–1910. For Frederick Douglass and other heirs of the radical-abolitionist tradition, these devices were so blatant in their discriminatory intent that they violated liberal conceptions of equal rights, to say nothing of radical-democratic conceptions of government by the people.[8]

The qualified "nonracial" suffrage that accompanied the grant of responsible government to the Cape Colony in 1854 represented the high point of mid-Victorian racial liberalism within the British empire. Following in the wake of slave emancipation and an effort to remove all legal disabilities from the indigenous Khoisan peoples of the Western Cape (known to whites as "Hottentots") it represented a sincere effort to apply the liberal political principles that were then triumphing in Britain to a multi-racial colonial environment. In recommending that the property or wage qualification be kept relatively low, the Crown's Attorney General, William Porter, called it "just and expedient to place the suffrage within the reach of the more intelligent and industrious of the men of colour, because it is a privilege which they deserve, and because by showing to all classes, those above and below them, that no man's station is, in this free country, determined by the accident of his colour, all ranks of men are stimulated to improve or maintain their relative positions." This conception of the vote as a privilege to be earned in a society of open social classes, and not to be denied because of accidents of birth, comes close to the heart of the mid-Victorian liberal faith. Social hierarchy was taken for granted in a more frank and open way that would have been likely in the United States with its more egalitarian mores, but the hierarchy had to be the "natural" consequence of acquired individual qualities, not the artificial product of racial and ethnic prejudice. The white settlers who approved the "color-blind" Cape franchise did not necessarily agree with the view that race was in principle irrelevant to social classification. Realizing that no explicit color bar would be countenanced, representatives of the

Dutch-speaking or Afrikaner majority had reasons of their own to prefer a low property qualification; without it many of their own people would be disfranchised, and they would have no chance of outvoting the more affluent English-speaking minority.[9]

In any case, there was no imminent danger of people of color exercising substantial power. Low as it was, the qualification was high enough to leave the Europeans firmly in control. When the franchise went into effect, most nonwhites who might be able to take advantage of its provisions were members of the diverse mixed-race group that was being lumped together under the designation "Colored." The great majority of these were farm workers who, if earning enough to qualify for the vote at all, were likely to vote the way their masters dictated. In the 1870s and '80s, however, the Cape incorporated within its borders the eastern frontier areas of the Ciskei and the Transkei, with their large African tribal populations. Although the suffrage was initially extended to these newly conquered Africans on the same basis that it was exercised by the Colored people of the Western Cape, the fear of having too many black voters led to new voting restrictions in 1887 and 1892. The first reduced the Eastern Cape African vote by ruling out communally held property as a basis for meeting the qualification, and the second by raising the qualification itself, which also had the effect of reducing the Colored vote in the Western Cape. Despite these limitations, the non-European voters continued to constitute between 16 and 17 percent of the Cape electorate at the turn of the century and held the balance of power in a number of key districts. Although there was no bar against Africans or Coloreds electing one of their own to the Cape Parliament, they never succeeded in doing so. In practice, the color-blind electoral system meant only that blacks had some choice over which white man would represent them. But the ballot was nevertheless highly prized by those who possessed it. Many believed they could influence colonial authorities to heed the interests and desires of their communities by using their leverage over the liberal politicians—known as "Friends of the Natives"—who depended on their votes. Furthermore the vote was a badge of honor implying that the holder was regarded as a "civilized" equal by the dominant whites. In the mind of covertly white supremacist liberals, the Cape franchise had value precisely because it constituted a safety valve for satisfying the aspirations of a new class of educated Africans and Coloreds who might otherwise take the lead in resisting the colonial order.[10]

A major difference between the Cape electoral system and the one that came into being in the South after Reconstruction is that the franchise restrictions in the Cape were at least honest ones that roughly conformed to nineteenth century liberal conceptions of fairness and equity. Although most whites could easily meet the property and wage qualifications, a few could not and no exception was made for them. Furthermore, blacks who met the formal criteria for the vote could actually exercise it, and there was reason to hope that the number of black voters would increase as more blacks acquired property. In the South, on the other hand, violence, intimidation, and fraud first reduced the black electorate, and the formal legal and constitutional provisions that completed the process provided loopholes allowing poor and illiterate whites to remain on the voter rolls and gave local registrars the authority to turn away blacks who actually met the property, educational, or "good character" requirements that were the official basis of inclusion. It was very difficult to make the facts of southern disfranchisement conform to any recognized liberal principles. In the eyes of blacks, certainly, nothing could be perceived except the arbitrary and tyrannical exercise of white power; for there was no room for a privileged elite that could realistically view suffrage as a reward for self-improvement. Finally, from the vantage point of 1900, the historical trajectories of black enfranchisement might have seemed very different in the South and the Cape. In the former political rights previously exercised were being denied and seemed likely to diminish further in the future; in the latter, recently conquered and incorporated Africans were apparently being offered a stake in the new political order that the white man had imposed—a modest one to be sure, but one that could grow in time—if one believed, as many Africans did, in the promise of future self-government implied by the Cape liberal slogan of "Equal rights for every civilized man south of the Zambezi."[11]

During the South African War of 1899 to 1902, Africans and Coloreds in the Cape were led to expect that a British victory over the Afrikaner republics of the Transvaal and the Orange Free State would lead to an extension of the Cape's nonracial electoral system to a larger South Africa unified under the British Crown. But these expectations were dashed when it became clear after 1902 that the imperial objective was accommodation of white settlers and not political rights for blacks. As the prospect of a self-governing white supremacist South Africa came into view, the Cape liberal idea became the basis of black protest and political mobilization. At approximately the same time, a group of black Americans met at Niagara to launch a militant civil rights struggle based mainly on protest against the evasion of constitutional guarantees of equal voting rights in the southern states.

The struggle to gain, preserve, and extend voting rights thus became a common preoccupation of the black protest politics that emerged in the United States and South Africa during the nineteenth century and persisted into the twentieth. Before examining, connecting, and comparing these suffrage-based struggles, it may be useful to draw out some of the implicit assumptions that underlay this emphasis on the ballot as source of racial equality and empowerment.

If a subordinate group defines its cause as equal political rights with a dominant group, it is apparently affirming the possibility of sharing a common polity with its current oppressors. It also appears to be acknowledging that the kind of political order that the dominant group has established for itself—in this case, Western-style representative democracy—is an acceptable basis for a more inclusive system. The political community to which African-American and black South African protest leaders were ostensibly seeking access was the one already constituted by whites. In demanding "equal rights," they were implicitly assenting to a conception of rights held by their conquerors or former masters. Their formal objection to the status quo, therefore, was to its exclusiveness—its explicit or implicit use of racial tests or color bars—and not to what were regarded as its essential or intrinsic characteristics. Protest movements emphasizing extension of the suffrage are therefore likely to be radical reformist rather than revolutionary, unless the franchise or agitation for it is viewed merely as an available means for creating a revolutionary group consciousness. If equal rights is indeed all that is being sought, and if there is some hope of achieving it, more radical objectives such as the dictatorship of the proletariat or full national self-determination would seem to be ruled out. In situations of racial inequality and conflict, agitation for equal voting rights seems to imply a desire for incorporation or integration into a color-blind democratic order. Hopes for vengeance, turning the tables, restoring an independent past, or establishing a new independent nationality—all potentially powerful emotions among oppressed racial or ethnic groups—have to be resisted, suppressed, or at least kept under wraps if a struggle aims at constitutional reform to eliminate racial exclusiveness in voting and holding office. For a social revolutionary or radical nationalist, demanding and using the vote is at best a device to raise consciousness and at worst a diversion from the true task of overthrowing the existing economic and social order, as well as the government that sustains it.

Even if one rules out revolutionary cynicism as a motive for equal-rights political struggles, the aim of such agitation is more ambiguous than it appears at first glance. Attaining the vote may be regarded as a step toward a genuinely color-blind society in which race becomes totally irrelevant and individuals identify themselves in other ways for political purposes. In such a polity, there would be no need for racially based parties, factions, or voting blocs because race had ceased to be an indicator of economic or cultural interests. But a less utopian and more ethnocentric viewpoint is possible. The vote might be viewed as a way of unifying and mobilizing a group that would increase its self-consciousness and solidarity, enabling it to pursue its collective needs and interests more effectively. The implicit model here is an ethnic political pluralism in which each distinct racial or ethnic group within the same society retains a strong sense of its identity and special interests, especially in the realm of culture and international affiliations, and uses its voting power to further group goals. According to this viewpoint, race or ethnicity is an essential and durable characteristic of human beings and not merely the artificial construction of racists. Equal rights does not mean the disappearance of race but rather the right to be a racially conscious person, loyal to one's own kind, without being penalized for it or inhibited in the pursuit of essential group interests.

Such a perspective could not be clearly articulated until well into the twentieth century, for previously there had been no pluralist social or cultural theory to sustain it. The universalistic conceptions of culture and civilization that predominated in the nineteenth century left little room for the political accommodation of enduring group differences. But the emphasis on race pride that often accompanied universalistic demands for individual equality and assimilation, as well as the recurring endorsement of separate action and group solidarity as a means to the attainment of equality and political incorporation, reveal the persistent tension between cosmopolitan, genuinely color-blind perspectives and those which viewed blackness as existentially inescapable and culturally invigorating. It was not so much a matter of choosing between integrationist cosmopolitanism and racially pluralistic ethnocentrism as a basis for political action, as finding a way to reconcile them so that blacks could find fulfillment in two ways at once—as the generic human beings of liberal theory and as a special people whose unique historical experience could be represented or symbolized by reference to skin color.

The Ballot in African-American Protest Thought, 1840–1905

The origins of black protest politics in the United States can be found in the activities and organizations of northern free blacks in the pre-Civil War period. The black population of the northern states was small, only about a quarter of a million in 1860, and the majority were impoverished laborers and servants whose full energies were required to keep their heads above water in a hostile environment, in which they were forced to compete on unequal terms with poor white immigrants (mostly Irish) for menial, low-paying jobs. But a small, educated elite emerged in cities like New York, Boston, Philadelphia, Pittsburgh, and Cleveland. As clergymen, teachers, journalists, and small-scale entrepreneurs, this vanguard of the black middle class was the only fraction of the black population of the United States that had the freedom, means, and leisure to engage in sustained political protest. Southern free blacks, approximately equal in number to those in the North, had greater freedom of movement and expression than the slaves—and a small number were relatively affluent—but they were denied most civil and political rights and were likely to be intimidated or ruthlessly repressed if they gave any sign of protesting against their condition or of allying themselves with their brothers and sisters in bondage. Northern free blacks, on the other hand, could express themselves freely, form associations for political purposes, and in some states could even vote on the same basis as whites (as in most of New England) or by meeting special property qualifications (as in New York after 1821).[12]

Northern free blacks founded newspapers, formed local associations, held a series of national conventions beginning in 1830, and espoused a variety of causes of importance to the black community. Their preeminent cause was of course the abolition of slavery. A complex and difficult relationship existed between black abolitionists and the white-led antislavery societies that agitated the nation in the 1830s and helped begin the process of sectional polarization that led to the Civil War. Most prominent black abolitionists were early supporters of William Lloyd Garrison (they were most of the original subscribers to his fiery antislavery organ *The Liberator*), and many of them remained loyal to this white champion of immediate emancipation throughout the antebellum period. Endorsing Garrison's Christian perfectionism, black abolitionists such as Charles Remond, William Wells Brown, Robert Purvis, James Forten, and William Whipper found little justification for racially specific organizations or activities. Other free black protest leaders, especially among those based outside of Garrisonian New England, had a greater tendency to act independently of whites. The Reverends Henry Highland Garnet of Troy, New York, Lewis Woodson of Pittsburgh, and Samuel Cornish of New York City carried the idea of separate action to the point where they anticipated some of the ideas later espoused by black nationalists. Frederick Douglass, the escaped slave who eventually made his home in Rochester, New York, and was the most conspicuous and influential of all the black abolitionists, began as a follower of Garrison, then broke with him over a number of issues, most notably Garrison's opposition to Douglass's founding of an antislavery journal of his own to rival *The Liberator*. But Douglass remained opposed to black separatism in principle, accepting it only when it could be justified on purely tactical grounds as promoting the ultimate objective of an integrated egalitarian society.[13]

In states where black suffrage was denied entirely, as in Pennsylvania after 1838, or was granted to blacks on a discriminatory basis, as in New York, the struggle for equal political rights rivaled or even at times eclipsed the battle against southern slavery. In the 1840s and '50s, Douglass and Garnet mobilized New York blacks around the suffrage issue. In 1856 Garnet addressed a state convention of representatives of New York's five or six thousand qualified black voters and called for them to use their precious ballots to support the new Republican party in the upcoming presidential election. Showing a pragmatic attitude toward use of the ballot, Garnet argued that the Republicans, with all their shortcomings, came "nearest to liberty and justice," for the party's "great principle was to stop the progress of slavery." The convention then devoted itself to its main business of organizing a petition campaign for extension of the franchise. In the 1850s, Garnet joined a number of black leaders in advocating emigration to Africa or the West Indies as part of an effort "to establish a grand center of Negro nationality." But his pan-African perspective did not prevent him from attaching great importance to black participation in the American political system.[14]

The educated and relatively affluent free blacks who expressed themselves in sermons, newspapers, pamphlets, and books sometimes complained that they were "leaders without followers." As victims of caste discrimination, they knew that their own attainment of equal rights with white Americans depended on the removal of the disabilities that American racial order imposed on all blacks; hence they had no difficulty in seeing themselves as representatives of the black community as a whole. But the thoroughly middle-class outlook and style of life that most of them manifested, as revealed by their characteristic preoccupation with temperance and "respectability" and sustained by their economic independence as professionals and small business owners, tended to cut them off from some of the most immediate concerns of working-class blacks. Their attitude toward the black masses was at times critical and condescending; lower-class blacks were objects of improvement and uplift more than comrades in a struggle. Martin Delany, one of the most militant and nationalistic of antebellum black leaders, was given to chastising lower class blacks for being willing to work as servants, forgetting apparently that most of them had no other choice.

Most of the free black leaders were self-made men, who had dragged themselves from poverty or even from slavery to economic independence and respectability. Like self-made men in most societies, they tended to lack patience with those who had been unable to overcome similar handicaps. Radical and militant on questions of racial injustice, they were also likely to have relatively conservative attitudes on other social questions. Implicitly or explicitly, they endorsed a less than egalitarian view of society in general. Along with most white abolitionists, they accorded legitimacy to what they viewed as just and normal social hierarchies, those based on character, education, and property acquired through honest effort. Consequently their concept of race leadership was of the vanguard type: their role was not to articulate the actual feelings and grievances of the masses; it was rather to elevate the masses so that they would have the proper feelings and understand their true and legitimate grievances. This style of leadership was to predominate in black politics for a long time to come.[15]

These pioneer protest leaders were attempting to develop and inculcate an ideology of liberation that would synthesize an idealized American nationalism with an incipient black nationalism. Much has been made of the allegedly irreconcilable conflict between integrationists and nationalists that arose in the antebellum period over whether blacks should establish racially exclusive bodies or seek to liberate themselves from American racism by emigrating to Africa or the West Indies. The extreme positions in these debates do indeed seem impossible to reconcile. William Whipper of Philadelphia, for example, was an unconditional integrationist who believed that blacks should never engage in independent reform activity or patronize segregated institutions. Martin Delany on the other hand argued in a notable book of 1852 that blacks had no future in the United States and should depart en masse for any destination that would afford them racial self-government. (At that point, he focused on the West Indies; later he would turn his attention to Africa.) In his 1854 report, "The Political Destiny of the Colored Race,"

Delany argued that the suffrage in the United States would not give blacks true citizenship, because it would not make them "an essential part of the *ruling element* of the country in which they live." As a disadvantaged minority, blacks could do no more with the ballot than give their "*approbation* to that which our *rulers may do*."[16]

Most free black activists struggled to find a middle ground between total integration and total separation that would allow them to maintain a dual identity. The black campaign against slavery and racial discrimination was based on identification with the promise of America as a land of liberty and equality, but to escape the burdens of racism and attain equality as American citizens, blacks had to develop racial pride and solidarity. Apart from Delany, no prominent black leaders of the antebellum period proposed that African-Americans secede en masse from the American republic and establish a separate state. The historian Leonard Sweet has captured the dominant attitude: "Blacks were nationalistic about their color and capabilities, indispensable ingredients in self-pride and creativity. Yet their nationalism did not preempt their demands for inclusion as Americans." Most antebellum emigrationists actually viewed the expatriation of some blacks to Africa or the West Indies as a supplement to the struggle for inclusion in American nationality and not as a substitute for it, the logic being that independent black achievements abroad would help to mitigate white American racial prejudices. Anti-emigrationists such as Frederick Douglass, who generally attacked such proposals as a distraction from the struggle at home, also called on blacks to develop pride of race and solidarity as a people. The basic objective was to achieve political integration without sacrificing a sense of black identity and peoplehood. "Blacks realized," according to Sweet, "that they had a common history of suffering and oppression that differentiated themselves from other groups and that, as a proscribed minority, they had special interests and special needs which required them to band together as a unit."[17]

Black consciousness and solidarity for the purpose of incorporation into an idealized American nation may not meet the requirements imposed by a strict standard of nationalism. What kind of nationalism is it, one might ask, that does not envision a separate national state or even manifest any clear conception of a distinctive national culture? But we lack another word to describe group identity and pride based on a common color, origin, and experience of oppression. For nineteenth century blacks, the America of their dreams and expectations was not the particular manifestation of a white or Anglo-Saxon race spirit. It was a universal nation based on color-blind principles of equality and human rights, a nation that did not require its citizens to surrender their ethnic or racial distinctiveness and the pride that went with it.

This nationalistic integrationism was normally sustained by the belief that racial prejudice was not innate or natural. "CONDITION and not *color*, is the chief cause of the prejudice under which we suffer" was one striking formulation of a doctrine accepted by most black abolitionists and antebellum proto-nationalists. One implication of this idea was that the abolition of slavery would go a long way toward eliminating racism; another was that self-improvement among free blacks would overcome much of the feeling against them. In urging their fellow African-Americans to educate themselves, acquire property, and adopt strict, puritanical moral standards, some members of the free black elite came close at times to blaming the victims for the discrimination against them. What perhaps pained them the most was that people like themselves who deserved equal treatment at the hands of the white middle class did not receive it; for whites lumped all blacks together and falsely attributed vices of the many to the virtuous few. Hence the only way to conquer prejudice was to raise the standards of the group as a whole, thus removing the stigma of inferiority from all blacks. This line of thought might have led to an emphasis on self-help at the expense of protest, had it not been for the existence of slavery, which no amount of self-help could conceivably eliminate. There was the further injustice of political discrimination in the free states, which meant that in states such as New York and Pennsylvania uneducated, disreputable whites, some of them just off the boat from Ireland, went freely to the polls, while many respectable, middle-class blacks were barred from exercising their rights of citizenship. Consequently the struggle had to be based on a dual strategy of self-improvement and organized protest—especially by the educated and respectable vanguard—against slavery and civic inequality.[18]

This ideology assumed that class is potentially more important than race in a capitalistic, free labor society. Since slavery was based on color, its defenders promoted the myth of innate black inferiority to justify their practice of human bondage; and the degraded condition of the slaves reinforced prejudiced attitudes in the minds of whites, including those who did not benefit directly from slavery. But if slavery could be abolished, prejudice and discrimination would disappear and blacks would be judged on their merits. This gross underestimation of the ability of racism to outlive slavery and even gain in strength after servitude was abolished was not unique to black antislavery thought. In fact it was a view shared by white abolitionists, who had often severed their ties with the colonizationist movement precisely on the question of whether prejudice was ineradicable or a result of "condition." One does not of course have to believe that prejudice is literally innate or instinctual to find an enduring racism in white American culture. Most black abolitionists and intellectuals of the antebellum period failed to perceive that more than two hundred years of slavery and discrimination had planted the notion of black "otherness" and inferiority so deeply into the white psyche that liberation of blacks from bondage could not remove it. Neither could strenuous efforts at self-help and self-improvement; for such efforts were bound to have limited results in the face of persistent prejudice and discrimination. Paradoxically, black thinkers of the mid-nineteenth century often appreciated the power of race consciousness in their own quest for pride and dignity but could not face up to the ways that it maintained and reinforced the positive self-image and status pretensions of white Americans.[19]

When the Civil War brought emancipation and the use of black troops to defend the Union, black leaders rejoiced in the impending fulfillment of their hope for full-fledged membership in the American nation. Frederick Douglass, in particular, viewed the war in millennialist terms as an apocalypse that would regenerate the nation and make it true to its promise as a land of freedom and equality. Martin Delany, the most thoroughgoing of prewar separatists, now accepted a commission in the Union army and for the time being abandoned the idea of a black exodus from the United States. In the South black slaves voted with their feet against the Confederacy and offered themselves en masse to the Union armies as laborers and soldiers, thus making a significant and perhaps decisive contribution to the Union victory. As the South was going down to defeat before an army and navy that included almost 200,000 blacks, most of whom were ex-slaves, northern black leaders were preparing a campaign for equal suffrage. They found a receptive white audience in the North when they claimed that blacks had earned the right to full citizenship and that they were the most "loyal" segment of the southern population. Without their votes, it was plausibly argued, the South could not be reconstructed in such a way as to assure the future safety of the Union. Furthermore, if American democracy was extended in a color-blind way to the southern states, there would be great opportunities for the black elite to find a constituency and exert political leadership.[20]

During Reconstruction the former slaves of the defeated South voted in massive numbers for the Republicans and constituted the backbone of the party in most states. But they did not dominate the politics of the region; white Republicans held a controlling share of the top leadership positions and elective offices except in South Carolina, which for a time came closer than any other state to achieving the "black domination" that southern white supremacist myth-makers would later claim was characteristic of Radical Reconstruction throughout the region. (South Carolina's combination of a substantial black majority and lack of a significant body of white Republican voters made it atypical of the ex-Confederate states.) Although deferring in most cases to the leadership of white "scalawags" and "carpetbaggers" substantial numbers of blacks were elected to local, state, and federal offices—16 served in the United States House of Representatives and two in the Senate during and immediately after the period of Republican ascendancy. Many self-educated ex-slaves were elected to local offices, but most of the blacks who gained state or federal office had been free before the Civil War and had enjoyed exceptional educational advantages, including college or university training for a significant number. Some of these were "black carpetbaggers," members of the northern free black elite who had come to the South as soldiers, missionaries, or Freedman's Bureau officers. In short, the freedmen were represented to a considerable extent by the same kind of leaders that had predominated in the northern-based agitation against slavery and racial discrimination in the antebellum period. This top political leadership was overwhelmingly middle-class or "bourgeois" in its outlook and did not generally favor radical social and economic policies. It did not, for the most part, press hard for land redistribution or other measures that would challenge established property rights.[21]

Black abolitionists who remained in the North likewise generally failed to give strenuous backing to land confiscation and redistribution schemes. Frederick Douglass, for example, favored a federal program to make small farms available to freedmen on relatively easy terms, but he balked at free homesteads. Middle-class black Republicans, like many white Republicans, were not so much fearful of the effect of confiscating the land of "traitors" as they were of the deleterious effect on the freedmen of "getting something for nothing." Douglass's standard answer to the question of "What shall we do with the Negro" was "Do Nothing with us! . . . if the Negro cannot stand on his own legs let him fall." By this he meant that once blacks had received full civil and political rights they should get no special advantages. They should not, in other words, become wards of the state or the objects of "class legislation." It was a legitimate application of the self-help doctrines of the prewar free black elite to make the freedmen primarily responsible for their own economic advancement. Education and political participation would enhance self-reliance, but free homesteads or government charity might undermine efforts to inculcate competitive efficiency.[22]

Douglass and other black Republican leaders obtained, on paper at least, essentially what they were asking for—equality under the law, as provided by the Civil Rights Act of 1866 and the Fourteenth Amendment to the Constitution and the right to vote as mandated for the South by the Reconstruction Acts and for the nation as a whole by the Fifteenth Amendment. The ballot was crucial, not because it would enable an impoverished black majority to redistribute the wealth of the South, but in order to give blacks the ability to resist de jure or de facto abridgment of their civil rights. These rights, if protected, would allegedly guarantee to the ex-slaves a fair chance to get ahead in a competitive capitalist society. What went wrong in their eyes was not that more fundamental reforms failed to pass Congress but rather that the federal government failed to enforce the right to vote and thus deprived blacks of the protection against discrimination and exploitation that the ballot was supposed to provide.

Contrary to what Booker T. Washington would later argue, the drive for black self-improvement was not a casualty of a preoccupation with politics during Reconstruction. What was being sought was a legal and political structure in which self-help could take place, free of blatant discrimination and artificial barriers to group advancement. But in practice, Gilded Age American politics was a poor school for inculcation of the Protestant virtues of industry, honesty, sobriety, and frugality. Given the spoils system that predominated in American politics and the emphasis on government aid for private economic activity (especially railroad construction) that dominated southern politics during the Reconstruction years, it was no surprise that the southern Republican governments of the period 1868–77 were less than models of honesty and efficiency. Even more than northern carpetbaggers, the middle-class blacks who entered politics full-time had no career alternatives outside of public service, for "civil society" in the South remained under white conservative control, and there was little patronage for professionals or business men of the wrong color or political persuasion. Government salaries were meager, standards of political ethics were low (the concept of "conflict of interest" hardly existed), and temptations for private enrichment through political activity were many, especially in the case of those who had little other opportunity to accumulate wealth. Less perhaps than white Republicans but still to a degree that was troubling to principled and idealistic supporters of black political participation, members of the black office-holding elite in the South engaged in the same kind of vote-selling and misuse of public funds that was endemic in the American politics of the period. This activity, as natural and inevitable as it may have been under the circumstances, had a heavy cost; it betrayed the high ideals of the conflict against slavery and gave an excuse to northern liberal reformers to abandon a radical-democratic experiment that was hard to justify in terms of their basic beliefs about race and class. But the Reconstruction governments were not overthrown primarily because they were corrupt or wasteful. The campaign to terrorize and intimidate black voters mounted by the Ku Klux Klan early in Reconstruction and by unmasked paramilitary groups in its later phases drew their main inspiration from the affront to white supremacist mores and traditions that was inherent in the mere fact of black voting and officeholding.[23]

The immediate post-Reconstruction period forced national black leaders and intellectuals to reassess their political situation. For the most part blacks remained loyal to the Republican party. Republicans may have failed in their efforts to reconstruct the South on the basis of black suffrage and by the late 1870s were clearly in the process of abandoning southern African-Americans to their fate. The party of emancipation, equal rights, and free labor was on its way to becoming the party of high tariffs, hard money, and big business. But the Democrats had the deserved reputation of being the party of blatant Negrophobia. Furthermore, the Republican politicians of the 1880s continued to protest at election time against the violence, intimidation, and fraud which were making it increasingly difficult for southern blacks to vote the Republican ticket, and the party continued to reward prominent black supporters with federal patronage jobs. Frederick Douglass, famous for his statement that for blacks "the Republican party is the ship, all else is the ocean," was one of those who benefited from federal patronage policies designed to preserve the loyalty of the black elite. On the state and local level in the North, Republicans sometimes put black loyalists at the bottom of the ticket or appointed them to minor offices. A more substantial indication of the party's continued concern for black rights, or at least for the difference black voters might make in close elections, was its role in the passage of state laws calling for equal access to public accommodations to replace the federal Civil Rights Act of 1875, which the Supreme Court declared unconstitutional in 1883.[24]

If a significant group of black politicians, many of whom depended for their livelihood on the Republican party, continued to put their faith in reviving the party's flagging commitment to equal rights, a group of black intellectuals and journalists began to explore political "independence" as an alternative to automatic support of the party of Lincoln and emancipation. T. Thomas Fortune, editor of the New York *Globe*, one of several black newspapers founded in the post-Reconstruction North, was the era's most conspicuous advocate of political independence. In a speech before the Colored Press Association of Washington in 1882, Fortune castigated the Republican party for forgetting its original principles and abandoning southern blacks to the not so tender mercies of white supremacist "redeemers." "The Republican party," he charged in his characteristically forceful fashion, "has degenerated into an ignoble scramble for place and power." Consequently it was no longer "binding upon colored men further to support the Republican party when other more advantageous affiliations" could be obtained. It was of course impossible for a black man to be a "Bourbon Democrat," but he could be "an independent, a progressive Democrat." More generally he called upon blacks to vote as individuals, concentrate on local rather than national issues, and divide on issues of public policy that did not involve their fundamental rights.[25]

The rise of Grover Cleveland as a reform-minded President who seemed less hostile to blacks than previous Democratic standard bearers caused a flurry of black interest in the Democratic party. After Cleveland appointed a few black Democrats to minor federal offices during his first term, interest increased, and Fortune supported him for reelection in 1888. But Douglass and most black opinion-makers, including the editors of most of the black journals, remained steadfast in their loyalty to the Republicans. The problem with independency was that neither party was prepared to make significant concessions to a black electorate that was diminishing and ceasing to be a factor in southern contests. Continuing support of the Republicans, despite the high level of frustration that this entailed, was based on the assumption that they at least had an incentive to resist the march toward disfranchisement that was insuring Democratic control of southern elections. When the Republicans gained control of both houses of Congress in 1888, they came close to fulfilling their longstanding pledge to pass a law providing for federal supervision of southern elections to prevent the intimidation of black voters. Henry

Cabot Lodge's "Force Bill" passed the House of Representatives in 1890 only to fail in the Senate. After that the party ceased even giving lip service to black voting rights, but African-Americans could still find no convincing reason to support the Democrats. The national party made no further overtures to blacks, and its southern supporters took the lead in revising state constitutions in such a way as to make black voting virtually impossible.[26]

In his notable book *Black and White: Land, Labor, and Capital in the South* (1884), T. Thomas Fortune provided the rationale for a new class-based politics that would replace older allegiances based on specifically racial concerns. Influenced by Henry George and the white anti-monopoly radicalism of the 1880s, he described the essential southern conflict as one of rich against poor, landed against landless, rather than black against white. He stated baldly that "*the condition of the black and white laborer is the same and . . . consequently their cause is common.*" But he failed to draw the seemingly obvious conclusion that blacks should join with working class whites to form an anti-monopoly third party. Instead, he advocated a flexible, nonpartisan independence. In 1886, Fortune supported the Democratic candidate for mayor of New York against Henry George and the United Labor party.[27]

A close reading of *Black and White* reveals that it contains two competing social and political philosophies. Fortune's assertions of an anti-monopoly, quasi-socialist radicalism did not prevent him from calling on southern blacks to adapt to the capitalist system and succeed on its terms. In accordance with the dogma that prejudice is the result of condition, he predicted that race conflict would end in the South "when the lowly condition of the black man has passed away"—when "he has successfully metamorphosed the condition which attaches to him as a badge of slavery and degradation, and made a reputation for himself as a financier, statesman, advocate and money-shark generally, his color will be swallowed up in his reputation, his bank-account, and his important money interests." As his ironic tone suggests, Fortune the social radical was critical of the roles blacks would have to play to overcome prejudice, yet he answered in the affirmative the question of whether such an adaptation to predatory capitalism was "the logical outgrowth of the Divine affirmation that of one blood he created all men to dwell upon the earth and of the Declaration of Independence that 'we hold these truths to be self-evident:—That all men are created equal. . . .' " He then went on to report approvingly on the extent to which blacks had already succeeded in business, the professions, and government service.[28]

Fortune's celebration of black people's ability to emulate the kind of acquisitive behavior that he condemned in whites was not necessarily inconsistent. It made sense in the context of a sophisticated historical determinism similar to the orthodox Marxist view that you have to achieve capitalism before you go can go on to socialism. Southern blacks, Fortune was implying, would have to modernize themselves through capitalist enterprise and develop their own class and ideological divisions before there could be a color-blind class conflict with blacks on both sides. In the short term, then, the main task facing blacks was to work the capitalistic system in order to compete successfully with whites. Anticipating almost all the principal doctrines of Booker T. Washington, with whom he would later be closely allied, Fortune included in his otherwise radical and quasi-Marxist book the advocacy of self-help, industrial education, cooperation with ruling-class southern whites, and encouragement of black capitalist enterprise. "All over the country," he concluded, "the colored man is coming to understand that if he is ever to have and enjoy a status at all commensurate with that of his fellow citizens, he must get a grip upon the elements of success which they enjoy with such effect, and boldly enter the lists, a competitor who must make a way for himself."[29]

Such an interpretation of Fortune's ideology puts his call for political independence and an end to bloc voting in a new light. Fortune was not advocating that blacks as a group put their weight behind a party other than the Republicans, whether it be a reformed Democracy or a laborite third party. Instead he was calling in effect to blacks to rely less heavily on parties and electoral politics in general as a basis for group advancement. In *Black and White*, he put forth a harshly negative view of Reconstruction and black participation in it that echoes the charges of many white critics and anticipates Booker T. Washington's view that blacks had tried to walk before they had learned to crawl. "Illiteracy," he wrote, ". . . has been the prime cause of more misgovernment in the South than any other one cause, not even the insatiable rapacity of the carpet-bag adventurers taking precedent over it." His account of what happened in South Carolina repeats the anti-Radical views of the Redeemers and their northern apologists in every particular except their justification of the deliberately provoked riots and massacres used to rectify the situation: "By the side of robbery, the embezzlement, the depletion of the treasury of South Carolina, and the imposition of ruinous and unnecessary taxation upon the people of the state by the carpet-bag harpies, aided and abetted by the ignorant negroes whom our government had not given time to shake the dust from their feet before it invited them to seats in the chambers of legislature, we must place the heartless butcheries of Hamburg and Ellerton."[30]

If Fortune implied, without quite saying it categorically, that the enfranchisement of "ignorant" blacks during Reconstruction had been a mistake and that more stress needed to be put on preparing for the ballot than on exercising it, other prominent black intellectuals of the 1880s were coming to this conclusion in a more straightforward fashion. John Bruce, a prominent black journalist who would have a long career as a "race man" and advocate of separatist activity, concluded after the Supreme Court decision of 1883 restricting federal action on behalf of civil rights that "we have already paid too much attention to politics." The focus should shift, he argued, to education.[31]

The same point was made more strongly in George Washington Williams's *History of the Negro Race in America* (1882), a book that would be quoted in suffrage debates in South Africa. Williams, an Ohio clergyman and the first black member of the state legislature, was, like Fortune, highly critical of Radical Reconstruction. The failure of the "Negro governments" was inevitable, he contended, "because they were not built on the granite foundation of intelligence and statesmanship." As a result of their fall, "a lesson was taught the colored people that is invaluable. Let them rejoice that they are out of politics. Let white men rule. Let *them* enjoy a political life to the exclusion of business and education, and they will sooner or later be driven out of their places by the same law that sent the Negro to the plantations and schools. And if the Negro is industrious, frugal, saving, diligent in labor, and laborious in study, there is another law that will quietly and peaceably, without a political shock, restore him to his normal relations in politics." Blacks will regain the ballot, he concluded, when they can pay taxes and read their own ballots, when they are "equal to all the exigencies of American citizenship."[32]

That education and economic self-help had a higher priority than political participation and even the exercise of political rights was still a minority viewpoint in the 1880s. Only after 1890, when the defeat of Lodge's Force Bill and the wholesale constitutional disfranchisement of Mississippi blacks made loss of southern voting rights appear irreversible, did this approach win wide acceptance among black leaders and intellectuals. Its foremost champion was Booker T. Washington, the principal of Tuskegee Institute and, after the death of Frederick Douglass in 1895, the most influential black man in the United States. Washington argued in 1900 that "we made a mistake at the beginning of our freedom in putting the emphasis on the wrong end. Politics and the holding of office were too largely emphasized, almost to the exclusion of every other interest." What had been forgotten was that "the individual or race that owns the property, pays these taxes, possesses the intelligence and substantial character, is the one which is going to exercise the greatest control on government, whether he lives in the North or whether he lives in the South." He made it clear that he did "not favor the Negro's giving up anything which is fundamental and which has been guaranteed to him by the Constitution of the United States." It was legitimate and indeed desirable for the South to restrict the franchise on the basis of property and education, but it should not do so, Washington warned, in a racially discriminatory fashion. Despite his reputation as an accommodationist, Washington protested openly on a number of occasions against the tendency of the disfranchising constitutional conventions to provide loopholes enabling poor and illiterate whites to meet the new voting qualifications. Like most nineteenth-century liberals, but unlike radical democrats such as Frederick Douglass, he believed that the ballot was a privilege to be reserved for the certifiably competent and intelligent members of the community. For blacks and whites alike, it should serve as a reward and incentive for the attainment of certain minimum standards of civilization and social efficiency. (He would have had no objection to the electoral system of the Cape Colony of South Africa.) Although Washington's powers of persuasion were not sufficient to prevent southern white supremacists from enacting a suffrage that was impartial in form only, it is less than fair to charge him with abandoning the cause of equal political rights.[33]

Whether applied fairly or unfairly, the establishment of new tests for voting would keep most southern blacks from the polls. But for Washington this was a small price to pay for a chance to concentrate on activities he deemed more fundamental for group progress than voting. His program of industrial education, for which he succeeded in getting substantial support from northern philanthropy and from southern state legislatures, was part of a larger plan to rehabilitate a "backward" black peasantry and make them competitive in a developing capitalist society. Like most Americans of the late nineteenth century, Washington believed that economics was a more important source of power and status than political participation. He also accepted the view that prejudice and discrimination were a response to poverty and immorality. Like T. Thomas Fortune, he apparently believed that southern blacks would be treated as equals and granted full political rights when they had produced a substantial and prosperous class of landowners and businessmen. Like advocates of black self-help in later periods, he regarded Jewish commercial success in the teeth of prejudice as a model for blacks to follow. In his estimation, it was incompatible with black pride to assume that blacks could make progress only through political activity aimed at influencing public policy in their favor.

Whether Washington's emphasis on self-help and autonomous economic action constitutes a kind of proto-nationalism that anticipates the doctrines of Marcus Garvey (as Garvey himself believed) or whether Washington was a long-term integrationist who was simply calculating what it would take to win full acceptance from white Americans remains in dispute, mainly because he was careful not to divulge his views on whether black Americans were destined to remain a separate people or be absorbed as individuals into a color-blind America. Whatever his real thoughts about race, it is clear that blacks in South Africa and throughout the world saw Washington as a model of black achievement and self reliance who refuted the racist canard that blacks could achieve nothing on their own. The view of Tuskegee as a thoroughly modern and progressive institution run by blacks for blacks was a powerful inspiration for those living under white colonial domination. Washington's legacy was therefore ambiguous; he was an accommodationist on the question of political rights and social segregation who also championed a black self-reliance and capacity for progress that contradicted white supremacist ideology. Du Bois' criticism would be directed mainly at the first aspect, and Garvey's praise would focus on the second.[34]

After 1900, a group of northern black intellectuals and professionals led by W. E. B. Du Bois and William Monroe Trotter charged that Washington's acceptance of a qualified suffrage had encouraged southern white supremacists to nullify the Fifteenth Amendment. Du Bois, Trotter, and the other black radicals who founded the Niagara Movement in 1905 made federal action to restore equal voting rights on the Reconstruction basis of universal manhood suffrage a centerpiece of their neo-abolitionist campaign for black equality. Washington was condemned not only for acceding to a qualified suffrage but also for leaving control of the franchise in the hands of southern white supremacists, which assured a discriminatory result. This was a valid criticism of Washington's policy. In his desire to accommodate southern whites and win their support for black education and economic opportunity, Washington opposed congressional legislation to enforce the Fifteenth Amendment; almost alone among prominent black leaders he had withheld his support from the Force Bill of 1890. Washington was not unintelligent or naïve; he must have realized that he was surrendering in practice what he persisted in affirming in principle—equal access to the ballot box. Like George Washington Williams, he believed that equal political rights would be restored when blacks had rehabilitated themselves economically and morally. His critics replied that political power was essential to equality of economic opportunity. Reverting to the logic of Reconstruction radicalism, they maintained that only through the ability to influence public policy by voting and holding office could blacks protect the fundamental civil rights that were essential to the improvement of their economic condition. According to Du Bois, writing in 1903, "The power of the ballot we need in sheer self-defense—else what will keep us from a second slavery?"[35]

The most persuasive defense of Washington's approach is that he was making the best of a bad situation. Federal interventionism had failed once and was unlikely to be revived in the foreseeable future. This being the case, conciliation of southern whites in the hope of getting them to acquiesce in the improvement of black education and the development of black enterprise, leading perhaps to a decline of race hostility and an eventual willingness to restore political rights, might be the best strategy available. It was a sad commentary on the strength of American racism at the turn of the century that this analysis could not be easily refuted.

The Cape Suffrage and the Origins of Black Protest Politics in South Africa

Black protest politics in South Africa could not begin until British troops and white settlers had overcome the violent resistance of independent Africans against European conquest. Furthermore, reformist activity of a kind that was analogous to the African-American struggle for equality could develop only in constitutional contexts that provided some mechanisms for African political participation and representation. Without a sanctioned way of expressing concerns or presenting grievances and the possibility that complaints would be addressed, the only politics conceivable would have been the politics of rebellion or revolution. The necessary conditions for lawful black protest activity of a significant kind in southern Africa first came into being in the eastern Cape Colony in the late nineteenth century. By the 1880s, the Xhosa chiefdoms on the eastern frontier had lost their independence but had not suffered enslavement or a total denial of political rights. In fact some of their former subjects were invited to participate as voters in what white liberals and humanitarians claimed was a "color-blind" political system.

For reasons peculiar to the history of the Cape colony, therefore, the first African political leaders who adjusted to the reality of European domination focused on issues involving the suffrage that many of them believed were analogous to those then facing African-Americans. Even if they had not been drawn to the comparison for reasons of their own, they could not have avoided it; for white enemies of the Cape African franchise were inclined to use it against them. In 1891, J. H. Hofmeyer, leader of an emergent political organization of Afrikaners called the Bond, argued for new suffrage qualifications designed to reduce the number of African voters by pointing to the alleged consequences of black voting in the United States. It had led to "fraud, violence, and bloodshed; to a systematic falsification of the register." According to Hofmeyer, the federal government had been forced to "stand by and wink," while southern whites used force to keep blacks from the polls. The only way to prevent such a disaster from befalling the Cape was to diminish the African electorate.[36]

The circumstances that gave the suffrage to some recently conquered indigenes in a self-governing British colony were unique to the Cape. When the colony was granted representative government in 1854, the majority of the white population were descendants of Dutch colonists, many of whom resented British hegemony and the forced Anglicanization previously undertaken by colonial authorities. Since English settlers were a substantial but distinct minority, the American system of white manhood suffrage would have led to Afrikaner predominance. One way of preventing Afrikaners from controlling the new representative institutions was to admit a proportion of the Colored population to the electorate, on the assumption that they could be induced to support English-speakers against their former slavemasters. Of course English hegemony could also have been assured, at least temporarily, by setting a very high property qualification for the franchise. But this would have dangerously embittered the relatively poor but fiercely proud Afrikaner farmers, who had earlier rebelled against British rule or trekked away to form independent republics in the interior of South Africa, and was also opposed by missionaries and liberal imperial officials, who believed that some people of color were capable of becoming "civilized" and should be rewarded for their efforts. Although the analogy is far from perfect, there is some resemblance between the mixture of idealistic and self-interested motives that led northern Republicans to impose black franchise on the South after the Civil War and those which impelled English authorities and settler representatives to extend the franchise to people of color in the mid-nineteenth century Cape. In both cases, adherence to liberal ideals of equal rights helped to enable one ethnic or sectional division of the white population to maintain its hegemony over another.[37]

In the decades immediately after their enfranchisement, the Colored voters of the Western Cape failed to organize independently and define an interest of their own. They did not, for example, mobilize against the draconian Master-Servants law of 1856, which subjected many of them to quasi-serfdom on white farms. An extraordinarily heterogeneous population of uprooted, detribalized indigenes and ex-slaves of diverse origin, they mostly lacked a group consciousness that transcended paternalistic dependence on the local white notables who monopolized office-holding.[38]

It was otherwise in the Eastern Cape, which was an expanding frontier until the 1880s. The Xhosa tribesman of the Ciskei and Transkei fought no less than ten wars against white expansion, beginning in the 1780s and lasting for a century. Few indigenous populations put up a more determined and effective resistance against white colonial conquest. The recurring pattern was for a conquered area to be first ruled directly by the British crown and then absorbed into the Cape Colony. In 1865 the Ciskei was attached to the Cape, and the Transkei was incorporated piecemeal during the 1870s and '80s. The result was a vast increase in the Bantu-speaking African population of the colony and a smaller, but still substantial, increase in the nonwhite electorate. Since most of the new African voters helped to send English-speaking representatives to the Cape Parliament, it is not surprising that it was the political spokesmen for an increasingly self-conscious Cape Afrikaner community, namely Hofmeyer and the Afrikaner Bond, who began the agitation for suffrage restriction.[39]

One distinct group of newly incorporated Africans adapted with alacrity and enthusiasm to the white man's electoral contests. The most eager voters tended to be those who had been converted to Christianity and attended mission schools. Beginning in the early nineteenth century, Christian missionaries established stations throughout the territory that would become South Africa, especially in areas under British control. As the region came under British rule in the mid to late nineteenth century, the Eastern Cape and the Transkei became a veritable hotbed of mission activity, and the location of notable boarding schools for Africans, such as Lovedale, a Presbyterian institution, and Healdtown, a Methodist training center. The missionaries believed that conversion and "civilization" went hand in hand; peoples born in barbarism could not become good Christians until they had changed all their habits and become culturally British. Most Africans in the Eastern Cape as elsewhere did not readily forsake all of their traditional customs to adopt the white man's ways, and the process of conversion was slow and uneven. Even those who responded with apparent enthusiasm to a mission education were not as thoroughly transfigured as their teachers liked to imagine; for they remained African in many of their deepest feeling and loyalties.[40]

The best opportunities for establishing Christian communities under missionary direction were among people whose way of life had already been severely disrupted. Refugees from the devastating series of wars resulting from the expansion of the Zulu Kingdom in southeastern Africa in the early nineteenth century—the *mfecane*—were particularly susceptible to missionary influence; defeat and the weakening of chiefs and traditional authorities left a political and cultural vacuum that missionaries were eager to fill. The Mfengu people—who fled from Natal to the Cape Colony in 1835, allied themselves with the British in subsequent wars with the Xhosa in the eastern Cape, and were eventually awarded with a substantial grant of land—were favorite targets of missionary endeavor; a substantial portion of them became Christians and peasant producers involved in the market economy. The first of the African politicians who urged their compatriots to vote and get involved in Cape politics came from this ethnic group.[41]

White missionaries viewed their charges as cultural inferiors—people living in the darkness of barbarism and heathenism. At best their attitudes toward Africans were benevolently paternalistic, at worst arrogantly disdainful. They were also of course enemies of African independence and contributed significantly to the expansion of European hegemony. But in their own ethnocentric way, missionaries were champions of African potential, at least in comparison with most white settlers with their belief in unalterable black subservience and inferiority. British missionaries of the mid-nineteenth century, especially those who came from dissenting churches or from the Church of Scotland, shared many attitudes with the general run of white American abolitionists; they were likely to support movements for humanitarian reform, including the abolition of slavery and caste inequality. Although they had little or no respect for traditional African culture, many of them nevertheless affirmed that properly nurtured Africans were eventually capable of reaching a level of civilization that would make them worthy of civic and political equality with white colonists. For the most part, the early- to mid-nineteenth century missionaries were not racist in the strict sense of biological determinism. In their view, Christian belief, Victorian sexual morality, the Protestant work ethic, literacy in the English language, and adaptation to a market economy were inseparable elements of a civilized standard that Africans were capable of attaining.[42]

Some Africans fulfilled their expectations. This converted and "civilized" minority became the nucleus of an elite of Western-educated blacks who sought to modernize their more tradition-minded compatriots and prepare them for participation in the Cape's theoretically nonracial polity. The missionaries also imparted economic attitudes and skills that helped to give this class a material foundation for meeting the suffrage qualifications and for making alliances with influential whites. Missions often provided individual families with small lots for cultivation and encouraged them to produce for the colonial market. The eastern Cape in late nineteenth century thus became a growth area for a new kind of African peasantry. The market for farm produce began expanding rapidly in the 1860s and then went on to greater heights as a result of the diamond and gold rushes of the '70s and '80s. Much land on the Eastern Cape frontier remained in African possession, and the demand for foodstuffs was so great that it could not be met exclusively by white farmers. Even after some Africans began trekking to the mines of Kimberley and the Witwatersrand to labor under conditions that quickly grew oppressive, others were prospering by growing foodstuffs to meet the new demand created by a burst of urbanization and industrialization. In the eastern Cape a community of interest developed between the emerging class of peasant producers and the white middlemen or commercial capitalists who handled their crops and supplied them with consumer goods. Historian Stanley Trapido has argued that this nexus provided the interracial liberalism of the Cape with a material base that gave it greater staying power than if it had been merely an afterglow of mid-nineteenth century British humanitarianism.[43]

But the most prominent and articulate of the early post-conquest African political leaders were not likely to be personally engaged in peasant production. They were more apt to be the sons of prosperous peasants who had acquired more education than was needed for farming and sought careers of a more sedentary and cerebral kind. By the 1880s and '90s a small number of converted Africans had graduated from mission high schools and were finding employment as clerks, interpreters, minor civil servants, journalists, or even clergymen. They were the vanguard of a Western-educated intelligentsia that would assume leadership in the struggle for African rights, especially equal rights for "civilized" blacks, in a colonial order that they hoped or imagined would live up to the highest ideals of Victorian liberalism.[44]

The forerunner or founding father of this new intellectual elite was Tiyo Soga, son of a Xhosa headman, who attended the Presbyterian school of Lovedale in the years immediately after its founding in 1841. When a frontier war forced the temporary closing of the school in 1846, he was sent to Scotland for further education at Glasgow University and the United Presbyterian Theological Seminary in Edinburgh. He returned to the Cape in 1857 with a degree in theology, ordination as a Presbyterian minister, and a pious Scottish wife. As the first black clergyman in South Africa, he confined himself for the most part to his duties as a missionary and generally avoided speaking out on political issues. Fully sharing the mission philosophy that no civilization was possible without Christianity, he was in no position to object to European domination and imperial expansion so long as exposure to the gospel went with it. But in 1865 he reacted angrily to an article in a missionary journal in which one of his white associates, discouraged by the slow progress of mission education, predicted the extinction of "the Kaffir race" because of its irredeemably "indolent habits." Soga replied publicly that God had promised Africa to "Ham and his descendants" and that the sheer numbers of blacks in Africa made their ultimate disappearance extremely unlikely. In an early anticipation of Pan-Africanism, he also noted that people of African descent had multiplied in the New World despite the most oppressive circumstances. He also made specific reference to the ordeal of newly emancipated blacks in the United States—possibly the first such reference by a black South African: "I find the negro in the present struggle in America looking forward—though still with chains on his hands and chains on his feet—yet looking forward to the dawn of a better day for himself and his brethren in Africa." This expression of black consciousness from the most "Westernized" black man in South Africa at the time reveals the danger of exaggerating the extent to which converted, educated, and "progressive" Africans of the nineteenth century turned their backs on the African masses and identified with Europeans. Soga sent his half-white sons to Scotland to be educated but also encouraged them to consider themselves Africans rather than Europeans. His youngest son, Alan Kirkland Soga, later became a political activist and a central figure in the organized African protest of the early twentieth century.[45]

Tiyo Soga's lack of interest or involvement in Cape politics was characteristic of the small Christianized elite before the 1880s. As clients of white missionaries, members of white-dominated mission churches, and (in some cases) collaborators with British authorities in the conquest of the last independent African societies of the Transkei, they accepted paternalistic white domination on the assumption that they would be rewarded for their loyalty to the cause of "Christian civilization" and find a place worthy of their talents in the new colonial order. But in the 1880s there were ominous signs that support for liberal, assimilationist policies among the British colonizers was declining. In the churches the impulse to ordain black ministers and give them independent authority began to weaken as the result of an increase in racist attitudes among missionaries. Also, a section of the white settler population that was viewed by Africans as peculiarly hostile to their rights began to organize themselves in an effort to exert more influence on the Cape parliament. During the early years of representative government in the Cape, Afrikaners had been almost as passive as people of color in their relations with a British-imposed and -dominated electoral system. But in the 1870s, there was an upsurge of ethnic consciousness among Cape Afrikaners culminating in 1879 in the founding of the Afrikaner Bond. In 1883 this ethnic political organization came under the astute and relatively pragmatic leadership of Jan Hofmeyer and became the first organized political party in the Cape. Africans viewed the Bond as a threat because it advocated a reform of the suffrage to reduce the electoral role of non-Europeans.[46]

Most of the black politicians operating within the late nineteenth-century colonial system showed a strong partiality for the British side in the developing Anglo-Boer conflict in southern Africa. Noting that the Afrikaners in their independent republics denied all political and civil rights to blacks, while the British in the Cape and Natal at least paid lip service to the idea of "equal rights for every civilized man" regardless of race, they tended to be on the lookout for any signs of an Afrikaner conspiracy to reduce them to servitude. The homes of most Christianized and educated Africans in the late nineteenth and early twentieth century had a picture of Queen Victoria on the wall, just as many black homes in post-Civil War America honored Abraham Lincoln. Analogous to the way many African-Americans of the Reconstruction era and afterwards identified Lincoln, the North, and the Republican party with freedom and the defense of their rights against the white supremacists of the South, most "Progressive" Africans viewed the Queen, the British empire, and the English-speaking liberal politicians of the Cape as their defenders against the unvarnished racism of the Afrikaners.

It was in response to the growth of the Afrikaner Bond that members of the educated elite of the Eastern Cape came together in Port Elizabeth in 1882 to form the first black political organization in South Africa—Imbumba Yama Nyama (literally "hard, solid sinew," or tightly unified group.) The announced purpose of the new organization was to unify Africans politically so that they could fight for their "national rights." Immediate concerns of Imbumba members included overcoming sectarian divisions among African Christians and protecting the franchise. Reacting to the move among Afrikaners to raise the qualification for the suffrage, an Imbumba conference of 1883 passed a resolution to be submitted to Parliament requesting that "the franchise should be as it is at present and not raised, so that browns [Africans and Coloreds] may always have the right of voting." This pioneer effort to organize Africans for political assertion was apparently a bit ahead of its time, for Imbumba disappears from the historical record after 1884. But the precedent would be remembered.[47]

A founder of Imbumba and the key figure in the birth of African protest politics in the Cape during the 1880s and '90s was John Tengo Jabavu, a second-generation Methodist convert of Mfengu ancestry. Born in 1859, Jabavu attended the Methodist school at Healdtown, qualified as a teacher, taught a few years in mission schools, and then embarked on a career in journalism as editor of the Xhosa-language edition of a missionary newspaper. In 1884, he became involved in electoral politics, serving as canvasser for a white parliamentary candidate who was seeking the African vote. Shortly thereafter, he accepted help from his white political patrons to found his own newspaper, *Imvo Zabatsundu* ("Native Opinion"), the first black-owned and -operated journal in South Africa. *Imvo* became a forum for African viewpoints and was quick to protest government actions viewed as contrary to black interests—although the most radical opinions were generally confined to the Xhosa columns of its bilingual format. In 1887 *Imvo* led the unsuccessful fight of politically active Africans against the Parliamentary Voters Registration Act, which curbed the growth of the African electorate by requiring individual rather than communal tenure to meet the property qualification for the suffrage. One of its editorials protested that "by eliminating the native factor," it "establishes the ascendancy of the Dutch in the Colony forever." In 1894, Jabavu opposed the Glen Gray Act, a measure which offered individual tenure to a minority of Africans in areas of the Transkei while effectively reducing the majority to a landless proletariat of potential migrant workers. But Jabavu's close alliance with a group of white liberal politicians who shifted their positions on African rights in response to the exigencies of the Cape's politics of faction and coalition, led him to waver in his commitment to African voting rights. In 1892, at the behest of his patrons, he failed to protest against the Franchise and Ballot Act, which further reduced the African role in Cape politics by tripling the property qualification for voting. In later years, Jabavu would continue to follow a zigzag course in deference to his white political allies.[48]

Opposition to Jabavu's less-than-consistent advocacy of African rights developed slowly during the 1890s. But when the white "friends of the natives" with whom he was closely associated dissented from the imperialist position in the confrontation between Britain and the Transvaal republic that preceded the War of 1899–1902 and formed an alliance with the Afrikaner Bond, Jabavu followed their lead and thus found himself at odds with the many educated Africans who strongly supported the cause of British hegemony over the Afrikaner republics. In 1898, his opponents, led by men such as Alan Kirkland Soga and the Rev. Walter Rubusana, founded a rival newspaper, *Izwi Labantu* ("Voice of the People"). This journal was financed initially by Cecil Rhodes, whose own political machinations at the time required wooing African voters. A two party system was emerging in the Cape in the late '90s, with the British imperialist Progressive Party of Rhodes opposing the South African Party (SAP) of Hofmeyer, now a coalition of Cape Afrikaners and most of the old-line English-speaking Cape liberals. Competition between the jingoistic Progressives and the SAP advocates of peaceful reconciliation between the British empire and the Afrikaner republics enhanced the value of the black vote but also divided it. Rhodes was at heart a white supremacist and had been generally hostile to the black suffrage, but in the election of 1898 the need for African and Colored votes to bring the Cape parliament behind his expansionist policies led him to coin the slogan that summed up the essence of Cape racial liberalism and provided a rallying cry for African progressives in the years to come—"Equal rights for every civilized man south of the Zambezi." (This was a self-conscious revision of his earlier demand, aimed at the Transvaal for its disfranchisement of English-speaking immigrants, for equal rights for all white men in southern Africa.)[49]

The founders of *Izwi* also formed an organization known as the South African Native Congress (SANC), which began in 1898 as an electoral convention supporting Rhodes's Progressive party but by 1902 had developed a permanent character and a broader set of aims. During the South African War, Jabavu held so firmly to a neutral position that the authorities closed down his paper for a time because of its alleged disloyalty to the British empire. *Izwi*, on the other hand, strongly supported the imperial cause, but did so on the assumption that a British victory would extend the Cape nonracial franchise to the Afrikaner republics, which had previously limited the vote to white men. This hope for the extension of equal rights to "civilized" blacks throughout a British-dominated South Africa reflected a general feeling among Africans and Coloreds that accounts for the substantial contribution—some of it military despite the pretense that blacks were not to be used as soldiers by either side—that they made to the British war effort. The establishment of the SANC on a permanent basis in 1902 may have been inspired by a fear that these expectations were being betrayed. Despite the fact that British war propaganda had attacked the republics for denying equal rights to nonwhite British subjects, the treaty of Vereiniging that ended the conflict provided that "the question of granting the franchise to the natives will not be decided until after the introduction of self-government," a provision that meant in effect that it was up to the white settlers rather than the imperial government to decide whether blacks would vote or not.[50]

In 1903, SANC petitioned the British Secretary of State for the Colonies on behalf of "the Natives and Colored People Resident in British South Africa" and set forth in detail their grievances on the suffrage and other matters. This document deserves close attention because it is the fullest statement of the political orientation of the Cape African elite in the wake of the South African War and set the general pattern for African protest politics in the early twentieth century. It began with a strong affirmation of loyalty to the British government and expressed the gratitude of African Christians to the British people for bringing "the Gospel of Salvation" to "the people that sit in darkness and the shadow of death." Reacting to white fears that the recent establishment of independent black churches portended a bid for political independence, it affirmed that "the black races are too conscious of their dependence upon the white missionaries, and of their obligations toward the British race and the benefits to be derived by their presence in the general control and guidance of the civil and religious affairs of the country to harbour foolish notions of political ascendancy." On the suffrage question, however, the SANC registered profound disappointment at the turn of events. Noting the promises made during the war for the extension of "equal rights to every civilized man," blacks had assumed that the British government was "bound to protect the rights of all classes in the reconstruction and admission of the newly-formed states into the Union." But the articles of peace suggested otherwise and constituted a surrender of Crown responsibilities and principles to a coalition of British and Afrikaner settlers which was "aiming at the elimination of the imperial factor to the ruin of the subject races of his gracious majesty." The petition was polite, almost obsequious, and did not charge deliberate betrayal; its strategy of persuasion was meant to suggest that the British government had made a mistake that it would surely be willing to rectify if it had all the facts. But a note of bitterness and frustration appeared in the way the petitioners' dampened hopes were described:

> We were of the opinion that conditions had undergone a change and that the Natives were no longer to be looked upon as a class for special and exclusive treatment, or to be governed by a policy of continued suspicion. We thought that they were now to be received with confidence within the imperial family circle as true citizens of the empire, and that the doors of the Temple of Peace would be thrown widely open so they might enter freely in with their white brethren to share in the coming prosperity which has been so eloquently described by the great advocates of the Commonwealth.[51]

When the Transvaal was granted responsible government in 1906 on the old republican basis of universal manhood suffrage for whites only, making it clear that the British government was firmly committed to ceding authority over the black population in South Africa to a self-governing white settler community with an Afrikaner majority, African disillusionment with British justice and benevolence increased. But it would take a long time for members of the educated black elite to give up all hope of inducing the British government to intervene on behalf of equal rights. For Africans in the Cape, the fact that their own qualified franchise survived the unification of South Africa took some of the edge off their discontent and gave their voteless compatriots in the other provinces some basis for hoping that the Cape liberal franchise would eventually be extended to the rest of South Africa. It was not until Cape Africans were removed from the common voters roll in 1936 that the Cape precedent and faith in the British egalitarianism that it allegedly embodied lost its power to define the goals of organized African political activity.[52]

This demand for equal political rights under British hegemony should not be mistaken for a cultural assimilationism that attached no importance to African distinctiveness and anticipated its eventual disappearance. Like Tiyo Soga, early Cape political leaders such as Jabavu, A. K. Soga, and Walter Rubusana were proud to be Africans and thought of themselves as showing their people the way to self-determination under the only conditions available. As historian André Odendaal has argued, "the so-called 'school people' or educated elite who participated in the new western forms of politics were more closely tied to their own communities, and much more concerned with traditional matters than has been realized." The minority of Africans who could vote did not usually do so as isolated individuals or as members of a privileged caste, but rather in accordance with the will of the community as registered by political meetings that included nonvoters. Being Christian, and therefore opposed to the traditional customs like polygamy that were condemned by missionaries, the modernizing elite did have to reckon with the distrust and antagonism of unconverted; and there was tension between the old authority of chiefs and the new authority that came from education and private property. But the usual attitude of the new elite toward African tradition was not simple rejection; more commonly the aspiration of the modernizers was to preserve those elements of traditional culture that were compatible with Christianity and liberalism, and they usually found much that could be salvaged and purified. There is no evidence whatever that they devalued African ethnicity as such; equal rights under the British Crown meant to them the prospect of political rights as a people or nation within a commonwealth composed of culturally diverse peoples and nations sharing a common set of political principles. Those principles may have been British in origin but were thought to be universal in application.[53]

In the Western Cape, the Colored population began to stir politically around the turn of the century in ways that paralleled the birth of African protest activity in the eastern part of the colony. In 1892, Coloreds in Kimberley, the center of diamond production, attempted to form a colony-wide organization to protest proposed legislation to raise the property requirement for voting, the first franchise restriction that directly affected Coloreds. They sent a petition with over ten thousand signatures to parliament, but it could not prevent passage of the Franchise and Ballot Act. They then attempted to organize Colored voters against candidates who had supported suffrage restriction, but after one notable success in 1894 the organization faded away. One special problem that inhibited the efforts of Coloreds to mobilize politically was the imprecise and permeable color line in the Western Cape. It was so easy for successful and relatively light-skinned people with nonwhite ancestry to pass over into the European population that it was difficult for a group-conscious leadership to emerge.[54]

But after 1900 there was an upsurge of Colored consciousness, especially in Cape Town, that led to the formation of the African Political Organization (APO) in 1902. It resulted in part from an increase in white discrimination against Coloreds, including the first efforts to segregate them in public facilities, and in part from the concern for the effect of the South African War on the rights of Coloreds who had migrated to the Afrikaner republics. As part of the propaganda campaign leading up to the war, British spokesmen had publicly condemned the denial of civil rights to Colored British subjects resident in the Transvaal and the Orange Free State. Like Africans, Coloreds felt betrayed by a peace settlement that failed to assure them that even the semblance of equality that they enjoyed in the Cape would be extended to the rest of what was now British South Africa. The resulting mobilization of the small Colored middle class under its own educated elite exhibited the tensions and contradictions created by the group's racially marginal status. Colored political leaders—like Abdul Abdurhahman, the Cape Malay physician who was the dominant figure in the APO from shortly after its founding until his death in 1940—were willing at times to cooperate with Africans in a common struggle for Cape liberal principles, but were more immediately concerned with the rights and privileges of their own group, which in South Africa's emerging three-tiered racial system gave them some advantages over Africans. Unlike the African organizations, the APO tended to embrace the goal of total assimilation into European society; in other words, Colored politicians and intellectuals usually lacked the sense of nationality or distinctive peoplehood that was a virtually inescapable component of African consciousness.[55]

Nevertheless, there was one tendency in the early twentieth-century Colored search for identity that reached out to a larger world of black and brown people—a kind of pan-Negroism that was stimulated by contact with black Americans or black American ideas. A key figure in the agitation that led to the founding of the APO was F. Z. S. Peregrino, a West African of apparently mixed racial origin who had been educated in England and had lived in the United States during the 1890s. After a brief career as a journalist and editor in Buffalo, New York, he attended the first Pan-African Congress in London in 1900 and then emigrated to Cape Town with the intention of spreading Pan-African ideas among the Colored population. Besides attempting to found race-conscious organizations, he established a newspaper, the *South African Spectator*, in which he reported on the struggles of black people throughout the world and especially in the United States. An admirer of both Booker T. Washington and W. E. B. Du Bois (he would be distressed when they had a falling out), he encouraged Coloreds to emulate their American cousins and organize themselves along race lines in order to press more effectively for equal rights. He also called for solidarity with Africans who were engaged in a similar struggle. Identifying fully with Washingtonian ideas of group progress, he regarded a qualified but genuinely impartial suffrage as an incentive for black self-improvement, a reward for advancing on the path to full equality with whites. According to a leading historian of Cape Colored politics, Peregrino's moderate Pan-Africanism and emphasis on racial pride and self-help "proved a catalyst in Coloured political organization in Cape Town," even though he did not—presumably because of his somewhat marginal status as a recent immigrant—play a major role in the founding of the APO and in its subsequent activities. His paper continued for several years to promulgate race pride and solidarity among the Colored population and thus helped to counter the eagerness of many members of the Colored elite to identify completely with Europeans and work merely for a downward shift of the color line to include them, but not Africans, among the ruling caste.[56]

Suffrage Struggles: Connections and Comparisons

Juxtaposing American and South African narratives of how blacks fought to protect the ballot against impending restriction or disfranchisement reveals some obvious differences in the context and character of these struggles. South Africa had not experienced a Radical Reconstruction and had no tradition of one man, one vote —although the Afrikaner republics had reproduced the antebellum American pattern of one white man, one vote. In the Cape, where some people of color had actually been enfranchised in the nineteenth century, the idea of a qualified suffrage based on education and property was well established and was in principle acceptable to the black political elite. Hence there was a radical-democratic thrust to the voting-rights campaign of African-American abolitionists such as Frederick Douglass and neo-abolitionists such as W. E. B. Du Bois and William Monroe Trotter that was lacking in the relatively more conservative efforts of Cape Africans and Coloreds to extend a qualified, color-blind franchise to the rest of South Africa after the defeat of the Afrikaner republics by the British. Unlike blacks in the abolitionist tradition, Booker T. Washington was willing to settle for a restricted, impartial franchise in the South, but in the American context such a suffrage policy represented a strategic retreat and an accommodation with white supremacy rather than defense of an equal-rights tradition.

Douglass and his successors in the American agitation for equal political rights believed that the mass of blacks would learn citizenship through the exercise of the ballot, while Jabavu, A. K. Soga, Abdurahman, and other black politicians in the Cape generally accepted the idea that the majority of their people needed to reach a higher level of "civilization" before they could be entrusted with the vote. They shared the view of a delegation of African clergymen from the Orange Free State who told the South African Native Affairs Commission in 1904 that the vote should not be given to "those natives who are still in darkness," but only "to those who are progressive in their minds and civilized in their modes of living."[57]

If one accepts the view of black leaders in both countries that there were universal standards of progress and civilization that Europeans had first attained but which blacks could and would reach in due time, this difference in attitudes toward the suffrage can be explained in part by perceptible differences between the black populations of the two countries. By 1900 a majority of American blacks were literate, a substantial and growing proportion possessed private property, and most were adherents of Christianity. Only a minority of Africans possessed one or more of these "civilized" attributes. This difference accounts for the common tendency of educated black leaders in South Africa to point to African-Americans as models of black progress. But they were also aware of the intense white racism aroused by the general enfranchisement of southern blacks, which was now threatening the achievement of equal rights in the United States. This knowledge made many of them tolerant of Booker T.

Washington's strategic retreat on the suffrage issue and also made them think twice about demanding more than the most liberal segments of British and white South African opinion seemed willing to grant. In other words, they resembled Washington in their reluctance to challenge directly the principle of white or European hegemony.[58]

As early as 1886, Jabavu's pioneer journal of African opinion, *Imvo Zabantsundu*, featured a debate on the American example. The Reverend P. J. Mzimba, a black Presbyterian missionary following the footsteps of Tiyo Soga, called on Africans to refrain from political activity and concentrate on education. Politics, he argued would offend white missionaries and impede their essential efforts to uplift Africans. To support his case, he cited the similar advice George Washington Williams had given to African-Americans in his *History of the Negro Race*. "Let the experience of Africans in America give warning in time to Africans in Africa to let politics alone at present," Mzimba advised. "Let us be content to be ruled by colonists. Let us only have to do with politics in order to encourage those white men who desire to give us schools and books." *Imvo* editorialized against Mzimba's analogy, arguing that the unsettled conditions that made black suffrage problematic in the United States did not exist in the Cape colony, where political involvement did not carry the same risks and in no way conflicted with an emphasis on education. In subsequent editorials of the late 1880s, Jabavu presented the African-American example of economic self-help and independence as a model for his own people to follow—without, however, subscribing to the view, soon to be fully articulated in the United States by Booker T. Washington, that they had to deemphasize politics in order to do it. Because the Cape African suffrage was restricted to "civilized natives" and did not threaten white domination of the colony, its exercise and preservation seemed to harmonize fully with a philosophy of gradual advancement through economic self-help and education that accommodated to white rule.[59]

When the possible extension of the Cape franchise to the rest of a unifying South Africa became the main preoccupation of African leaders around the turn of the century, American franchise struggles assumed greater relevance. The most incisive comparative analysis of the black suffrage question in the two societies came from the pen of Alan Kirkland Soga, a leader of the South African Native Congress and editor of *Izwi Labantu*. An unjustly neglected figure in the history of black politics in South Africa, Soga was perhaps the most significant and influential African intellectual of the first decade of the twentieth century. In a sense, he played W. E. B. Du Bois to Jabavu's Booker T. Washington. The youngest son of the Reverend Tiyo Soga, he was educated in Scotland under the supervision of his then-widowed Scottish mother, studying law and humanities at Glasgow University. Returning to South Africa, he identified with his father's people and embarked on a career in the civil service of the Cape Colony at a time when there was no official bar against blacks in responsible government jobs. But Soga soon discovered that there was an unofficial color bar that blocked his advancement to positions for which he would have qualified had he been white. He resigned angrily from the civil service to take up the editorship of *Izwi* in 1898. He was the principal author of the previously cited SANC petition of 1903 protesting the treaty of Vereininging and was undoubtedly responsible for the allusions to American developments that it contained. In 1903 and 1904, he wrote a series of articles for his journal that were reprinted in the *Colored American Magazine* under the general title "Call the Black Man to Conference." In these early examples of Pan-African discourse, he reflected on the black experience in the United States and South Africa, with particular attention to suffrage issues, and searched for general principles that would apply to both peoples.[60]

The SANC petition condemned the failure of juries to convict whites accused of murdering Africans, warning that such injustices would "encourage a bitter and lasting hatred such as exists between White and Black in the United States of America." It also supported a complaint against discriminatory treatment of blacks in the civil service (such as Soga himself had apparently experienced) by citing President Theodore Roosevelt's recent defense of his policy of appointing blacks to federal offices in the southern states. Roosevelt (who would soon backtrack from these principles) affirmed that "it is a good thing to let the colored man know that if he shows in marked degree the qualities of good citizenship—the qualities that in a white man we feel are entitled to reward—that he will not be cut off from all hope of future reward." On the suffrage question itself the petition quoted black Republican politician P. B. S. Pinchback's warning that "whenever colored men have been deprived of the ballot, unjust class legislation has speedily followed, race antagonism has been intensified, and lawlessness and outrage against the race increased." The lesson for South Africa was that a denial of the suffrage to Africans in the Transvaal and the Orange Free State would lead to a continuation of the ill-treatment and flagrant oppression characteristic of the prewar Afrikaner regimes. Implicit here was the conception of the ballot as a means of group protection against whites with a heritage of slavery and blatant racism—the same argument that Douglass and other black Radicals had used to justify the extension of suffrage to southern freedmen after emancipation.[61]

Soga's information about American affairs seems to have come in large part from his reading of the *New York Age*, a prominent African-American newspaper, which was then being edited by the redoubtable T. Thomas Fortune. The articles from *Izwi* that Soga contributed to the *Colored American Magazine* quoted Fortune extensively, referred to him in the most admiring terms ("among race leaders Mr. Fortune occupies a distinguished place"), and generally endorsed his perspective on what was happening in black America.[62]

Fortune at this time was allied with Booker T. Washington and was defending the principal of Tuskegee against black critics of his accommodationist policies. But Fortune's position was not identical to Washington's.

Although he strongly supported an emphasis on industrial education and believed that blacks in the South had to cultivate good relationships with local whites—views that he had held long before he became associated with Washington—he condemned lynching and disfranchisement in the kind of strong language that Washington generally avoided. He politely dissented from Washington's endorsement of a qualified suffrage and in fact supported Washington's opponents on this issue: he wanted federal legislation to guarantee a fair and equal suffrage for all Americans. But, partly because he was beholden to Washington for financial support and partly because he genuinely admired him and accepted his educational philosophy, he had no use for those who disrupted Washington's speeches and accused him of betraying the race. Fortune was a strong proponent of race unity. In 1890, he had founded the Afro-American League, which he hoped would incorporate state and local organizations into a national effort to defend black political and civil rights. After that organization failed to get off the ground, he joined with other influential black leaders in 1898 to form a successor group with similar objectives called the Afro-American Council.

The Council had greater influence and staying power than the League, mainly because it had the patronage and support of Booker T. Washington. But Washington did not back organizations that he could not control, and the Council soon became his creature, which made it less militant and protest-oriented than Fortune's original League. In Fortune's view it was better to be unified behind a moderate program than for the race to be divided. His *modus vivendi* was in effect a dual strategy that prescribed Washingtonian accommodationism for blacks in the South but permitted northern blacks to protest vigorously against southern developments that threatened the basic rights of blacks as set forth in the Reconstruction amendments. The *New York Age* denounced lynching, disfranchisement, and Jim Crow laws, while at the same time lavishing praise on Washington and his educational program. (Washington did not in fact require his northern allies and clients to adopt his own, cautious accommodating stance). When radical northern blacks began to attack Washington and question his southern strategy, Fortune condemned them for undermining a prestigious black leader who was doing valuable work under difficult circumstances; from his point of view, they were endangering the unity and tactical flexibility essential to the progress of the race.[63]

The first of Soga's articles in the *Colored American Magazine* reviewed for South African readers the emerging controversy between Washington and the group of protest-oriented northern blacks who would soon be led by W. E. B. Du Bois. Although recognizing some merit in the criticism of Washington's educational philosophy as narrow and philistine, Soga nevertheless found Washington to be a benefactor to the black race: "Mr. Washington has no claim to a monopoly of greatness among the many bright stars in the Afro-American firmament, who are renowned in their own spheres; but in the realm of industrial education he is peerless, and in the inculcation of duty and the dignity of labor, and the practiced application of brains to manual work, he has no equals." He also praised Washington for his "attempts to conciliate black and white, to appease angry passions, and to lead men's thoughts away from the turmoil and strife fostered by continual political agitation." Not surprisingly, therefore, Soga condemned in the strongest terms those African-Americans who had recently disrupted an address of Washington in Boston and deprived him of the right to speak. "Unity for common objects, and the common safety, is the only hope of the black man's salvation," he wrote, and this need for solidarity meant that "the black man cannot afford to see the leaders of the race subjected to insult, or degraded by members of the race without the most solemn protest." He then quoted at length another black South African editor, F. S. Z. Peregrino, who was dispensing similar judgments to the Coloreds of the Western Cape. Peregrino compared the squabbles among black Americans to "the vagaries of the mythical kilkenny cats" and described the attack on Washington as an indication that black Americans had failed to develop " 'Race Pride.' "[64]

The reactions of Soga and Peregrino reflect the enormous international prestige of Washington as an exemplar of black celebrity and achievement. But when Soga turned in subsequent articles to the question of the franchise, he distanced himself from Washington and embraced Fortune's position, which he paraphrased as realization that "the franchise is too important to be left to the whim of the states." He then quoted the editor of the *New York Age* at length on why he disagreed with Washington's willingness to allow state governments to set property and educational qualifications for voting. Fortune's view was "that the suffrage question is the basic principle of national citizenship, that it should therefore be controlled absolutely by the Federal Government," and that "in a democracy like ours," all men who are liable for taxation and military service "should have an equal voice in the election of those who spend his taxes and levy wars that he must fight, and make laws regulating his conduct in the social and civil compact."[65]

Soga attached enormous importance to the suffrage. "The ballot," he wrote, "is a guarantee of protection, and in that sense is the Palladium of the People's liberties. Its denial or withdrawal from any class may be regarded as opening the door to oppression, crime, and virtual slavery." Nevertheless, he had some difficulty applying Fortune's radical-democratic principle to South Africa, because all that he and SANC were calling for was an extension of the Cape's qualified suffrage to the other British colonies. Acknowledging "the weakness of the Cape Colony franchise," because it did not even give the vote to all blacks who paid taxes, Soga nevertheless refrained from calling for a lowering of the qualification and invoked Washington to justify such prudence. At a time when white liberals in the Cape were advocating a nonracial franchise to the rest of South Africa but were also willing to consider *raising* the qualifications to make it more palatable to the white settler communities, Soga found "no call for black men to agitate the question on the matter of rights, for there are some things that may be right in principle, but it is not always expedient to demand them. The force of this truth is more apparent to Booker Washington than to some of his countrymen. . . ." Where Fortune's doctrine was immediately and directly applicable was to the question of *who* should control the suffrage. Using the language of American federalism, Soga reported that the SANC suffrage petition had "urged that the question of the enfranchisement of His Majesty's native and Colored subjects should not be left to the decision of the States." Language from the petition itself referred to the former republics as "proposed Federal States" in the process of "reconstruction and admission . . . into the Union" under a "British Constitution" that provided equal rights to its black subjects. In Soga's thinking, Britain was equivalent to the federal government of the United States and the newly incorporated colonies were analogues of the southern states that had failed in their bid for secession and then had to be reconstructed under the egalitarian political principles that allegedly set the terms for unification. The danger was that Britain would follow the American example and fail to enforce its constitutional principles in recalcitrant states or provinces: ". . . surrender of a principle in the Transvaal, hitherto recognized by the oldest colony—the Cape of Good Hope,—will be to reproduce, in course of time, similar conditions to those existing in the Southern States. . . ."[66]

Soga's views of 1903 and 1904, and those of the South African Native Congress for which he was principal spokesman, closely paralleled those of relatively pragmatic northern African-American leaders who tried to support Booker T. Washington's work in the South without forgoing their right to protest or conceding the principle of federal protection of the ballot. Fortune was the most conspicuous of these, but others included Bishop Alexander Walters of the African Methodist Episcopal Zion Church, who was first president of the Afro-American Council, and Kelly Miller, dean of Howard University and the nation's most prominent black academic. Before he began his direct attack on Washington in 1903, W. E. B. Du Bois had shared this basic perspective. Men like Fortune, Walters, and Miller did not follow Du Bois into the neo-abolitionist camp of William Monroe Trotter, partly because they attached a high value to race unity and strategic flexibility, believing that both accommodationist and protest strategies had their place in the larger black movement.[67]

This inclusive or "all-in" approach was the theory behind the Afro-American Council. Such hopes for a federated unity of northern and southern black leaders would soon be a casualty of the Washington-Du Bois controversy. But before the militant opposition to Washington crystallized in the Niagara Movement of 1905, the Fortune-Walters-Miller concept of a diversified but coordinated struggle could serve as a model for Soga, Peregrino, and like-minded black South Africans. A biographical sketch of Soga in the *Colored American Magazine* of February 1904, which appears to have been written by Soga himself despite its nominal authorship by a member of the magazine's staff, likened the aims of the South African Native Congress to those of the Afro-American Council: "Unfortunately, the higher interests of the natives are suffering by the divisions over politics and the native press, and it is sincerely hoped that for the protection of these larger interests the Native Congress, *which is a body akin to the Afro-American Council*, will be able to draw the most intelligent classes together in unity and social cooperation."[68]

From Soga's perspective, the disrupter of unity among Cape Africans—the equivalent of the Trotter and the Boston rebels in the United States—was John Tengo Jabavu, who in 1903 and 1904 stood aloof from the Congress, repudiated in principle the idea of separate black political organizations, and supported the white political party that sought unity between English and Afrikaners rather than the one that stood for the British hegemony that Soga believed was the only hope for extension of the Cape liberal suffrage throughout South Africa. Jabavu did protest against the treaty of Vereininging and defend the Cape franchise, but his strategy was to curry favor with prominent white politicians rather than join in a separate black political movement that he believed would serve only to alienate whites. One is tempted to compare his position on black politics with that of Frederick Douglass toward the end of his career. Douglass had refused to join the Afro-American League because of a similar opposition to forming separate national organizations for quasi-political purposes. Both Douglass and Jabavu favored a political integrationism that precluded independent action of the kind favored by Fortune and Soga. But there were differences between the political contexts in which the two men operated that explain why Douglass usually projected the image of a protest leader while Jabavu appeared at this time to be playing an accommodationist role. Douglass backed a political party that at one time had fought for the legal and political equality of blacks, and he never gave up hope of reviving the spirit of Reconstruction-era Radical Republicanism. Jabavu worked with white politicians who were liberal paternalists and pragmatic defenders of white supremacy. Both were leaning on weak structures, but Douglass at least had some claim on a tradition within his party that promised more than a token suffrage for a "civilized" black minority.[69]

From the vantage point of historical hindsight, Soga's broader analogy between the circumstances and possibilities of black politics in the United States and South Africa appears to have been an unstable compound of insight and illusion. His likening of the British Empire to the American Union as an arena within which blacks could gain political rights was, time would show, seriously defective. The British empire was moving toward decentralization; London would take progressively less responsibility for the rights of people of color in the commonwealths to which it was according self-government. Once South Africa was defined as an autonomous white settler society like Australia or Canada, the hopes of blacks for equal rights as royal subjects were doomed. Furthermore, the South African Union that came into being in 1910 was not the kind of federal union under the dominance of English-speakers and British traditions that Soga anticipated. Rather it was a unified state with an Afrikaner majority and an English minority that had its own fiercely white supremacist component. The Cape franchise survived for a quarter of a century after union, but only as a local exception to the general practice of excluding blacks from the electorate. The United States on the other hand was moving toward centralization; federal power was increasing at the expense of the states. It took more than half a century for the centralization of government to manifest itself in the enforcement of the equal suffrage rights for all citizens, regardless of race, as prescribed by the Fifteenth Amendment; but the growth of federal activism that began in the Progressive era and accelerated during the New Deal was a precondition for the success of the voting rights movement of the 1960s. Hence Soga's assumption that the political-constitutional contexts of the two struggles were comparable in the possibilities that they offered for black enfranchisement was wishful thinking.[70]

Soga showed more prescience when he warned that if the suffrage was denied to Africans and people of color in a South African union they would suffer a further erosion of their rights and be reduced to "virtual slavery," as appeared to be happening to blacks in the American South in the wake of disfranchisement. The steady expansion of segregation and legalized discrimination after 1910 were a realization of his worst fears.

Soga's belief in the ballot as the bedrock of black progress was what made African-American discourse seem to him so relevant to his assessment of the black situation in South Africa. Like black American leaders from Frederick Douglass to W. E. B. Du Bois and beyond, he did not conceive of the right to vote as simply the chance for blacks as individual citizens to participate in governing themselves. In a context of white prejudice and discrimination, suffrage might provide the race as a whole with a weapon to resist oppression and the means to create an environment in which collective efforts for social, economic, and cultural betterment would have some chance of success. This perspective enabled the oppressed group to develop a sense of pride, patriotism, and self-interest, while at the same time looking forward to participating with whites as equal citizens of a common polity. It was thus an integrationist nationalism—to be distinguished from pure assimilationism on the one hand and separatist nationalism on the other—that dominated the thinking of protest leaders in the United States and South Africa at the turn of the century.

The hopes for a color-blind democracy in which distinctive groups could remain different without being unequal was linked to an optimistic vision of human progress and enlightenment. Differences of opinion on political tactics and strategy may obscure the fact that black leaders at the beginning of the twentieth century—Douglass, Washington, Fortune, Trotter, and the young Du Bois in the United States; Soga and Jabavu in

South Africa—shared not only a sense of race pride and loyalty but also a liberal Victorian view of human history as the march of humanity toward a higher level of "civilization." The movement of the West from feudalism, aristocracy, and superstition toward democratic capitalism, civic equality, and an enlightened Christianity compatible with modern science and technology was seen as the proper and inevitable path for black people and for multi-racial societies with a history of slavery and white supremacy. Such values were not regarded as "Western" in any racial or ethnic sense but as universal, reflecting the God-given capacity and destiny of all the nations and races of the world.[71]

Events of the twentieth century would soon challenge this liberal-progressive faith and make it harder to sustain. As the status of black people in the United States and South Africa deteriorated in the early years of the new century, liberal progressivism began to seem less compelling as a political ideology on which blacks could base their hopes for freedom and self-determination. Both Fortune and Soga later became disillusioned with liberal values and modes of action. Fortune's growing sense of the strength of American racism led him by the early 1920s to an association with Marcus Garvey's Universal Negro Improvement Association and the editorship of the Garveyite organ *The Negro World*. Although Fortune continued to urge African-Americans to fight for their constitutional rights, this affiliation signified a deemphasis on liberal reformism in an American political context and the willingness to collaborate with a separatist form of nationalism that he would found have anathema in earlier years.[72]

As white settler hegemony over South Africa became more and more inescapable, Soga began to wonder if the British surrender of African rights might be due to the desire of capitalists to exploit African labor without the interference of a state committed to the protection of basic human rights, and he simultaneously registered serious doubts about whether Western civilization was an entirely desirable model for blacks to follow. By 1908, *Izwi* was making favorable references to socialism and to traditional African customs. "All that the blackman need do," Soga wrote in that year, "is to borrow the best that Western civilization offers. But we must get rid of the educated black serf who will blindly tie his race to the juggernaut of Western civilization, through sheer ignorance abandoning the social qualities that have preserved them hitherto from the social decay which is unavoidable under the present effete systems of European civilization." In addition to his other more-or-less forgotten achievements, A. K Soga may also have been one of the forerunners of that special blend of Western economic radicalism and African communalism that would later be known as "African socialism."[73]

· 2 ·

"Ethiopia Shall Stretch Forth Her Hands": Black Christianity and the Politics of Liberation

The Problem of Religion and Resistance

The black Americans and South Africans who struggled for the suffrage in the late nineteenth and early twentieth century did not rest their appeal primarily on an eighteenth-century notion of humanity's inalienable right to self-government. More often they claimed equal political rights as people who had attained, or were capable of attaining, a level of "civilization" that made them qualified to perform the duties of democratic citizenship. For many, the most important element in this acquired competence was their belief in Christianity, which not only attested to their moral enlightenment and sense of responsibility, but also allowed them to claim membership in a spiritual community composed of those, of whatever race or previous condition of servitude, who had accepted Christ and were willing to obey his commandments. The abolitionist movement in the United States and missionary-led campaigns against forced labor in early- to mid nineteenth-century South Africa had been based on the assumption that the Golden Rule forbade the enslavement or enserfment of human beings who had souls to be saved. Blacks who had themselves become Christians were therefore able to condemn white oppressors and discriminators for violating a color-blind system of ethics to which whites themselves supposedly subscribed. Biblical passages such as Acts 18:26—"He created of one blood all the nations of men to dwell on the face of the earth"—were interpreted as providing divine sanction for interracial struggles against slavery and legalized white supremacy in the nineteenth century.[1]

The gospel of human solidarity preached by white missionaries and abolitionists was egalitarian in an ultimate theological sense, but in practice it normally placed whites in a position of cultural superiority and validated paternalistic attitudes toward the blacks who were allegedly being rescued from heathenism and oppression. The tendency to identify Christianity with European or Euro-American civilization made whites the teachers and blacks the pupils—a hierarchical relationship that was supposed to be temporary but that tended to become, certainly in the missionary thinking of the late nineteenth century, less of a sudden liberation from sin and unbelief through the miracle of divine grace and more of a long-term guardianship. It is not surprising, therefore, that many black Christians, believing themselves to be fully reborn, rebelled against a white tutelage they no longer viewed as necessary and against being treated within the churches as second class or probationary members. Even more painful was the realization that, despite their professions of brotherhood and sisterhood, many white Christian humanitarians, including some of the reformers and philanthropists who took a special interest in the improvement of blacks, revealed through words and deeds that they actually regarded Africans or people of African descent as inherently inferior to whites.

Out of such tensions and disillusionments came separate or independent black churches. In the United States, separate Methodist denominations, based mainly among urban free blacks, emerged during the late eighteenth and early nineteenth century. After emancipation the freedmen withdrew en masse from the white-controlled Baptist and Methodist congregations that had ministered to them under slavery to become members of new or preexisting black denominations. By the turn of the century only a small minority of African-American church members remained in white-dominated churches. In South Africa, the secessions began in the 1880s and led in the twentieth century to a proliferation of independent churches. But in contrast to what happened in the United States a substantial majority of black Christians in South Africa have remained in historically white-dominated mission churches. Differences in the extent and character of religious independency may correlate quite closely with contrasts betweeen the more general circumstances and prospects of blacks in the two societies. African-Americans left churches that they could never hope to control on a denominational level because they were never likely to become a majority of the membership. A large proportion of black South Africans remained in mission churches under a form of white dominance that might be overcome in the future if church governance could be made more democratic.[2]

Independent church organizations did not necessarily mean that the gospel preached therein differed in any substantial way from what was offered in their white churches of origin. In ways crucial to believers, black Methodists, Baptists, and Pentecostals had more in common with the churches from which they had seceded than with other black denominations or sects. But for some black churchmen independence provided an opportunity to express a distinct and particularistic black version of Christianity. A further extension of the impulse behind religious separatism was to make it the ideological basis for an assertion of black autonomy and distinctiveness that had more than a narrowly religious meaning. Some historians of twentieth century black nationalism and Pan-Africanism have traced the roots of these liberationist ideologies to the Christian Pan-Negroism or "Ethiopianism" that emerged out of the religious separatism of the nineteenth century.[3]

It would be a mistake, however, to make too sharp a distinction between universalistic and particularistic varieties of black Christianity and to identify the former with political accommodation to white hegemony and the latter with militant opposition to it. According to C. Eric Lincoln and Lawrence H. Mamiya, the black church in the United States since emancipation has been characterized by a number of ambivalences or "dialectical tensions," among which are the tensions between nonracial universalism and ethnic particularism, as well as between resistance and accommodation to white domination. It also seems clear that one cannot predict the degree of political militancy from positions on a scale running from cosmopolitan universalism to racial chauvinism. African Methodist Bishop Henry M. Turner combined sharp denunciations of white injustice with a thoroughgoing racial separatism, but some of the most radical and activist black clergymen have been based in predominantly white denominations and have agitated among their white co-religionists rather than walking out on them. The most celebrated and effective of all activist black ministers, Martin Luther King, Jr., was simultaneously the product of separatist black Christianity and the eloquent exponent of a Christian universalism that sought reconciliation with whites in an interracial "beloved community." The black Christian experience in sub-Saharan Africa reveals that separation from whites might lead either to other-worldly escapism or to millennialist uprisings against colonial authority, while remaining within the mission churches might signify acquiescence to white control or the assertion of black power within the ecclesiastical structures that Europeans had established.[4]

Historians and sociologists of both the United States and South Africa have long debated the issue of whether black Christianity was an obstacle or a stimulus in the struggle for liberation. On one side are those who argue that religious independence and distinctiveness cradled black nationalism or liberationism and nurtured political resistance to white supremacy and racial inequality. At the other extreme are those who view the usual role of black Christianity as politically barren or even accommodationist, a diversion of psychic energies from the terrain of popular struggle for tangible objectives to that of spiritual or psychological satisfactions and compensations. But the issue requires greater clarification than it has normally received. The relationship between a generalized religious intensity or piety and political activism, one way that the question has been posed, has led to disputes over how to measure religiosity that have not been satisfactorily resolved and do not seem likely to be. A broad acquaintance with human history ought to tell us, however, that intense religiosity can be associated with virtually any point on a spectrum of political orientations running from other-worldly passivity or quietism to revolutionary millennialism. A better predictor than sheer piety would be the attitudes toward the exercise of physical power found in the doctrines of a particular religion group. An indifference to who governs and how—as with the early Christians and some Buddhist groups—normally results in political passivity, or the effort to achieve freedom from politics rather than through it. A belief that state must enforce godliness or be resisted by the godly, as espoused by John Calvin, John Winthrop, and the rulers of modern Iran, is likely to involve religion deeply and violently in politics. But the extent to which any set of potentially political religious beliefs expresses itself as direct involvement in the struggles for earthly power that characterize the domain of political action as it is usually conceived would also seem to depend on the specific political, social, and economic context in which a group of believers find themselves.[5]

African-American and black South African Christians have been historically ambivalent on whether their faith mandates fellowship with whites and the obliteration of racial distinctions in a flood of Christian love or the development of a distinct national or ethnic religion that will support claims to a special God-given destiny apart from that of their white oppressors. Impetus to move in either direction might come, therefore, from a perception of whether the white Christian world was receptive to interracial brotherhood or adamantly opposed to it. An integrationist or separatist tendency in black religion would to some extent be determined by whether, at a particular point in history, the Christianity that whites professed appeared to be racially inclusive or exclusive. But a religious belief with political implications does not lead automatically to political movements and programs. The human needs that religion seeks to satisfy are not the same as those addressed by politics; one must still explain how and why an obligation to love all of God's children *or* a belief that one's own nation or ethnic group has a special relationship with God sometimes remains a purely religious consolation or hope for redemption "some day" and at other times becomes an effort to change the world here and now. To explain why one thing happens and not the other, the historian is required to examine and analyze the specific circumstances surrounding any encounter between ultimate spiritual concerns and struggles for power and justice within the physical world.

The crusade against slavery in the United States provides a relatively clear-cut example of how the universalistic side of black religion promoted a certain kind of political action. The mainstream of the black abolitionist movement, as represented by Frederick Douglass and the majority of black antislavery activists, condemned whites for two acts of hypocrisy: one was betrayal of the political egalitarianism of the Declaration of Independence; the other was violation of Christ's injunction to love one's neighbors. A common platform of Christian fraternity made possible the collaboration of white and black abolitionists, at least to the extent that radical white Christians such as William Lloyd Garrison practiced what they preached. The great evangelical revival that swept the United States in the early nineteenth century affected free blacks and slaves with pious masters, bringing many into the Christian fold for the first time and intensifying the faith of others. When the revival spawned a movement for radical reform that in the North encouraged the belief that slavery was a sin and had no place in a Christian republic, the side of black Christianity that affirmed an ideal of interracial fellowship with whites was powerfully energized. History seemed in the process of validating a belief shared with some whites that obviously served the cause of black emancipation from servitude.[6]

Nevertheless, the opposite tendency—toward a Christianity that affirmed difference from whites rather than identity with them and found a special place for blacks in God's providential design—also had deep roots in the antebellum black experience. As several scholars have demonstrated, the Christianity of the slaves was far from being a carbon copy of that of their masters. Differing from the white prototype in modes of worship, choice of central Biblical texts, and ethical priorities, it laid the foundations for an African-American cultural nationalism. Whether it also implied a political nationalism is more problematic. But a strain of thought among antebellum free blacks made an explicit connection between black religious distinctiveness and a future of political self-determination. This early separatist thinking was encouraged by the setbacks and disappointments that beset the antislavery cause, as southern whites geared for a militant defense of slavery and many northern whites excoriated and mobbed the abolitionists for advocating racial equality and threatening the peace of the union. These pioneer black nationalists put forth a prophetic view of black redemption that would inspire African-American emigrationists, the first Pan-Africanists, and leaders of the "Ethiopian" movement among black Christians in South Africa. St. Clair Drake called this pattern of thought and imagery "Ethiopianism" or "the Ethiopian myth." The extent to which this world view became a major source for twentieth century black protest and resistance is a question that can profitably be addressed only after the genealogy of this black religious nationalism has been explored.[7]

Ethiopianist Thought in the Nineteenth Century

The biblical source of the Ethiopian vision was a somewhat obscure passage in Psalms 68:31, which prophesied that "Princes shall come out of Egypt; Ethiopia shall soon stretch forth her hands unto God." The context is provided by the next line—"Sing unto God, ye Kingdoms of the Earth"—but readers could well ask why Egypt and Ethiopia are singled out while other kingdoms or parts of the world that were known in biblical times are not named. The use of "Ethiopia" as a synonym for black Africa as a whole, and not merely for the actual Christian kingdom also known as Abyssinia, has a remote origin in the English language and was commonplace during the era of the Renaissance and Reformation. In Elizabethan drama, for example, Africans are often referred to as "Ethiops." During the Middle Ages, a positive image of blacks had arisen from the hazy knowledge that there were black Christians to the south of Islamic North Africa. The legend of Prester John, the saintly black Christian monarch of Ethiopia, had fueled the hopes of Crusaders that blacks would someday join Europeans in a successful holy war against Islam.[8]

The association of black Africans with Christian triumphalism receded after the Crusades and would have found relatively few European adherents once slave traders began to transport massive numbers of heathen West Africans to the New World in the sixteenth century. But the notion that Africans or Ethiopians had a starring role in the drama of Christian redemption reemerged in Western thought during the eighteenth and early nineteenth centuries. The eighteenth-century Swedish philosopher and mystic Emmanuel Swedenborg believed in an historical sequence of "true churches," the most recent of which was the European Christian Church. But Europe's day was coming to an end, he prophesied, and a new and purer church would soon emerge elsewhere. Although Swedenborg did not clearly specify that the new church would arise in Africa, his descriptions of blacks in the interior of that continent as more spiritual than other people led some of his followers to believe that Africa was the destined site of his "New Jerusalem."

During the nineteenth century, British and American Swedenborgians were influential in promulgating a millenarian view of African destiny. In fact they made a creative synthesis of Christian eschatology with an emerging ethnology based on the belief in permanent differences in temperament and capacity among human races, a doctrine that in other hands would justify white superiority and dominance over blacks. According to some Swedenborgians, blacks were the race that God and nature had endowed with the greatest aptitude for Christianity. Whites were naturally too cerebral, self-seeking, and aggressive to meet the standards of the Sermon on the Mount; only Africans had the believing, affectionate, and altruistic temperament that was the right soil for the full flowering of Christian faith and virtue. Hence the prophecy of Ethiopia stretching forth her hands unto God meant that the redemption of Africa would realize the Kingdom of God on earth. This religious version of "romantic racialism" had a considerable impact on white American abolitionists, colonizationists, and advocates of missions to Africa. It found its most eloquent statement and largest audience in Harriet Beecher Stowe's epoch-making antislavery novel *Uncle Tom's Cabin*.[9]

No evidence has been uncovered to show that Swedenborgian eschatology directly influenced the thought of free black Christians in early-nineteenth century America, but black abolitionists certainly became aware of such ideas when antislavery whites began to invoke them in the 1840s. At that point, however, their function would have been to give further confirmation to pre-existing views on what holy scripture revealed or implied about what God had in store for black people. For antebellum black Christians, the Ethiopian prophecy had two clear meanings: that black people throughout the world would soon be converted to Christianity and that God would deliver New World Africans from slavery. Within the earliest and most rudimentary formulations was the germ of the nationalistic belief that blacks were a chosen people with a special and distinctive destiny—a providential role similar to that of the Jews of the Old Testament. The suffering of captivity and slavery, a miraculous emancipation (first prophesied and then actually realized), the wandering in the wilderness, and the return to the Promised Land (which might in this case be Africa rather than Canaan) provided an intellectually and emotionally satisfying narrative structure for black hopes and aspirations. It also planted the seeds of Pan-Negroism, or Pan-Africanism.[10]

The idiom of Ethiopianism was central to the rise of a literature of black political protest in Jacksonian America. In 1829 Robert Alexander Young, a black New Yorker, published at his own expense *The Ethiopian Manifesto, Issued in Defense of the Black Man's Rights in the Scale of Universal Freedom.* Addressed not simply to American blacks but to "all those proceeding in descent from the Ethiopian or African people," it paraphrased the biblical prophecy to make it an explicit affirmation of black nationality. "God . . . hath said 'surely hath the cries of the black, a most persecuted people, ascended to my throne and craved my mercy; now behold! I will stretch forth mine hand and gather them to the palm, that they become unto me a people, and I unto them their God.' " Young predicted the coming of a messiah who would "call together black people as a nation in themselves."[11]

Later the same year, a more elaborate and incendiary call for black liberation was published in Boston. David Walker's *Appeal . . . to the Colored Citizens of the World* sent shock waves through the South because of its call for slaves to prove their manhood by rebelling against their masters. This rich and complex document was also a jeremiad directed at whites who had betrayed their Christian principles by oppressing blacks—but who could still repent in time to prevent divine retribution—and at free blacks who were not doing their utmost to elevate themselves and disprove racist charges of inferiority to whites. The *Appeal* was animated throughout by a spirit of apocalyptic Christianity. In the introduction to the third edition (1830) Walker announced that his main purpose was to convey to "all coloured men, women, and children over every nation, language, and tongue under heaven," a hope for divine deliverance from oppression, but only on condition that blacks struggle for their own liberation: ". . . the God of the Etheopeans [*sic*], has been pleased to hear our moans . . . and the day of our redemption from abject wretchedness draweth near, when we shall be enabled . . . to stretch forth our hands to the LORD Our GOD, but there must be a willingness on our part for GOD to do these things for us, for we may be assured that he will not take us by the hairs of our head against our will and desire, and drag us from our very, mean, low, and abject condition." The final clause on the necessity of self-help is of crucial importance; what could make Ethiopianism an activist, potentially political faith was a recognition that the prophecy was conditional, that God would help blacks only if they helped themselves.[12]

Walker's *Appeal* also featured his own version of the romantic racialist conception of black moral superiority and messianic destiny: "I know," he wrote, "that the blacks, take them half enlightened and ignorant, are more humane and merciful than the most enlightened and refined European that can be found in all the earth." He applied this racial contrast to the history of Christianity in the course of refuting Thomas Jefferson's negative assessment of black capacities in *Notes on Virginia*. If one were to judge whites by their behavior, Walker wrote, one would have to conclude that they "have always been an unjust, jealous, unmerciful, avaricious and bloodthirsty set of beings, always seeking after power and authority." Conversion to Christianity had not changed their essential nature, for they had been incapable of following its moral precepts. "The blacks of Africa and the mulattoes of Asia," however, "have never been half so avaricious, deceitful and ummerciful as the whites." Ironically reversing Jefferson's "suspicion" of black inferiority, Walker was impelled to "advance my suspicion of them, whether they are as *good by nature* as we are." Whites had long been exposed to the Gospel but had been unchanged by it; "the Ethiopians" had not, but they were now "to have it in its meridian splendor—the Lord will give it to them to their satisfaction." In the footnote that concluded this comparison of the races, Walker proclaimed as his "solemn belief, that if ever the world becomes Christianized (which must certainly take place before long) it will be through the means under God of the *Blacks*." The prophecy that Africans would be redeemers of Christendom and harbingers of the salvation of the world had been put to a new use—the mobilization of blacks against slavery and oppression.[13]

As a fervent opponent of the African colonization movement of the 1820s—which he regarded as a proslavery plot—Walker did not support an immediate campaign to redeem Africa through the agency of converted American blacks. He believed that Americans, black and white, must redeem themselves from the corruption and demoralization of slavery and racism before they sent missionaries to Africa. But the failure of the abolitionist movement to mount an effective challenge to the slave power in the 1830s and '40s, which became painfully obvious with the passage of the Fugitive Slave Act of 1850, brought some of the black abolitionists who followed in Walker's footsteps to the sad conclusion that slavery was becoming more entrenched. The Fugitive Slave Act dampened black hopes for abolishing slavery in the foreseeable future, posed a direct threat to the liberty of northern free Negroes, and caused an upsurge of interest in emigration from the United States. At first the new emigrationist movement of the 1850s shunned Africa as a destination and focused its attention on potential havens that were closer to home and easier to reach, such as Canada and the West Indies. The ill-repute of the American Colonization Society and its Liberian colony continued to deter blacks from contemplating a return to the continent of their ancestors, even though Liberia became independent in 1847. For the emigrationists of the early 1850s, the Promised Land for blacks fleeing white racial tyranny—the American equivalent of Pharaoh—was any place on earth where they might hope to establish an independent nation of their own. By the late 1850s, however, a small group of prominent black intellectuals overcame their repugnance at the African colonization idea and made an explicit connection between their desire to provide a refuge for oppressed American blacks and their hopes for the redemption of Africa.[14]

One of these was the Reverend Henry Highland Garnet, the pastor of black congregations within the predominantly white Presbyterian church and the most conspicuous representative of the black activist clergy. In his notable 1848 address, *The Past and Present Condition, and the Destiny of the Colored Race*, he provided Africa with a glorious past, describing the achievements of ancient Egypt and Ethiopia, and invoked the Ethiopian prophecy as a sign of greater things to come: "It is said that 'Princes shall come out of Egypt, and Ethiopia shall soon stretch out her hands unto God.' It is thought by some that this divine declaration was fulfilled when Philip baptised the converted eunuch of the household of Candes, the Queen of the Ethiopians. In this transaction, a part of the prophecy may have been fulfilled, and only a part." In the same address, however, he explicitly rejected the notion that African-Americans had a separate destiny from white Americans. "It is too late," he said, "to separate the black and white people in the New World," and he went so far as to predict that "*this western world is destined to be filled with a mixed race.*" Blacks in the United States faced great obstacles in their struggle for inclusion; echoing David Walker's "suspicion" of white moral depravity, Garnet noted that "the besetting sins of the Anglo-Saxon race are, the love of gain and the love of power." But he nevertheless proclaimed himself an American patriot: "America is my home, my country, and I have no other."[15]

This affirmation did not impel Garnet to discourage all blacks from emigrating, as Frederick Douglass normally did. Even before the passage of the Fugitive Slave Act of 1850, Garnet endorsed a limited and selective emigration, not only within the Americas, but also to Liberia. As might be expected, he was prominent in the emigration movement of the 1850s; after 1858 his attention focused on Africa, and he became the president of the African Civilization Society, a white-supported effort to build up an independent black Christian nation in West Africa to serve as a base for the evangelization of the continent. The further aim was "to establish a grand center of Negro nationality, from which shall flow streams of commercial, intellectual, and political power which shall make colored people respected everywhere." But Garnet did not advocate the mass emigration of American blacks to Africa. The scheme he endorsed contemplated sending only a few thousand a year, and he did not rule out the possibility that his "grand center of Negro nationality" would be in the Western hemisphere, perhaps even in the American South, rather than in Africa. Furthermore, he never abandoned his hope of equality within the American republic; the triumph of black nationhood elsewhere in the world was seen as a contribution to the black struggle for liberation and equal rights in the United States rather than as an alternative to it.[16]

A more radical and thoroughgoing advocate of black emigration was Martin R. Delany, a physician and journalist who had been a close associate of Frederick Douglass during the 1840s. Delany broke with Douglass and mainstream black abolitionism after receiving two rude shocks in 1850: he was forced to withdraw from the Harvard Medical School because white students objected to his presence, and Congress passed the Fugitive Slave Act. For Delany that infamous piece of legislation sounded the death knell of black hopes for equality on American soil and pointed to the conclusion that blacks could exercise their civil and political rights only in a nation of their own. As a political nationalist in the strict sense, he had no patience with those softer forms of nationalistic thinking that sought a reconciliation of black ethnicity and American republicanism. After exploring various emigration possibilities in the Americas, Delany went to Africa in 1859 and led an expedition to the Niger valley in search of land on which to colonize African-Americans. In the "Official Report" of the expedition, he anticipated the slogan of twentieth-century African nationalist movements: "Our policy must be . . . *Africa for the African race and black men to rule them.*"[17]

Delany's thinking was rooted in his own somewhat secularized interpretation of the Ethiopian prophecy. On one occasion in 1852 he reportedly made the text (and Garnet's gloss on it) the explicit basis for a God-given assurance that his work for black emigration was furthering "the grandest prospect for the regeneration of a people that ever was presented in the history of the world." In his great work *The Condition, Elevation, Emigration, and Destiny of the Colored People of the United States* (1852), Delany emphasized the special mission of African-American emigrants in the regeneration process: "Our race is to be redeemed; it is a great and glorious work, and we are the instrumentalities by which it is to be done. But we must go away from our oppressors; it never can be done by staying among them." But at this time Delany was still committed to a destination within the Americas: "God has, as certain as he has designed anything, has [*sic*] designed this great portion of the New World for us, the colored race; and as certain as we stubborn our hearts, and stiffen our necks against it, his protecting arm and sustaining care will be withdrawn from us." Like David Walker, Delany made the fulfillment of God's promise of black redemption, wherever it might occur, dependent on the initiative and exertions of blacks themselves.[18]

But Delany did not often invoke God and biblical prophecy to support his black nationalist ideology. For the most part, he stressed practical and worldly forms of self-help rather than divine support and inspiration. Indeed he was critical of the more intense varieties of black religiosity and viewed them as a hindrance to the cause of black liberation. Accepting the romantic racialist view of blacks as naturally pious, he saw this trait as part of the problem rather than the key to a solution. "The colored races are highly susceptible of religion," he wrote; "it is a constituent principle of their nature and an excellent trait in their character. But, unfortunately for them, they carry it too far. Their hope is largely developed, and consequently, they usually stand still—hope in God, and really expect Him to do that for them, which it is necessary they should do themselves." Delany's critical comments on black religious excesses did not mean, as some historians have suggested, that he was anti-religious. Read in context his remarks show that he was no atheist or agnostic: he was rather a liberal, nonsectarian Christian who believed that God's purposes must be achieved by men working through the "fixed laws of nature" that God had laid down rather than merely praying and waiting for a miraculous intervention.

This attitude explains the ambivalence toward missions and missionaries that he manifested after he became actively interested in African colonization. His concern was that mere Christianization of Africa was not enough to redeem the continent if religious conversion was not accompanied by the reconstruction of society, the development of natural resources, and a full application of "the improved arts of civilized life." Delany was an admirer of the economic and technological achievements of Western capitalist society, and he feared that black people's reputed emotionalism, supernaturalism, and susceptibility to religious fervor might impede their efforts to become equal to whites—and independent from them—by preventing them from developing the practical skills and physical resources that their "elevation" demanded. He thus laid bare a tension within Ethiopianism, and within black nationalism generally, between the mystical, romantic racialist, black chauvinist aspect, and the stress on self-help endeavors that were modeled to a considerable extent on white or European examples of material achievement and "civilization."[19]

Nineteenth-century Ethiopianist thought attained its highest level of complexity and sophistication in the writings and sermons of Edward Blyden and Alexander Crummell. Both emerged from the African diaspora in the New World to become Protestant missionaries and educators in Liberia. Their trans-Atlantic perspective on the black experience made them early and effective exponents of what was called Pan-Negroism in the nineteenth century and Pan-Africanism in the twentieth. Blyden was born and raised in the Danish West Indies and spent eight months in the United States in 1850 vainly seeking admission to a theological seminary before emigrating to the West African republic, where he completed his education and was ordained as a Presbyterian minister in 1858. Active in Liberian politics and education for more than thirty years, he also served as a promoter of the persistent efforts of the American Colonization Society to encourage African-American emigration to Liberia.[20]

Crummell grew up in New York as a close friend of Henry Highland

Garnet and other future members of the city's African-American intellectual elite. Like Garnet, he was ordained in a predominantly white denomination, but he encountered even more discrimination from the Episcopal hierarchy than Garnet had suffered at the hands of white Presbyterian authorities, and in 1848, while on a fundraising trip to England, he accepted a fellowship to Cambridge and stayed there long enough to take a degree in 1853. Instead of returning to the United States and engaging in the domestic struggle against slavery and racism, as he had originally intended, he accepted the commission of the Episcopal Church of the United States to serve as a missionary to Liberia, where he remained for twenty years. During that time, he was often closely associated with Blyden. Both men labored mightily to raise the religious, educational, and cultural level of Africa's only independent black republic and to increase its "civilized" population by attracting the right kind of black immigrants from the New World. Blyden ended his career in Africa, but political difficulties forced him to move from Liberia to Sierra Leone in 1885; Crummell returned to the United States in 1873, where he became a prominent black intellectual figure, the principal founder of the American Negro Academy in 1897, and a major influence on the Pan-Negro thought of the young W. E. B. Du Bois.[21]

The thought of Blyden and Crummell during their Liberian years elaborated the essentials of an Ethiopianist world-view. They differed somewhat in emphasis, with Crummell stressing the religious or spiritual aspects of African redemption more than Blyden, who paid greater attention to the political and economic prerequisites for black self-determination. According to the providential view of black history that they shared with other nineteenth century nationalists, the Africans of antiquity, especially the black inhabitants of Egypt and Ethiopia, had attained a high level of civilization and were in fact the progenitors of the civilization that later flowered in Europe. Subsequently, however, Africa had fallen into degeneracy and barbarism—according to Crummell this was because they had incurred the wrath of God for lapsing into the practice of idolatry. Europe had progressed mainly because of its early embrace of Christianity, which was essential to the highest form of civilization. But an officially Christian Europe had flagrantly violated the ethical precepts of its own religion by enslaving Africans and taking them to labor on the plantations of the New World. This apparent catastrophe was in fact part of God's design for the redemption of Africa and the black race. People of African descent in the New World had benefited from their exposure to the religion of their masters. The natural characteristics of blacks as a gentle and affectionate race enabled them to respond to the altruistic ideals of their European religious mentors without being infected by their actual greed and hypocrisy. It was the mission of some converted and civilized American blacks to return to Africa as missionaries or settlers and redeem the land of their ancestors from heathenism and barbarism. In this way the Ethiopian prophecy would be fulfilled. For Blyden, especially, the building up of "Negro states" was an essential part of this process. "Nationality is an ordinance of nature," he wrote in 1862. "The heart of every true Negro yearns after a distinct and separate nationality."[22]

Blyden and Crummell both had very long careers, and their thought was far from static. Both eventually retreated somewhat from their earlier stress on African-American agency in Africa. After his return to the United States, Crummell focused his attention on building "character" and "civilization" among domestic blacks in the hope that they could achieve equality within an American context without sacrificing the distinctiveness, pride, and solidarity that he associated with a heightened consciousness of racial heritage and identity. Blyden, for his part, became more Africanized, developing an increased respect for traditional African cultures and even for Islam as a positive force for improvement in the sub-Saharan regions that it had penetrated. He eventually concluded that Africans themselves, rather than black emigrants from the New World, would play the leading role in the redemption of the continent.[23]

The intellectualized Ethiopianism of Blyden and Crummell combined two ways of thinking about race and culture that appear difficult to reconcile—what Wilson Jeremiah Moses has called "civilizationism" and what I have elsewhere labeled "romantic racialism." Civilizationism derived from a universalist theory of human social and cultural progress that emerged in eighteenth- and nineteenth-century Europe and was often used as a rationale for imperialism and colonialism. Their adherence to it sometimes led Blyden and Crummell to condone or even endorse the expansion of European, and especially British, rule in Africa as an essential phase in the civilizing of the continent. Carried to its logical conclusion, it meant that there was a single standard of civilization, which could be found in nineteenth-century Europe and some of its overseas extensions. Accepting this doctrine did not have to mean endorsing a fixed hierarchy of races. Many white Americans and Europeans of the late nineteenth century were of course convinced that a capacity for civilization was an inborn racial trait limited to Caucasians, but black civilizationists maintained on the contrary that all human beings had a capacity for civilization and that the current superiority of the West was an historical accident or part of a providential design which pointed toward the eventual uplifting of non-European peoples to a comparable or even higher level of enlightenment, achievement, and physical well-being. If civilizationism was a universalist progressivism with its roots in the European Enlightenment, romantic racialism—as the term implies—was a product of the nineteenth-century reaction against the Enlightenment. It held that each "race" or "nation" had its own inherent peculiarities of mind and temperament and would develop according to its own special "genius" rather than follow some model based on the experience of peoples with different inherited or inbred characteristics. A reconciliation or synthesis of these two views of human progress is difficult, and a tension between them characterizes the thinking of both Blyden and Crummell.[24]

In the case of Blyden, it appears that a commitment to racial particularism eventually predominated over universalist conceptions of progress toward a world civilization. Blyden's views on race, as ably summarized by his biographer Hollis Lynch, did not crystallize until the 1870s, but they quickly hardened into a rigid ideology. In his hands the standard romantic racialist notion that each race or nation had distinctive innate characteristics and that blacks and whites were opposites in temperament and aptitude was combined with some of the doctrines of nineteenth-century scientific racism. Blyden did not acccept the racist conception of a hierarchy of superior and inferior races, but he did come to believe in the superiority of pure races over mixed ones and to view miscegenation with as much disdain as any white supremacist. Of apparently unmixed African ancestry himself, he viewed mulattoes as inferior to pure blacks and as an undesirable, disruptive element in any "black" community that incorporated them. To some extent his prejudices derived from Liberian politics. Colonists from the United States and the West Indies had divided into mulatto and Negro parties, with the browns establishing an early ascendency over the blacks; Blyden and the equally dark-skinned Crummell had been fervent champions of the darker party. In their view the mulatto elite had adopted the exclusiveness, arrogance, and highhandedness of their southern white ancestors and was especially culpable for its cruel and disdainful treatment of the colony's indigenous African population. Crummell, for the most part, kept his anti-mulatto biases to himself after he returned to the United States, but Blyden cultivated his color prejudices until they became almost an obsession. In his efforts on behalf of the American Colonization Society, he made strenuous efforts to prevent mulattoes from joining the exodus to Liberia and openly proclaimed that he wanted "pure Negroes" only. Late in his life, he told a friend that he wanted nothing on his tombstone but "He hated mulattoes." His view that only those of unmixed ancestry could be trusted to embody the black genius anticipated some of the rhetoric of Marcus Garvey after World War I.[25]

In his comparisons of the innate characteristics of Europeans and blacks, as expertly summed up by Hollis Lynch, Blyden repeated the familiar romantic racialist litany of black virtues and white vices but gave them a twist of his own:

> The European character, according to Blyden, was harsh, individualistic, competitive and combative; European civilization was highly materialistic; the worship of science and industry was replacing that of God. In the character of the African, averred Blyden, was to be found "the softer aspects of human nature": cheerfulness, sympathy, willingness to serve, were some of its marked attributes. The special contribution of the African to civilization would be a spiritual one. Africa did not need to participate in the mad and headlong rush for scientific and industrial progress which had left Europe with little time or inclination to cultivate the spiritual side of life, which was ultimately the most important one.[26]

Blyden, therefore, ended up rejecting orthodox civilizationism with its celebration of material progress on the European model in favor of a romantic racialist conception of African spiritual superiority. His formulation of the difference between a soulless Europe and a soulful Africa provided a link between nineteenth-century Ethiopianism and the cultural Pan-Africanism of the twentieth century. Whether they ascribed the difference to race in the genetic sense, as Blyden did, or merely to the weight of culture and historical experience, later Pan-Africanists would often contrast the selfish materialism of Europe with the altruistic spirituality of Africa. Blyden's own commitment to Ethiopianism in its narrow and original sense as the redemption of Africa through the exclusive agency of Protestant Christianity became attenuated in his later years. He began to find convincing evidence of African spirituality in Islam and even in traditional animistic religions.[27]

For Crummell, who was more deeply attached to orthodox Protestantism than Blyden, a reconciliation of racialism and universal civilizationism was possible in the supra-rational realm of Christian faith. His notable paper on "The Race Problem in America"—delivered in 1882, almost a decade after he had returned to the United States and reimmersed himself in African-American affairs—provided one of his strongest statements on the fundamental importance of race in human affairs. "Races, like families," he wrote, "are the organisms and ordinances of God. The extinction of race feeling is just as impossible as the extinction of family feeling." He went on to show how races and nations had kept their essential characteristics through all recorded history. This was even true, he argued, of the Europeans of various nationalities who had emigrated to the United States. The Irish remained Irish, the Germans German, and the Jews Jewish. Implicitly taking issue with assimilationist black progressives such as Frederick Douglass and T. Thomas Fortune—who viewed the ultimate destiny of Afro-Americans as biological amalgamation with whites—he saw no reason to expect "the future dissolution of this race and its final loss." But racial or ethnic diversity, including differences of color, and the natural desire of each group to associate mainly with its own kind did not constitute an insuperable barrier to peace and equality among races.

At this point he reverted to his belief in universal progress. In his view, as in that of most other civilizationists, there were two essential components of human advancement—the spread of Christianity and the growth of "civil freedom." Neither, according to Crummell, had "yet gained its fullest triumphs." But such triumphs were inevitable in the future because the progress and ultimate victory of true religion and personal freedom were divinely ordained: "It is God's hand in history. It is the providence of the Almighty and no earthly power can stay it." Hence racial antagonism would disappear and "every person in the land," without any loss of racial or ethnic identity, would be "guaranteed fully every civil and political right and prerogative." By linking the progress of civil rights to Christian redemption and the maintenance of racial integrity to hopes for an egalitarian ethnic pluralism, Crummell anticipated the Social Gospel in American Protestantism and also later conceptions of how democracy could be reconciled with ethnic and cultural diversity. By all objective standards, Crummell's optimism was a dubious response to the race relations of post-Reconstruction America. It would not be the last time that faith in God's promise of human solidarity would be the recourse of those facing a world so torn and disfigured by racial conflict and injustice that changing it through mere human agency appeared impossible.[28]

Crummell's domesticated and reformist Ethiopianism of the 1880s was a far cry from David Walker's apocalyptic and potentially revolutionary version of 1829 and 1830. Walker had seen some chance that "Americans yet under God, will become a happy and united people"; for "nothing is impossible with God." But it was far from certain that whites would repent of their sins against blacks; it was at least as likely, if not more so, that they would continue in their merciless, avaricious ways and be punished by God through the agency of rebellious blacks. Walker's negative assessment of white character and his view of how little Christianity had thus far been able to improve it would seem to support the cataclysmic alternative.[29]

Crummell, unlike Blyden and some other black proponents of romantic racialism, did not follow Walker's example and dwell on the harsh and cruel aspects of the Caucasian's innate character. He was content to exalt his own race without demeaning the other; hence he could more readily adopt an attitude of Christian benevolence toward whites and conceive of a future of racial harmony and equality within the United States. His late nineteenth-century liberal view of history as the progressive unfolding of God's design for a peaceful, fraternal, and egalitarian world may have lacked the emotional depth and tragic sense of Walker's more Calvinistic sense of history as divine judgment and potential apocalypse, but it did provide an antidote to despair at a time when racism and discrimination were intensifying in the United States.

Crummell's emphasis on race consciousness and cultural revitalization had a significant influence on the young W. E. B. Du Bois, who also shared his mentor's hope for the growth of freedom within an American context. But Du Bois lacked Crummell's secure faith that a benevolent deity would reconcile all conflicts and contradictions in the fullness of time. His was a more troubled and restless spirit that would seek first in Hegelian idealism and later in Marxist materialism the kind of assurance for black equality and liberation that Crummell had derived from his Christian post-millennialism. An application of Hegelian dialectics to the Ethiopian prophecy was featured in one of Du Bois's first public addresses. In his 1890 Harvard commencement oration on Jefferson Davis, the twenty-two-year-old Du Bois made the standard romantic racialist contrast between "Teutonic" strength and ruthlessness, on one hand, and the Negro idea of "submission apart from cowardice, laziness, and stupidity" on the other. According to biographer David Levering Lewis, "Du

Bois foretold a better world emerging from the dialectical struggle between the Strong Man and the Submissive Man, ending his commencement speech with the supplication, 'You owe a debt to humanity for this Ethiopia of the Outstretched Arm.' "[30]

Du Bois's first major effort to formulate a philosophy of race revealed his debt to Crummell and might be regarded as an effort to translate Ethiopianism into an idiom compatible with the secular philosophy and social science that Du Bois had learned at Harvard and the University of Berlin. His paper on "The Conservation of Races," delivered in 1897 to the National Negro Academy (the organization of the black cultural elite that Crummell and others had founded), was essentially a revision of Ethiopianism that dispensed for the most party with its specifically Christian and biblical underpinnings. It differed from Blyden's recent reformulation of the Ethiopian tradition to make it more Africa-centered by retaining a central role for African-Americans in the redemption of the race. Like Crummell and Blyden, Du Bois in 1897 regarded race as a fundamental and irreducible fact of enormous consequence. There could be no doubt, he affirmed, "of the widespread, nay universal, prevalence of the race idea, the race spirit, the race ideal, and as to its efficiency as the vastest and most ingenious invention of human progress."

Du Bois at this time was able to combine racialism and civilizationism by viewing progress toward modernity as the achievement of particular nations and races, a notion that he derived in part from his graduate work in Germany. But all races or nations, and not just Germans or Anglo-Saxons, were capable of attaining the highest levels of development. If blacks were to become modern and progressive, they had to develop their own race consciousness and sense of their unique gifts and potentialities. This meant that African-Americans should not aim at assimilation into white society, as Frederick Douglass had proposed, but should cherish and nurture their sense of racial distinctiveness. "For the development of Negro genius, of Negro literature and art, of Negro spirit," Du Bois wrote, "only Negroes bounded and welded together, Negroes bounded by one vast ideal, can work out in its fulness the great message we have for humanity." Consequently, "the advance guard of the Negro people—the eight million people of Negro blood in the United States of America—must come to realize that if they are to take their just place in the vanguard of Pan-Negroism, then their destiny is not absorption by the white Americans." African-Americans should regard themselves rather as "the first fruits of this new nation, the harbinger of that black tomorrow which is yet destined to soften the whiteness of the Teutonic society."[31]

Believing that black Americans were members of a great historic race and should provide leadership for the race's struggle to realize its "genius," Du Bois at the outset of his career took a direct interest in Africa and encouraged his fellow African-Americans to do likewise. In the same year that he exhorted the Negro Academy to conserve and develop the race, he wrote to Belgian authorities proposing a limited African-

American emigration to the Congo: Although "the American Negro . . . feels himself an American of Americans . . . and rightly claims that no constituent element of the country has a better right to citizenship than he, . . . he recognizes the fact that his cause after all is the cause of the historic Negro race; and that Africa in truth is his greater fatherland; . . . he feels as the advance guard of Pan-Negroism that the future development of Africa will depend more or less on his efforts." Apparently the Belgians were not interested, and Du Bois did not become involved in emigrationist activities; but he did serve as one of the convenors of the first Pan-African Congress in 1900. Throughout his long subsequent career, he would seek to correlate the struggle for equality in the United States with efforts for the advancement and liberation of Africa. But for him, in contrast to many of his clerical Ethiopianist predecessors, African redemption was viewed in forthrightly political rather than ostensibly religious terms—as freedom from European domination and exploitation rather than liberation from the darkness of heathenism. At no time did Du Bois support massive emigration to Africa or (until the very end of his long life) abandon the struggle for racial equality in the United States. He could not rely, like Crummell, on the hand of God; but in his early years, he used his own brand of philosophical idealism—a dynamic, historical idealism inspired to some extent by Hegel—to support a hopeful view of the future of blacks in the United States. Blacks could, he believed, be true to their own *Volkgeist*, be an inspiration for blacks in Africa and throughout the world, and still gain equality in American society. Indeed their peculiar racial gifts might contribute to the redemption of America from soulless materialism.[32]

Popular Ethiopianism and African-American Missions to Africa

The sophisticated Ethiopianism or Pan-Negroism of the nineteenth-century black intellectuals left behind a body of thought and literature that would influence the black nationalist and Pan-Africanist thinkers of the twentieth century, even though most of them followed the example of Du Bois and substituted some form of humanistic philosophy for divine revelation as the basis for their hope that all blacks throughout the world could be redeemed or liberated through a common effort. But the belief that blacks could struggle for collective redemption knowing that their ultimate triumph was part of God's design for the salvation of the world had never been confined to the theological and metahistorical speculations of an educated elite, and it would survive the process of secularization that made some intellectuals seek a new frame of reference for black liberation. A less intellectualized, more populist version of Ethiopianism became the creed of religious and social movements that some people, especially South African whites, believed had radical, even revolutionary, political implications.

As St. Clair Drake has pointed out, the Ethiopianist thought of exceptionally articulate and well-educated black spokesmen such as Garnet, Delaney, Crummell, and Blyden arose in conjunction with the efforts of "Negro folk theologians" to find a place for blacks in the Bible and in the drama of human salvation that it foretold. A basic version of the Ethiopian myth was undoubtedly familar to many of the oppressed, impoverished, and uneducated blacks who worshipped in the covert or "invisible" plantation churches of the slave era and in the ramshackle rural churches of the post-emancipation period. Ironically, most of the black clergymen who published the sermons, pamphlets, and books that provide the historical sources for an understanding of mid-nineteenth-century Ethiopianism were employed by mainstream white Protestant denominations—they were Presbyterians, Congregationalists, and Episcopalians. Reflecting the segregated character of Protestant religious life, they mostly pastored all-black congregations, but their parishioners were unlikely to be a cross-section of the black community; only the educated members of the black middle class were likely to be receptive to their learned discourses. In the post-Civil War period, when the southern freedmen had a right to join any church they chose, most African-Americans became affiliated with separate black churches, especially the independent Baptist congregations and those of the most highly organized and visible of black denominations—the African Methodist Episcopal (AME). We have no way of knowing how often the Ethiopian prophecy was the text of sermons in these churches, but it was in all likelihood a favorite passage. According to historian Albert Raboteau, it was "without doubt the most quoted verse in black religious history." There was a natural, almost inevitable, association between the flight of the Israelites from bondage in Egypt and their return to the Promised Land and God's pledge to the Ethiopians, some of whom had also found themselves freed from bondage in a foreign country.[33]

The Ethiopian prophecy was a source of inspiration for the upsurge of interest within the independent black denominations of the late nineteenth century in working directly for the redemption of Africa. Between 1877 and 1900, independent black churches sponsored at least 76 missionaries of their own in Africa. Justifying this endeavor was the Ethiopianist theory of Providential Design, the belief that God's purpose in permitting the agony of enslavement in the New World was the ultimate salvation of Africa. Having received Christianity from the whites and having made it their own, black Americans were commissioned by God to return to their fatherland and save it from the darkness of heathenism. For most advocates of an independent black mission enterprise, this interest in Africa did not imply that most or even very large numbers of African-Americans should go there. In fact the elevation and empowerment of Africa that a missionary vanguard would help to bring about was more often viewed as a means to black advancement in the United States. As the AME *Church Review* put it in 1883, "Never will Africa's sons [in the United States] be honored until Africa herself sits among the civilized powers."[34]

Whether they thought of emigrating or not, whether it was Africa or

America that they considered their fatherland or ultimate home, those African-Americans who joined independent churches were implicitly affirming the most fundamental and durable Ethiopianist assumption—that blacks were a distinctive people who needed to control their own spiritual and cultural lives, if not their economic and political affairs as well. Frederick Douglass—a whole-hearted proponent of ultimate black assimilation who regarded separate organizations as concessions to the American racial caste system that could be justified, if at all, only as a temporary expedient in the struggle for equal rights—understood this implication of religious separatism very well and consistently opposed independent denominations, arguing that blacks should remain in the white churches and fight for equality there, as in society as a whole. The very existence of an independent black Christianity showed that there was an embryonic cultural nationalism among the masses of blacks that would become a political nationalism as well if black people concluded that achieving their spiritual destiny required them to govern themselves in secular matters as well.[35]

The most visible and well-endowed of the independent black denominations, the African Methodist Episcopal Church became a conspicuous battleground in the late nineteenth century between an interpretation of Ethiopianism that would strictly limit its nationalism to the religious sphere and a militant variety that sought to give practical political meaning to the ideal of black independence and self-government. The conservative faction, which dominated the church's hierarchy and controlled its relatively affluent and "respectable" northern urban congregations, was dedicated to self-improvement and character building; its leaders assumed that the American race problem would be solved when blacks demonstrated through their upright behavior and internalization of the Protestant work ethic that they were qualified to exercise all the rights of citizenship. Like the black abolitionists of the antebellum era, they tended to view white prejudice as a relatively superficial response to current black degradation and not as a deeply ingrained attitude that frustrated self-improvement at its source and had to be neutralized before blacks could even begin to make progress. As the AME's missionary activities among the southern freedmen demonstrated, church leaders saw their essential task as the moral and spiritual rehabilitation of a people degraded by slavery. Interest in African missions normally reflected the same Victorian emphasis on building character and stimulating self-help among people whose unfortunate historical experience and lack of civilization required guidance from those who had enjoyed greater advantages. No distinctively black or African-American values were involved in this conception of black redemption; it was nationalistic only in its insistence that blacks uplift other blacks rather than rely on white Christians to do it for them.[36]

In opposition to the AME establishment, Bishop Henry M. Turner advocated a more radical form of black separatism and a more vital association between blacks in the United States and Africa. Turner is mainly remembered as an outspoken proponent of emigration to Africa; but he is also significant for having helped forge a link between independent black Christianity in the United States and the separatist Ethiopian movement in South Africa. More than that, he exemplified the political potential of church-based Ethiopianism more effectively than any African-American leader of the nineteenth century.[37]

Born free in South Carolina, Turner became a preacher before the Civil War, first for the white southern Methodists and then for the AME Church. While serving during the war as pastor of an important congregation in Washington, D.C., Turner fervently supported the Union cause, led in the agitation for the enlistment of black troops, and when the use of blacks became government policy, played a major role in recruitment. As a reward for his efforts, he was made a chaplain to black soldiers, the first African-American minister to hold such a position in the United States Army. Appointed an official of the Freedmen's Bureau in Georgia after the war, Turner proselytized successfully for his church among the ex-slaves and was one of a number of black clergymen to enter Republican politics, winning election to the state legislature in 1868. But Turner's deep disillusionment with black prospects for gaining political equality in the United States began when the white-dominated Georgia state legislature refused to seat blacks. As a leading black Republican in Georgia, Turner did manage to hold minor federal offices during the Reconstruction era, but neither his personal ambitions nor those he held for his race were fulfilled. Subsequently he emerged as the era's leading proponent of black emigration from the United States to Africa, and its most consistent and thoroughgoing supporter of black separatism.[38]

Turner had a relatively simple message that he presented in a blunt and salty fashion. The failure of Reconstruction convinced him that whites were unalterably opposed to black equality in the United States and that African-American liberation could be achieved only among other blacks in Africa. Any doubts he may have had on the subject were put to rest when the Supreme Court in 1883 declared the Civil Rights Act of 1875 unconstitutional. His conviction that hopes for black advancement and civil rights in the United States were futile seemed borne out by the wave of lynchings that swept the South beginning in the late 1880s. Addressing an emigration convention in 1893, Turner put the issue in stark, unvarnished terms: "To passively remain here and occupy our present ignoble status, with the possibility of being shot, hung, or burnt," was clearly unacceptable. "To do so would be to declare ourselves unfit to be free men. . . . For God hates the submission of cowardice. But on the other hand to talk about physical resistance is literally madness. . . . The idea of eight or ten million of ex-slaves contending with sixty million people of the most powerful race under heaven!" Obviously the only alternative was a mass exodus.[39]

Turner cannot be dismissed as a voice in the wilderness futilely trying to wean blacks from their blind devotion to America. His emigrationism attracted little support from established African-American leaders, but it did speak to the panic and desperation of many southern blacks. Reconstruction had raised hopes for a better life that had been dashed by terror, poverty, and discrimination. During the period between the late 1870s and the early years of the twentieth century, the impulse to emigrate from the rural South expressed itself in a variety of movements and schemes, beginning with the efforts of Martin Delany and other South Carolina blacks to establish a steamship connection with Liberia in 1878. The Liberian Exodus Joint Stock Company sent only one shipload of settlers, but thousands of others, responding to the "Africa fever" sweeping the South after the fall of the last Reconstruction governments, would probably have embarked if the enterprise had not gone bankrupt. The impulse to escape from the post-Reconstruction South, if not from the United States itself, was expresssed most dramatically in the mass migration of the "Exodusters" from Mississippi, Louisiana, and East Texas to Kansas in 1879. During the 1880s and '90s, Turner and other African emigrationists responded to the demonstrated willingness of masses of southern blacks to pull up stakes and go elsewhere in search of a better life by cooperating with the American Colonization Society's persistent efforts to find colonists for Liberia and supporting the bills for government-subsidized expatriation of American blacks to Africa that white supremacist southern senators introduced in Congress. These efforts had little practical consequence, but there is evidence that they had considerable popular support, especially in the rural South, and that large numbers of desperate and disillusioned blacks would have emigrated if the means had existed.[40]

Turner, however, did not actually advocate or anticipate the wholesale removal of America's black population to Africa. His populist appeal to the masses of blacks was to some extent qualified by his own version of the elitism and "civilizationism" that was characteristic of nineteenth-century black nationalist thought. He often referred disdainfully to the majority of blacks and made it clear that only a superior minority would join the exodus. In 1884, he predicted that only "the better class of colored men of this country will go to Africa and build up a mighty nation, while the riff-raffs of our race will remain here." "I have said forty times over," he wrote in 1893, "that all I advocated was that five hundred thousand or a million of us should go to Africa and build a civil government that would serve as an asylum for the oppressed and degraded portion of our race. I have never advocated all the colored people going to Africa, for I am well aware that the bulk of them are lacking in common sense and are too fond of worshipping white Gods." Like David Walker and some other nineteenth-century black nationalists, Turner could condemn the majority of his fellow African-Americans for their cowardice and degradation in such a sweeping and unforgiving way that he came close to conceding the fact of African-American inferiority to the white supremacists—although of course he rejected their racist explanation.[41]

Turner seems to have regarded Africans, even heathen Africans, as morally and culturally superior to the "bulk" of African-Americans. When he first traveled to West Africa in 1891, he sent back reports that were overwhelmingly favorable; he not only admired the physical appearance and demeanor of the natives but even praised many aspects of their culture. Carrying his tendency to criticize African-Americans to an uncharacteristic extreme, he "noted the tact, taste, genius, and manly bearing of the higher grade of the natives" and concluded that "we poor American Negroes were the tail-end of the African races." Because the dregs of Africa had been sold into slavery, Turner argued, their descendants had ugly or grotesque physical characteristics lacking in the descendents of those who had stayed at home. But Turner could scarcely have been as unqualified in his admiration of African culture as he was of the African physique. He was, after all, a Christian minister and advocate of the redemption of Africa through the agency of black missionaries and emigrants from the United States. More sympathetic to traditional customs than Alexander Crummell and most African-American missionaries, even Turner had to draw the line at polygamy and other practices that clearly conflicted with Christian ethics and social practices. He adhered to the essential Ethiopianist view that enslavement in America had been part of God's design for the redemption of Africa, because it had engendered a converted and civilized vanguard for the race—although he stressed saving black Americans from racism as well as redeeming Africa from heathenism. Consequently it is difficult to avoid the conclusion that "the better class of colored men" in the United States, those whom Turner regarded as suitable candidates for emigration, were believed by him to be culturally superior to most native Africans, even if the latter could be viewed as superior to the African-American "riff-raffs."[42]

Turner was, like Alexander Crummell, essentially an advocate of "civilizationism." He took many of the characteristics and achievements of the West as a model, because to his mind they reflected universal values, not racially specific traits. According to historian James Campbell, Turner denied to "Europeans a monopoly on Christian civilization," which enabled him "to reconcile race pride with equally firm commitments to Christianity and western-style progress." In the tradition of romantic racialism, he attributed to Africans a greater sensitivity to "invisible forces" than earth-bound Europeans could demonstrate, and he found evidence for this spirituality in animistic religions. Provided access to the higher and more sublime truths of Christianity, the African would, he wrote, easily "transfer his faith from superstition to Christ Jesus the Lord." Given his commitment to the evangelization of the continent, his judgments on Western imperial expansion in Africa could not be entirely negative. He attacked many aspects of imperialism, especially the introduction of alcohol. But he also praised the British in particular for suppressing slavery and polygamy in their colonies. Like Crummell, Blyden, and most educated Anglophone Africans of the time, he saw British influence as benign and essential to the progress of Africa. After a visit to

Sierra Leone, he gave strong expression to this Anglophilia: "God save old England, is my prayer."[43]

Ethiopianism in South Africa

Bishop Turner paid an historic visit to South Africa in 1898 and cemented an enduring relationship between African Methodism in the United States and what was called Ethiopianism in South Africa. At about the same time, African-American Baptists were establishing missions in the southern part of the African continent. But it would be wrong to see independent black Christianity as an export from the United States. Impulses of purely indigenous origin, as well as other foreign influences, contributed to the rise of a Christian alternative to white-dominated missions. A version of Ethiopianist religious nationalism that was more sharply focused on Africa and that appeared to present a more serious challenge to white domination than the American variant came into prominence in South Africa between the 1870s and the 1910s. Before detailing the sustenance that it received from African-American sources, we must place it firmly in its South African context.

The immediate cause of the secession of Africans from white mission churches was a growing pattern of discrimination against black ministers and evangelists. Beginning with the education and ordination of Tiyo Soga by the Presbyterians in the 1850s, Methodists, Presbyterians, and Congregationalists in southern Africa began to prepare selected African converts for evangelical work and eventual ordination. Some mid-nineteenth century white missionaries anticipated that Africans would gradually take over many of their responsibities and that a "native clergy" would eventually assume a central role in the redemption of black South Africa from heathenism. But the rise of racist and imperialist ideas in the late nineteenth century penetrated even the missionary establishment and raised new doubts about the capacity of even the most pious and educated blacks for the independent exercise of religious authority. The condescending paternalism that had always characterized the mission enterprise tended to harden into a caste system with white ministers fully in charge for the foreseeable future and the black clergy fixed in the role of perpetual auxiliaries.[44]

The resulting tensions were felt in all the major missionary denominations but with some differences in the timing and results. The Scottish Presbyterians in the Cape Colony had a high educational requirement and ordained relatively few Africans, but appeared willing to accord more respect and responsibility to the black clergy than denominations with less exacting standards. Nevertheless, the most prominent black Presbyterian divine after the death of Tiyo Soga, the same P. J. Mzimba who had advised Africans to eschew politics in 1886, seceded from the Free Church of Scotland in 1898 to form the independent Presbyterian Church of Africa, taking with him most of the congregation at Lovedale where the Presbyterians had their school and headquarters. Mzimba complained of racial slights and asserted that the time had come for blacks to govern themselves in ecclesiastical matters. The American Congregationalists in Natal were obligated by their system of church government to grant considerable autonomy to local congregations; but, as early as the 1880s, one black minister went further than was allowable when he turned an important mission into a totally independent community church. The problem grew worse in the 1890s, when the independence-minded factions of two congregations, one in Natal and one composed of Zulu migrants to Johannesburg, seceded and joined together to form a new "Ethiopian" demonination—the Zulu Congregational Church. More than most other denominations, however, the American Congregationalists learned their lesson and later attracted many of the secessionists back to the fold by granting greater autonomy to black ministers and congregations. Small-scale schisms also took place during the 1890s among Lutheran and Anglican converts in the Transvaal.[45]

The most dramatic and far-reaching secessions took place, as earlier in the United States, among the Methodists. The combination of hierarchical structure and a willingness to license large numbers of relatively uneducated exhorters or evangelists meant that Methodism had an almost built-in tendency to involve substantial numbers of blacks in active religious work without granting them independent authority. The evangelical populism characteristic of the Methodists attracted Africans in search of an emotionally sustaining faith somewhat more readily than the more staid and elitist approach of the Anglicans, Presbyterians, and Congregationalists, but its episcopal form of governance kept the converts under close supervision. The first assertion of black Methodist independency in South Africa took place in 1883 in Thembuland in the Transkei. Nehemiah Tile, a black evangelist and probationary minister at odds with his white supervisor, cooperated with the chief of this still quasi–independent tribe to form the Tembu National Church—the first completely autonomous black Christian church in South Africa. This incident demonstrates how religious independency could sometimes play an important role in the struggles of traditionalist African communities to resist white domination and, after the consolidation of white rule, in the internal politics of "native reserves."[46]

The Methodist secession that had the greatest impact and aroused the most concern among whites throughout South Africa involved a relatively urbanized and detribalized group of Africans on the industrializing Witwatersrand in the early 1890s. The most prominent of the few African Methodist ministers who had been ordained in the 1880s, the Reverend Mangena Mokone of Kilnerton near Pretoria reacted to racial discrimination and segregation within the Methodist mission community of the Transvaal by resigning from the Methodist clergy in 1892 and founding the independent Ethiopian Church in 1893. No discussion of precisely why the name was chosen survives, but it undoubtedly reflected the same broad biblical use of the term as a synonym for African that was current among black Americans in the nineteenth century and was in all likelihood inspired in part by the prophecy of Ethiopia stretching forth her hands unto God. Shortly after the church was founded, Mokone reportedly proposed that it send missionaries throughout black Africa, which would have been a strong indication that he conceived of it as the vehicle for a Pan-African Christianity. He also recommended dispatching envoys to Abyssinia, showing that he wished to establish links with Ethiopia (in the narrower sense) as the source of an indigenous African Christian tradition. There is no evidence, however, that he had any conception at this time of African-Americans playing a special role in the redemption of Africa.[47]

The chain of events leading to the forging of a connection between South African Ethiopianism and the African Methodism in the United States began with the visit of an African-American singing group, the Virginia Jubilee Singers, to South Africa in 1890. Led by Orpheus M. McAdoo, the group followed the example of the more celebrated Fisk Jubilee Singers and performed a combination of spirituals and lighter fare for racially mixed audiences in Kimberley and other cities; especially appreciative, it appears, were Cape Coloreds and mission-educated Africans literate in English. What these auditors found enormously stimulating was the singers' ability to combine a Western-influenced sophistication with a distinctive and putatively black style and sensibility. The group's achievement helped to make African-Americans a model for African advancement and "civilization" that avoided a staightforward imitation of white or European cultural styles, and its immediate effect was to inspire members of the Christian African elite to put on musical and dramatic shows of their own that featured new combinations of European, African-American, and traditional African material. Shortly after the 1890 visit of the Virginia Jubilee Singers, white businessmen in Kimberley agreed to finance a tour of Britain by a newly recruited African Jubilee Choir, which would give half of its performance in native African languages and traditional costumes and the other half in English and modern European dress. Although it performed for Queen Victoria and members of Parliament, the troup ran out of money in England, and its members had to find their own way back to South Africa. But it was soon reconstituted and dispatched on a second tour, this one climaxed by a visit to the United States. The choir ran out of funds again in Cleveland, and this time they lacked the wherewithal to return home. Six of the marooned singers were rescued by the African Methodist Episcopal Church and given fellowships to the denomination's Wilberforce University in Ohio, where they were trained to be AME missionaries.[48]

One of the African singers enrolled at Wilberforce was Charlotte Manye, whose family was acquainted with Mangena Mokone, the founder of the Ethiopian Church. When Charlotte wrote to her sister in Johannesburg in 1895 to tell her the news, she also provided a description of the AME Church. The sister showed the letter to Mokone, who quickly saw the possibility of some kind of affiliation between independent black Methodism in South Africa and the United States. He wrote to Bishop Turner for more information about his denomination, and the negotiations began that resulted in the incorporation of the Ethiopian church into the AME Church in 1898. The Africans, it appears, hoped that their American co-religionists could provide financial aid and educational opportunities that would replace the assistance provided by white Methodists before the schism—and without the discrimination and authoritarianism that was the price of European sponsorship.[49]

Mokone's inquiry was not the only appeal from South Africa that Turner received in 1895. In a letter that revealed much about the value Africans saw in independent black missionaries, John Tule, a Tembu who had migrated to Cape Town in search of work, wrote to Turner's journal *The Voice of Missions* that "our people at home in the Transkei are in a bad state; needing two principal modes of life; . . . Christianity and civilization." But they could not rely on white missionaries to meet these needs, because the latter in fact acted as agents of white oppression, doing only "what the government wants them to do," rather than following the will of God and promulgating "the good news of salvation." White missionaries, Tule argued, caused dissension and factional conflict within African communities to serve the aim of divide and conquer. He urged the AME to "send us ministers": "You are born of God (as Moses in Egypt). Brothers consider this clearly. Don't put those talents in safes . . . use them . . . to purchase the freedom of your brothers in South Africa, or in the whole of Africa." Remarkable for its understanding of how an AME presence in South Africa would further the cause of African autonomy and encourage a Pan-African vision, Tule's communication was exactly what Turner wanted to hear. He called it "a heavy indictment against the white missionaries in South Africa" and "a strong appeal to the A.M.E. Church to come and deliver them from the chains of sin, and the treachery of our brothers in white."[50]

In 1896, the Ethiopian Church voted to seek an affiliation with the AME Church and dispatched the Reverend James Dwane to the United States to make the arrangments. With little hesitation, the black American Methodists agreed to make the Ethiopians part of their denomination. In 1898, Turner came to South Africa to cement the relationship and encourage the spread of African Methodism. He ordained 59 African ministers, consecrated Dwane as "vicar bishop" of the new district, and toured the country for six weeks, speaking with calculated moderation and emphasizing safe religious themes when whites were present; he even earned the respect of President Paul Kruger of the Transvaal, who claimed that Turner was the first black man with whom he had ever shaken hands. (Kruger apparently had some notion that AME missionaries could serve as a counterweight to the pro-imperialist British missionary establishment.) But to all-black audiences, Turner may have given somewhat freer expression to his black nationalist political views. According to an account of his visit published in 1935, he was on at least one occasion quite outspoken in his advocacy of Africa for the Africans.

> The white man does not appreciate our values, [Turner is alleged to have said] because he believes himself by divine right to be the dominant race. . . . The black is the race of the future, and one day the black man will wake up and shake off the white man's yoke. He is already rubbing his eyes and feeling his muscles. . . . The time has now come to replace them with their antiquated methods and superannuated principles. Our new doctrine is more suited to the African awakening, and only the sons of New Africa may be trusted to propagate it, not any aliens.[51]

If Turner actually made such a direct connection between the struggle for religious independence and the future rejection of white political hegemony in South Africa, he would not have been the first to do so. Joseph Booth, a radical white missionary, had made the same linkage two years earlier. Booth, who was born in England and lived in New Zealand before becoming a freelance Baptist missionary in Nyasaland in 1892, turned up in Natal in 1896 with a scheme for African solidarity and self-help that he called "The African Christian Union." Its aim was to "pursue steadily and unswervingly the policy of AFRICA FOR THE AFRICAN and look for and hasten by prayer and united effort the forming of the AFRICAN CHRISTIAN NATION by God's power and in his own time and way." Booth's thought clearly demonstated how the Ethiopianist tradition could be harnessed to a political Pan-Africanism. Having previously traveled to the United States and established contact with black emigrationists, he endorsed their view that African-Americans would play a leading role in the redemption of Africa—a process that he saw as a fusion of evangelization and political nationalism. Part of his grand scheme was the same proposal for government-assisted Negro emigration to Africa that Bishop Turner endorsed. Although he aroused some interest in Natal, Booth did not succeed in starting a movement in 1896. The assembly of 120 educated Africans to whom he presented his scheme rejected it after prolonged and serious deliberation. The fact that it came from a white man may have aroused suspicion, but a more likely reason for Booth's failure was the generally cautious and accommodationist attitude of the Kholwa (Christianized African) community of Natal. Nevertheless colonial authorities were alarmed that prominent Africans attended such a meeting, conveyed informally their unmistakable approval of much that was said, and seriously considered endorsing the scheme.[52]

Unlike Booth's Christian Unionism, the AME Church did get a permanent foothold in South Africa, but to do so it muted the radicalism associated at times with Turner and played down the possible political implications of religious separatism. Between the time of Turner's triumphal visit and the report of the South African Native Affairs Commission in 1905, the AME and other independent black churches were the object of intense white suspicion. Many settlers took Booth's doctrines to be the fixed convictions of all religious separatists, and Ethiopian congregations were widely viewed as nests of sedition, the loci for agitation against European rule and possible preparation for armed rebellion. Panic about Ethiopianism was particularly intense in Natal, where AME missionaries were explicitly banned. It was possible for the African Methodists to evangelize in the other provinces, and churches were founded in the Cape, the Orange Free State, and the Transvaal. But they often had difficulty getting permission to perform legal marriages or acquire property. Church authorities, especially the African-American bishops and missionaries, gave high priority to reducing white fears and obtaining legal recognition. In their testimony before the Native Affairs Commission that met between 1903 and 1905, AME clergymen insisted that their church had no hidden political agenda and in fact accepted British colonial domination as a civilizing influence beneficial to Africans. This accommodationist posture was successful in getting the commission to report that fears of a politicized Ethiopianism had been exaggerated and that separate churches posed no threat to European rule. As a result, the panic subsided and the repression of black religious independency was eased considerably.[53]

Another problem that the AME Church's South African mission had to overcome was the tension created by the paternalistic attitudes of African-American churchmen and missionaries toward native Africans. The black Americans had no doubt that they were culturally more advanced than their African co-religionists, and some of them had a very low opinion of traditional African cultures. The Africans, on the other hand, were attempting to mediate between traditional African beliefs and what they took to be the values and practices of Christian civilization. Some of them deeply resented the condescension of the black Americans, who at times seemed almost as intolerant and autocratic as the white missionaries from whom the Africans had separated. Schism and secession began almost immediately after the incorporation of the Ethiopian Church into the AME Church in 1898. Bishop James Dwane, reacting to what he perceived as the failure of the American church to give full recognition to his consecration and to fulfill its promises of financial assistance for the building of a school, withdrew from the AME Church in 1899 to head an African order within the Anglican Church. Subsequent bishops were African-American clergymen who did not share the militant Pan-Africanism of Bishop Turner and were less likely to conceal their feeling of cultural superiority to Africans. They were generally capable administrators, but they did little more than white missionaries would have done to encourage African self-determination. Consequently, there were further secessions, leading up to the major schism of 1908 that reestablished the Ethiopian Church as an autonomous entity. Africans were seeking independence from what some regarded as a new form of ecclesiastical imperialism, and the number of distinct Ethiopian denominations proliferated. What gave the AME Church staying power, more than anything else, was the return of the African students who were educated at Wilberforce and other colleges in the United States. The church provided fellowships to about 30 such students at the beginning of the twentieth century, and several of them became ministers who came back to South Africa to serve in AME pulpits. Thoroughly imbued with the values of the African-American middle class but more sensitive to African needs and expectations than American-born blacks were likely to be, they served as an effective bridge between the American mother church and its South African adherents.[54]

If an accommodationist avoidance of political assertiveness and the promotion of self-help in the spirit of Booker T. Washington became the keynotes of official AME policy in South Africa, it is doubtful that religious independency in general can be made to fit this mold. Although separatist church leaders frequently disavowed political concerns and proclaimed that their members were obligated to obey the "authorities established by God," it was difficult to prevent members from making their own connections between religious and political independence. White fears that religious autonomy would lead directly to violent rebellion may have been exaggerated, but the danger that a charismatic prophet or messiah could turn popular religious enthusiasm into a divinely inspired mass uprising was a real one—as the revolt in neighboring Nyasaland in 1915 demonstrated. Led by John Chilembwe, an African protégé of Joseph Booth who was affiliated with the American black Baptists, it demonstrated that a visionary Ethiopianist preacher could arouse Africans to violent resistance against white domination. Nothing equivalent occurred in South Africa. (The nearest approximation was in 1921 when the African prophet Enoch Mgijima's refusal to disperse his encamped followers at Bulhoek resulted in their being massacred by white troops.) The Zulu rebellion that took place in Natal in 1906 was essentially a last stand of Zulu traditionalists against European political and cultural hegemony. Some members of the independent African churches supported the mainly non-Christian rebels, but others remained loyal to the colonial authorities. South African Ethiopianism before the 1920s could not be accused of inspiring revolutionary millennialism; as whites became aware of its essentially nonviolent character, they became more tolerant of the independent churches.[55]

But a failure to rise up did not mean that members of separatist churches were politically inactive. The history of African resistance to colonial rule in rural districts and African reserves is just beginning to be written, but it is already evident that Ethiopian churches were sometimes at the root of organized opposition to government schemes for the systematic control and exploitation of African peasants. In the Qumbu district of the Transkei, for example, members of Mzimba's African Presbyterian Church led the effective campaign of 1902 for noncooperation with the Cape government's effort to set up appointed councils of Africans to rubber-stamp the decisions of government officials and take the heat for imposing unpopular taxes. A recurring pattern that developed in rural areas saw separatist Christians cooperating with traditionalists against "progressive" government programs that were sometimes supported by Africans in mission churches.[56]

Despite the lack of an explicit political program, the independent churches undoubtedly provided an outlet for feelings of resentment of white domination that went beyond a desire to be left alone to practice their religion as they saw fit. The Rev. James Dwane, who had broken with the AME Church to form a separate black "Order of Ethiopia" within the Anglican Church, testified to the South African Native Affairs Commission of 1903–5 that the Ethiopians did not officially oppose white rule, "but in the practical teaching and training the tendency is to set the black race against the white race." Other commission informants testified that Ethiopians condemned those who remained in the white-dominated mission churches as toadies of the white man and traitors to their race. Respect for Europeans and acceptance of their supremacy was apparently being eroded in the African independent churches.[57]

Clergymen of the Ethiopian churches took an active role in the protest organizations that were formed in the South African colonies between the Anglo-Boer war and the emergence of the Union of South Africa in 1910. Mzimba, Dwane, and James Goduka—leader of the African National Mission Church—were all active in the South African Native Congress of the Cape (the organization for which A. K. Soga was a leading spokesman); F. M. Gow of the Capetown AME church served for a time as president of the African People's Organization. In the Transvaal, the most visible minister of an independent church who engaged in agitation for African political rights was H. R. Ngcayiya, leader of the Ethiopian Church, founded in 1908 as the result of a secession from the AME Church that may have been a reaction to its political moderation. In 1912 Ngcayiya helped to found the South African National Native Congress, later the African National Congress, and served for many years as one of its chaplains. In Natal, Abner Mtimkulu, minister of the Bantu Methodist Church, was for decades a leading figure in Natal's black politics and eventually became a senior chaplain of the ANC. Obviously political passivity was not the general reaction of religious separatists in the early twentieth century.[58]

It would be misleading, however, to describe the Ethiopian ministers who were politically active in the early twentieth century as radical black nationalists. Joseph Booth's goal of black political self-determination, as distinguished from purely religious separatism, was not part of the public agenda of the politically active clergy and laity who were based in the independent African churches. In fact they normally advocated a liberal reformism that differed in no essential way from the stances of African protest leaders such as A. K. Soga, the Reverend Walter L. Rubusana, and the Reverend John R. Dube, who had remained clergymen or prominent laymen in white-dominated mission churches. Mtimkulu, for example, was a follower and close associate of Dube, a non-separating Congregationalist and one of the most moderate of the African political leaders of the period between 1900 and 1940. It is therefore difficult to make direct and unambiguous correlations between Ethiopianist religious separatism and the mainstream movement for African rights that coalesced in the National Congress. Ethiopianist clergy may have been prominent in the founding of the movement, but it is difficult to detect any distinctive contribution that they made to its program or ideology. In other words, early religious separatism did not straightforwardly or inevitably translate into the Africanist or "orthodox nationalist" view of the South African future that was suggested by the slogan "Africa for the Africans." In the 1920s, Pan-Africanist rhetoric occasionally appeared in the sermons and speeches of prominent clergymen, but it was at least as likely to issue from African ministers in the white-dominated denominations as from those in the independent churches. Of course private attitudes may have differed from public positions. The independent churches were not well situated to carry on radical political agitation. They existed on the sufferance of the government and could not minister to the souls of their adherents if suppressed or denied recognition.[59]

The linkage that some historians would like to make between religious independency and the developing liberation struggle in South Africa is thus very difficult to substantiate if one concentrates on religious differences among an African elite that included clergy and socially prominent members from both white-controlled and independent churches. From the beginning, a majority of the leaders of the National Congress remained within the mission churches and this proportion grew over time; contrary to what might have been anticipated, the correlation between religious independency and mainstream political activism became weaker rather than stronger as the twentieth century progressed. After a rapid expansion in the 1920s, the Ethiopian churches themselves went into decline, losing members to separate churches of a different type—the resolutely other-worldly Zionist or Pentecostal churches that eschewed political activity and concentrated on faith healing and religious ecstasy. Furthermore, as some of the post-World War I mission churches became relatively more liberal in their attitude toward African authority and participation in church governance, the tendency of missionized Africans to secede and form their own churches of the same denominational type declined. The educated elite that dominated the African political organizations of the interwar period were often pious Christians, but they rarely made any explicit connection between their reformist political program and divine prophecies concerning the special destiny and mission of the black race.[60]

On the surface, therefore, Ethiopianism merely contributed to a reformist black politics in South Africa, helping to inaugurate a liberation movement but not adding anything essential to its spirit and ideology. On a deeper level, however, the beliefs that it nurtured may have helped to set in motion an Africanist countercurrent to the dominant tradition of protest for equal rights with Europeans in a common society. In 1908, the New Kleinfontein and Boksburg Native Vigilance Association of the Transvaal responded to the failure of political agitation to prevent the growth of discriminatory legislation in the province by declaring a day of prayer and humiliation and offering a "petition to God" rather than to the unresponsive white authorities. The God to whom they addressed their appeal was the God of the Ethiopians:

> We feel that our sympathies should be broad enough to include the whole of the African races when we approach our Maker. For instance, the atrocities which our brethren are suffering under the administration of the Congo Free State should appeal to us with a loud voice for our sympathy and prayers. Undoubtedly the time has come for the sons of Africa to stretch forth their hands unitedly, as was prophesised [*sic*] of us in Psalm 68 verse 31. . . . [God] will bless us and send a wave of his spirit, which will pass through the whole of Africa from the Cape to Egypt, the effects of which will be felt even by our brethren in America, who come from the same original stock as ourselves.[61]

So stated, the Ethiopian idea called for the fraternal union of blacks throughout the world in opposition to white oppression. Noticeably absent were millennialist expectations for the triumph of brotherly love between blacks and whites in South Africa—the hope of mainstream Christians and liberal reformists. Race apparently provided the natural limit of human sympathy. An appeal for divine help and not an incitement to revolution, this "petition to God" revealed a way of thinking about the world that could readily be turned to militant or revolutionary purposes. It contained seeds of the radical Africanism of later decades—the claim that South Africa was a black man's country and not a multi-racial one, which meant that Africans alone had the right to determine its destiny.[62]

The prayer also shows what could happen to the Ethiopian myth when it passed from African-American to African hands. The American version had made New World blacks the providential redeemers of Africa; the Transvaal Vigilance Associations saw the "wave of His spirit" originating in Africa itself, conquering the continent, and later being felt "even by our brethren in America." Such Africanism was undoubtedly widespread among supporters of Ethiopianism in South Africa and helps to explain much of the resistence to the AME Church's emphasis on African-American agency and guidance. But African Methodist congregations in remote rural areas and native reserves enjoyed considerable autonomy from the church's cautious hierarchy and could themselves serve as centers of Africanist opposition to white hegemony. In the 1920s, a women's boycott of white traders and government-sponsored schools in the Herschel district in the Northeastern Cape was associated with a shift of membership from the regular Methodist to the African Methodist Church. According to historian William Beinart: "Religious separatism was being linked to a new political radicalism." As we will see later in a discussion of black populism in the 1920s, ideas of black redemption that had a distinct Ethiopianist flavor and were sometimes associated with independent black churches were influential at the grass roots of black South African life in the years immediately after the First World War.[63]

There is no simple answer from the evidence presented to the question of how and to what extent religious separatism and ethnocentric conceptions of Christianity contributed to black political resistance to white supremacy in South Africa. A leading authority on African separatist churches, Bengt Sundkler, failed in his classic study of *Bantu Prophets* to find "sufficient proof of any definite political trend." Although there was "anti-white" propaganda in some of these churches, their leaders and members had generally proved "loyal" to the South African government. Yet he could not deny that they were somehow "nationalistic." Sundkler is noted for establishing a distinction between two types of separatist churches. Ethiopian churches proper were the result of secessions from white denominations, and they retained much of the theology and ritual of their churches of origin. "Zionist" churches originated from the mission work at the beginning of the century by white American missionaries from the apocalyptic religious community in Zion, Illinois, and were thus initially inspired by the Holiness and Pentecostal movements in American Protestantism. Waiting for the end of the world, faith healing, abstinence from alcohol and tobacco, adult baptism, and (among some groups) speaking in tongues became earmarks of the Zionist churches. But they did not remain long under white influence. Proliferating into a bewildering variety of sects, they prospered as a result of their ability to synthesize elements of traditional African culture with Holiness or Pentecostal practices. Spirit possession, magical healing, and other pre-Christian practices (including in some cases polygamy) were reconciled with evangelical Christian conceptions of miraculous conversion and the power of prayer. Many of these sects made use of Ethiopian mythology, a tendency that was accentuated during the international crisis resulting from Italy's invasion of Ethiopia in the 1930s. Some of them even claimed to be in apostolic succession as the result of a fictive relationship with the Ethiopian Orthodox Church. Notably lacking, however, was any inclination to go from a narrowly religious interpretation of the Ethiopian vision to one that could serve as a foundation of political nationalism.[64]

Anthropologist Jean Comaroff has found in the ostensibly apolitical and otherworldly Zionist churches a form of cultural resistance to white domination. Comaroff analyzes the decline of the churches Sundkler classified as Ethiopian in an insightful and provocative fashion:

> The dwindling of the African Independent movement has been due to the fact that it occupies the middle ground in an increasingly polarized community. Its reformism prevents it from offering the radical alternative presented by the Zionists, but it also lacks the emblems of established status—the

material trappings, educated leadership, and rational bureaucratic structure favored by the liberal bourgeoisie. In the latter half of the twentieth century, as black political consciousness developed its own secular models, both liberal and radical, the embryonic nationalism of a biblical Ethiopia lost all salience and was quite simply eclipsed.[65]

Comaroff is convincing when she attributes the decline of Ethiopianism to the modernization and partial secularization of the African bourgeoisie that dominated the mainstream protest movement and perceptive in her recognition that Zionism was primarily a lower-class or proletarian phenomenon. But her attribution of political radicalism to the Zionists depends on a questionable conception of the political. Most historians and social scientists assume that politics involves the conflict of groups for power over the state. Max Weber's classic discussions of power and domination have allowed sociologists to delineate a political sphere that can be distinguished from other realms of human thought and action. Ultimately, politics in this sense arises from the efforts of some people to make other people *act* in a certain way. Culture, on the other hand, is about symbols and representations, which may directly legitimize power or, conversely, provide an impetus for challenging established authority. But a third possibity also exists; it may give rise to a world-view that conflicts in essential ways with that of the ruling group but nevertheless has the practical effect of encouraging political acquiescence or passivity. Alienation from a dominant culture may therefore be either political or apolitical, and it is the task of the historian to determine which.[66]

As Sundkler recognized, the Zionist churches in South Africa have normally preached loyalty to the state and enjoined their members to avoid any involvement in political protest. In 1985, the largest Zionist group in South Africa publicly and ceremoniously described the white supremacist regime as a government instituted by God and thus to be obeyed as a matter of religious duty. For its part, the South African government has never felt threatened by Zionism or worried about its deviations from white cultural norms. It is in fact doubtful that the power of the apartheid regime ever depended to any significant degree on cultural hegemony in the sense of persuading the oppressed to believe in the world view of the dominant group. Enough force to induce a sense of powerlessness is the essence of minority rule in plural societies, not the search for cultural consensus. Cultural diversity has in fact been encouraged by the regime because of its ability to divide the black population and divert black energy from the struggle for physical power to the realm of purely psychic satisfactions.[67]

The contribution that Ethiopianism in its heyday made to the African liberation struggle was to give an early direction to black anger at being deprived of self-government. Its ability to reinforce the protest movement depended more on its capacity to absorb and reshape elements of the oppressor's culture than on its ability to project an antithetical Africanist view of the world—which is another way of saying that its Africanism was more political than cultural. Because they had absorbed from Western secular thought the idea that peoples have a right to self-determination, and from the scriptures brought by European missionaries a special sanction for their own particular struggle, Ethiopianists had been able to begin challenging the West in more direct ways than if they had resisted Western cultural influences more successfully. By the 1930s and '40s, Ethiopianism would be less effective in sustaining opposition to white domination than Marxism, democratic liberalism, or a de-Christianized Pan-Africanism. One lesson that might be drawn from this aspect of South African history is that cultural insularity in the face of oppression can be a prescription for political impotence.

What Happened to American Ethiopianism?

In the United States, as in South Africa, a politicized Ethiopianism got a new lease on life in the years immediately after World War I; but by the late 1920s this scripturally based black nationalism sank beneath the surface of political struggle into the realms of a mostly symbolic or cultural creativity, surviving as one element of an essentially escapist lower class black religiosity. St. Clair Drake has described the process well. As a result of "the gradual secularization of black leadership" beginning at the end of the nineteenth century, he argues, " 'vindication of the race' passed from the hands of those who believed in Providential Design and Biblically sanctioned Ethiopianism into the hands of trained historians, anthropologists, and archaeologists." Ethiopianism lingered as "a basic subsidiary reinforcing myth in Marcus Garvey's Universal Negro Improvement Association (UNIA)," but the Pan-Africanism that survived into the era of decolonization was mainly under the guidance of secular intellectuals. Nevertheless, according to Drake, Ethiopianism survived at the grass roots:

> By the mid 1920's Ethiopianist ideas were so deeply embedded in the urban subcultures of the Afro-American urban communities, from the impact of the Negro church and the UNIA, that an apperceptive mass was present to which founders of cults and social movements could appeal. All of these cults provide meaningful schemes of living to the identity quest for a small fraction of the black population in the United States. Their political influence as organized groups is minimal. . . .[68]

Randall Burkett has described in detail how Ethiopian mythology was incorporated in the "black civil religion" of the Garvey movement of the 1920s. He notes that Psalms 68:31—"Ethiopia shall stretch forth her hands to God"—was the most popular text for the sermons preached to UNIA meetings and that the official catechism for members of the movement interpreted the passage as proving "That Negroes will set up their own government in Africa with rulers of their own race." Not surprisingly, the black clergymen who supported the UNIA were also likely to be enamored of this passage. Garveyism will receive detailed treatment elsewhere in this study. In the present context, however, it suffices to say that Garveyism was transitional between nineteenth century Ethiopianism and the largely secularized Pan-Africanism that succeeded it. The rise of the UNIA was the first and only time that an association of Ethiopianist religious ideas with a militant form of political nationalism became the wellspring of a mass movement.[69]

The black churches of the United States, as we have seen, were characterized by a "dialectical tension" between accommodation and resistance. But the churches of the Holiness-Pentecostal type—which were the most likely after the 1920s to invoke Ethiopianist mythology—were also the least likely to encourage political activism or militancy. The main loci of political engagement by black Christians, as would become clear in the era of Martin Luther King, Jr., would be the mainstream black Baptist and Methodist churches (especially the former). Although Ethiopianism had been an element in the thinking of these churches since their formation, the main emphasis was increasingly on the apparently nonracial interpretation of the Gospel that was reflected in the AME Church's motto "Christ our Redeemer, Man our Brother"—the universalist message that King expressed eloquently and translated into militant politics as nonviolent direct action in the 1950s and '60s. St. Clair Drake described this version of black Christian thought as a moderate alternative to Ethiopianist nationalism, one that managed to retain a special role for blacks in the drama of human redemption: "Since black nationalism seemed unrealistic to moderate black leaders as a goal for American Negroes, the concept [of 'black messianism'] as developed in the United States from Booker T. Washington to Martin Luther King, Jr., (with overtones even in Stokely Carmichael) was that their suffering would 'save' America and the world."[70]

The portrayal of blacks as divinely inspired agents for the redemption of America from the sins of racism and selfish materialism, which King brought to full consciousness, did have a subtle relationship to the Ethiopianist tradition. The belief, originating in the black nationalism and romantic racialism of the early to mid-nineteeth century, that blacks had a natural aptitude for Christian self-sacrifice became in the 1950s and early '60s the implicit basis for a crusade "to redeem the soul of America" and realize its promise as an egalitarian interracial society. King's dream substituted an integrated "beloved community" for the original Ethiopianist image of a black Christian utopia, but it shared with more overtly nationalistic versions of black Christianity the conviction that the African-American experience of enduring the agony of slavery and segregation without losing faith in a God of love and justice meant that the freedom struggle of black people had divine approbation and would someday bring full liberation from white oppression. But to achieve that goal without the separation from whites that Turner and Garvey had advocated would require the conversion and redemption of the oppressors. Only as truly exemplary Christians could blacks hope to accomplish that enormous task.[71]

· 3 ·

Protest of "The Talented Tenth": Black Elites and the Rise of Segregation

The Making of Segregation

During the first two decades of the twentieth century, the relatively new term "segregation" was at the center of public discourse about race relations in both the United States and South Africa. Its principal use was to describe new laws and public policies designed to increase the physical separation of blacks and whites. In the United States, the word was first used in this way during the 1890s when southern states began passing laws that required separate accommodation for blacks and whites in railroads, passenger boats, and streetcars. In South Africa, it originated shortly after the turn of the century, and from the beginning referred mainly to proposals for controlling interactions between whites and Africans by dividing the country up between "native reserves" and the much larger area where white interests and prerogatives would be firmly entrenched. The common element of both forms of segregation was the use of law and governmental action, in a more systematic and comprehensive way than had been attempted previously, to prevent any mixing of the races that might conceivably engender conflict and threaten white dominance or social superiority.[1]

If segregation means simply the enforced social or territorial separation of dominant and subaltern "races," the practice (if not the term) had a long history in both countries. As Alexis de Tocqueville noted in the 1830s, black and white Americans lived thoroughly separate social lives in cities like Philadelphia, a division that prevailed even among incarcerated criminals and beyond the grave in the form of separate burial grounds. Historians of race relations in antebellum cities have found that separate-and-unequal was the norm in public contacts between whites and free blacks in both the North and the South. After emancipation, there was a brief challenge to some forms of public discrimination during the Reconstruction years, but the older pattern quickly reasserted itself. When the Radicals opened new kinds of opportunities and facilities to the mass of freedmen in the South—public schools, orphanages, hospitals, insane asylums, and so on—they rarely tried to do so on an integrated basis. Their efforts were controversial enough, they realized, without incurring the added burden of directly challenging the racial mores of a region (and a nation) that remained dead-set against the "social equality" of blacks and whites. Blacks themselves contributed to the post-Civil War pattern of separation; they used their new freedom to join or establish separate black churches, fraternal organizations, and other kinds of associations. If one includes in the term segregation all forms of de facto, customary, and voluntary social separation of the races, the South was already a generally segregated society before the passage of the notorious Jim Crow laws of the 1890s and early 1900s. This circumstance has led some historians to argue that the legislation merely confirmed a preexisting pattern rather than constitute a new departure in racial policy.[2]

In South Africa, too, there were policies that could have been described as segregationist well before the word became shorthand for a comprehensive "solution to the native problem" after the war between Great Britain and the Afrikaner republics. The Dutch Reformed Church in the Cape instituted separate services for Europeans and people of color in 1857. Applying the principle of "no equality in church or state," the Afrikaner republics simply excluded Africans and Coloreds from virtually all institutions and facilities open to whites, thus guaranteeing that whites and blacks would inhabit different worlds except to the extent that they came together as masters and servants. Finally, the English colony in Natal pioneered formal geographical or territorial segregation by assigning most Africans to extensive reserves that were governed by a set of laws and regulations different from those that applied to white settlements. It can be argued with some persuasiveness that all the ingredients of a segregationist policy were in place before 1902 and thereafter merely had to be elaborated and made more systematic.[3]

If segregation is viewed broadly as any form of contrived separation between groups defined as racially distinct and inherently unequal, one would have to conclude that it flows directly and irresistibly from deeply rooted and persistent patterns of unequal ethnic status. From this perspective, the "growth" of segregation is merely the emergence of new contexts or situations to which the logic of racism can be applied. As historians have often pointed out, segregation is mainly an urban phenomenon; in rural situations where the distinctions between masters and servants (or slaves) are firmly and clearly established, there is no need to take conscious action to separate the races in order to maintain social distance and racial hierarchy. But in cities—places where people who do not know each other bump and jostle in public places—there is no way to indicate superiority or inferiority without customs or laws requiring differential or separate use of facilities. Similarly, railroad and streetcar segregation makes no sense before such means of transportation come into general use and become essential means of locomotion for both blacks and whites. Or, to take a specifically South African example, there is no reason to establish separate locations or townships for Africans until the demand for their labor brings large numbers of them to the cities. Segregation might therefore be viewed as the normal, virtually inevitable adjustment of a society with an established pattern of racial inequality to the physical necessities of social and economic modernization.[4]

But the emergence and extensive usage of the term "segregation" around the turn of the century reflected changes in the way race relations were conceived and discussed, as well as in the settings of racial interaction. One way to view this ideological transformation is to see it as reflecting a shift from "paternalist" to "competitive" relationships. When blacks or Africans were slaves or the serf-like dependents of white landowners, a form of direct domination was possible that was paternalist or patriarchal in form if not necessarily in spirit. But the abolition of slavery and the growth of an industrial capitalist economy with a white working class meant that the blacks within the modernizing sector of society were actual or potential competitors of lower-class whites, many of whom clamored for protection from employers who might to be tempted to hire black or other nonwhite workers in the belief that they could be paid less because they had lower expectations and were less likely to be members of unions. Real or incipient class conflict among whites could breed intense racial antagonism in an industrializing society. The relationship between class divisions among whites and status differences between white and black "castes" changed radically when the United States and South Africa underwent the great transformation from predominantly agrarian to urban-industrial societies. The emergence of systematic segregation as the policy of governments with authority over large areas—as opposed to segregation by local customs or regulations—was in some sense a response to the forces released by these massive social and economic changes and a reflection of the new configurations of race and class that they engendered.[5]

But it would be unwise to apply this structural explanation for the rise of legalized segregation in a mechanical way. Industrializing capitalism does not inevitably promote legalized racial segmentation. Indeed, the response of nineteenth-century liberals to slavery and official racial discrimination suggests the opposite. Radical Reconstruction in the United States and, in a more equivocal way, the career of Cape Liberalism in South Africa would seem to show that liberal adherents of modernization have greater affinity for official color-blindness than for legally sanctioned color lines. Furthermore, class-conscious white workers found no mandate for color bars in their radical-republican or socialistic ideologies. Overwhelmingly, the modernist social and economic doctrines associated with capitalist economic development proclaimed the ascendancy of class over ethnic or racial identities. It was, as we have seen, precisely their adherence to these doctrines that gave heart to the emerging middle class of educated and "Westernized" blacks in the nineteenth century. The simplest plausible explanation of how we derive Jim Crow or "Native Segregation" from such unpromising ideological terrain is that inherited or ingrained white supremacist attitudes were so powerful that they could shape—or deform—the new order to reflect the prejudices derived from the earlier agrarian pattern of racial domination.[6]

But this cultural explanation, like the class analysis associated with Marxism, risks becoming mechanical and deterministic. A closer examination of the rise of segregation in the United States and South Africa shows complex interactions of race, class, and politics that make it arbitrary and ahistorical to single out one element and call it efficacious. C. Vann Woodward's multi-causal view of "the strange career of Jim Crow" in the American South retains considerable cogency. Although his critics have been correct in showing that there was much de facto and customary racial separation before the 1890s, they have not demonstrated that it attained anything like the rigor or comprehensiveness that was achieved after the Jim Crow laws were passed. What still seems remarkable, at least from the standpoint of those who view racism as a consuming, all-powerful force, is that it took some time for segregation to be implemented in new facilities for mass transportation that were the center of white supremacist attention. Blacks, for example, rode the streetcars in an unsegregated fashion in many cities until after 1900. Although railroad and steamboat segregation began in the 1890s or earlier, it was not until the first decade of the twentieth century that the formal Jim-Crowing of all public facilities took place throughout the South and showed signs of spreading to the North.[7]

The most obvious explanation for the timing of legalized segregation was the changing policy of the federal government. Although active enforcement of the Fourteenth and Fifteenth amendments ended in the 1870s, the Republican party continued to protest the treatment of southern blacks and threaten new legislation to protect their rights until the 1890s, when it became clear that southern white supremacists no longer had anything to fear from the party responsible for emancipation and Reconstruction. Meanwhile, a Supreme Court dominated by Republican appointees was undermining the constitutional foundation of federal action against segregation laws. In 1883, it declared unconstitutional the Civil Rights Act of 1875—a law designed to provide equal and unsegregated access to most public facilities. In *Plessy v. Ferguson* (1896) the court found that laws recently passed mandating segregation in public transportation did not violate the "equal protection" clause of the Fourteenth Amendment if the separate facilities provided for whites and blacks could be regarded as equal. From the 1890s on, therefore, no serious political or legal obstacles stood in the path of southern segregationists.

A pragmatic devotion to party interests—the willingness to abandon old positions that have become unpopular and change course in ways that betray previous ideological commitments in order to win or retain office—is a recurring feature of American politics. The shift in Republican race policy after Reconstruction that came to fruition in the 1890s was an example of such an adjustment. Republicans, who during Reconstruction had believed that they needed southern and black votes in order to control the federal government, had learned by the 1890s that they could win without blacks and the South and without the political risk of alienating white supporters by vigorously espousing the cause of an increasingly unpopular racial minority. By this time, the party's most influential constituency was big business and its camp followers rather than the small entrepreneurs, farmers, and artisans whose democratic-republican idealism had given the party its radical and reformist edge during the era of sectional controversy and the War for the Union. In the age of McKinley and Republican hegemony in the Midwest and parts of the Northeast there seemed no profit in resisting Jim Crow, particularly since the African-American voters who remained important to the party in some parts of the North had nowhere to go except to a Democratic party that harbored active southern segregationists.[8]

On the state level in the South, *Democratic* political partisanship played a major role in the movement to legalize segregation. The Populist movement of the 1890s briefly threatened Democratic hegemony, especially on those occasions when Populists appealed to African-Americans to vote their class interests as poor farmers rather than wasting their ballot on Republicans or succumbing to planter pressure to vote Democratic. The Populists failed in their efforts to reorient southern politics on a class basis, and one-party Democratic rule was reestablished by all the former Confederate states by 1898. But the politics of racial antagonism and white solidarity that the Democratic establishment had used to defeat the Populists survived the political challenge that had called it forth. Race-baiting became an avenue to power for personalities and factions within the Democratic party and encouraged a climate of opinion in which it was good politics to propose new ways to humiliate blacks and block the path to "social equality." In ways that no one explicitly acknowledged, a kind of bargain was struck between the economic elite of planters, merchants, and manufacturers and the mass of white farmers and mill workers. Although it was hardly a fair deal, lower class whites at least enhanced their social status at the expense of blacks (the dictum that the lowest white was superior to the highest black was now enshrined in law), while the elite, which did not need segregation laws to ensure its position on top of the heap, retained its dominance over the South's exploitative, low-wage economy. Legalized segregation was thus a key element in a southern settlement based on one-party government, low wages, and official white supremacy that would last for nearly half a century.[9]

But this Woodwardian explanation for the rise of Jim Crow ignores one crucial element. If African-Americans had accommodated without protest to customary segregation and discrimination there would have been no need to pass laws to sustain the color-caste system. Historians have only recently begun to uncover the record of black assertiveness after Reconstruction that made state action necessary to guarantee white prerogatives. Southern African-Americans were far from docile and compliant to white prejudices in the 1880s and '90s. To some extent, perhaps, adult freedmen of the Reconstruction period had needed prompting from those blacks who had known freedom prior to the war before they could begin to shed the deferential habits of slavery. But the postwar generation that had scarcely known slavery and had grown up hearing the egalitarian rhetoric of the Radical Republicans was primed to claim the equal rights supposedly sanctioned by the Fourteenth and Fifteenth amendments. And it was in fact much better prepared to exercise these rights than its forebears had been.

Between emancipation and 1900, the black illiteracy rate declined from over 90 percent to slightly more than 50 percent; from 1880 to 1900 the percentage of black farmers in the Deep South who owned their own land increased from between 3 and 8 percent (depending on the state) to an overall figure of around 25 percent. In southern towns and cities, a small but rapidly growing black middle class of small businessmen, professionals, and educators came into existence. The new class of educated, aspiring, and modestly achieving southern African-Americans collided with the de facto color line at points where it denied them the social status that they had earned by their self-improving efforts. When railroads acting on their own authority in deference to the wishes of white passengers denied well-dressed, well-behaved blacks access to the first-class, or "ladies'," car and tried to seat them in the second-class, or "smoking," car where rough, tobacco-spitting males of both races congregated, incidents of protest and even physical resistance occurred and lawsuits often resulted. The earliest of all the railroad segregations laws, the Tennessee law of 1881, was an effort to accommodate protesting middle-class blacks while respecting the racial sensitivities of genteel whites; it simply provided for separate first-class cars for blacks without requiring the segregation of second-class facilities.[10]

Opposition to racial discrimination was not limited to an emerging black elite. Many times in the 1880s and '90s, black crowds formed in southern cities to resist the arrest or brutal treatment of blacks by white policemen. When streetcar segregation was introduced around 1900, it was the elite that led the boycotts of the transit lines that often resulted, but the short term effectiveness of these ultimately unsuccessful campaigns would not have been possible without massive support from the rank and file of black laborers and servants. Black militancy in the late nineteenth- and early-twentieth-century South was diffuse, localized, and almost totally devoid of support from the North or the federal government. It provoked segregationist legislation by helping to persuade white supremacists that informal methods would not suffice to keep an increasingly assertive black community in its place. It also called forth futile boycotts and other protests against such laws once they were passed. In the end southern blacks accommodated to the Jim Crow system because, absent federal intervention on their behalf, they lacked the power to do anything else. That state of forced acquiescence would last until the 1950s and '60s.[11]

The rise of what was called segregation in South Africa in the early twentieth century has many distinctive features that would seem to make direct comparison with what happened in the United States tenuous and difficult. It was above all a response of white authorities to the unsettled relations between Europeans and Africans in the wake of the South African War of 1899–1902. As the former colonies and republics of the region moved toward unification under British auspices between 1902 and 1909, the question that was uppermost in the minds of influential Europeans was how to define and develop a common "native policy" that would prove satisfactory to major white interest groups. The fact that the issue was formulated in such a fashion indicated that a strong consensus existed behind white supremacy and the belief that Africans had only such rights as Europeans found to be compatible with their own security and interests.

Now that independent African societies had ceased to offer military opposition to white expansion, blacks were of interest to whites mainly as a source of labor. But "native policy" could not be based purely and simply on economic advantage. The British government and the missionary or humanitarian lobbies to which it listened had to be persuaded that a self-governing South Africa would meet some minimal standard of fairness to "subject races." Segregation emerged as the magic solution to the problem of reconciling British traditions of "fair play," settlers' aspirations for self-government within the British empire, and the desire of white farmers and mine owners to have ready access to cheap and exploitable African labor.[12]

The South African Native Affairs Commission, which was dominated by British colonial officials, met and took testimony between 1903 and 1905. In its report it set the pattern for future racial policy when it recommended that the country be divided into white and African territories. The basis for such a separation already existed in the heavy concentration of African population in areas such as the Transkei, Zululand, and the northern Transvaal, where white farmers were few and most land remained in the possession of Bantu-speaking peasant communities. What could be more humanitarian and in accordance with enlightened principles of "aborigines' protection" than to declare these areas "native reserves" and thus halt white expansion on their borders? Of course these areas were not economically self-sufficient, even in 1905. Large numbers of Africans would have to leave them to work in white mines, farms, and households for extended periods in order to sustain themselves. But if they did so, they could be regarded as transients or temporary sojourners in white areas rather than as candidates for citizenship. Their political rights, the commission concluded, could best be expressed by a measure of local self-government in the reserves and through some form of separate representation in the central political bodies of a South African union.[13]

The beauty of this scheme was that it seemed to offer something to almost everybody. To whites in general it gave privileged access to the approximately 90 percent of the country that they already controlled, which included the best farmlands and richest mineral deposits. To white employers it offered a migratory labor force that did not have to be paid a family wage because, in theory at least, African workers' families could still be supported in part from agricultural activities in the reserves. To African chiefs and their traditionalist followers, it offered a chance to preserve ancestral ways and some measure of their pre-conquest authority and autonomy. Indeed it seemed to be a recognition of the enduring importance of their ethnic identities and cultures. No wonder that many liberal "friends of the natives" found the segregation idea seductive.

Putting segregation into effect was a long drawn-out process that did not reach fruition until the 1930s. Rather than being imposed by a British imperial authority that had some concern for African as well as settler interests, it was done piecemeal by an all-white South African parliament. Consequently it was limited to separatist measures that served the immediate needs of white interest groups or would be attractive to a white electorate that was periodically warned of a "black danger" if Africans were not firmly controlled. African leaders had no say in how segregation was implemented; quickly dashed were the original hopes of some of them that territorial separation could be carried out in an equitable and mutually advantageous way.[14]

The constitution of 1909 that unified South Africa and made it self-governing contained no explicit provision for segregation. A white monopoly of political power in the new state—only Europeans were eligible for election to the Westminster-style parliament—was its key provision. It did, however, make a token gesture in the direction of political segregation by providing that Africans outside the Cape would be "represented" in the relatively powerless upper house by four appointed white senators. (Cape Africans and Coloreds retained the franchise but lost the right to elect one of their own to the highest legislative assembly.)[15]

In 1913, the Union parliament formally initiated the segregation process by passing the Natives' Land Act, which served as the cornerstone of all subsequent provisions for racial separation. Following the lead of the South African Native Affairs Commission, it divided the country into African and European areas and limited the right to own or lease land in each area to the designated racial group. Sometimes described as "territorial segregation," the principle at work was really "possessory segregation," since it did not in fact prohibit Africans and whites from being present in each other's areas. Particularly galling to Africans was the grossly inequitable division of territory; when the initial lines were drawn in 1916, Africans were assigned less than 8 percent of the total despite the fact that they outnumbered whites by more than four to one. They were scarcely mollified by a provision that future surveys and adjustments might lead to a slight increase in the percentage of land set aside for African occupancy.[16]

The Land Act did not spring primarily from abstract theorizing about racial adjustment; it derived mainly from the desire of capitalistic white farmers to gain easier access to black labor and eliminate the competition from the large number of African peasants who occupied white-owned land as renters or sharecroppers. Under the legislation, quasi-independent African farming in areas designated as "European" was prohibited, leaving the Africans there with no choice except to become wage workers or "labor tenants." (Labor tenancy meant working for several months of the year directly for white farmers in return for the right to occupy a bit of land on which subsistence crops and a few head of livestock might be raised.) The Land Act was unevenly enforced and sharecropping persisted in some areas as late as the 1930s, but historians are generally agreed that the legislation was the death knell of the African peasantry that had prospered from the expanding market opportunities of the late nineteenth century. Large scale capitalistic farming characterized by the direct supervision of low-paid and essentially rightless black contract laborers by white farm owners or managers was clearly the wave of the future in South Africa.[17]

Industrial employers of black labor also benefited from the act because the African reserves that it designated were already overcrowded; family subsistence there depended on the opportunities of male family members to go forth to work in the mines and other white-owned industrial enterprises under short-term labor contracts (usually nine months in the mines). Such a labor force was "ultra-exploitable." Its wages could be fixed at a low level, and labor contracts precluded the right to organize and strike for higher wages. The one major white interest group that derived no direct benefit from the Land Act and might even be disadvantaged by it was the working class. As low-paid African workers became increasingly available, employers would be tempted, especially in hard times, to substitute cheap blacks for expensive whites. Only with the economic "color bar" legislation of the 1920s did white labor gain a protected and privileged position within the framework of segregation.[18]

Other than use of the same word to describe the process that was occurring, what did legislation to require separate public facilities in the American South have in common with the kind of separation proposed by the South African Native Affairs Commission and mandated by the Natives' Land Act? The principal motivation—to insure white domination—was clearly similar, although one might conclude that social ascendancy was the center of concern in the United States, while the impulse to dominate and manipulate black labor played a more obvious role in the South African case. Also analogous was the desire to assure white dominance through legislation designed to prevent the kind of race mixing and group interaction that a laissez-faire policy would have permitted. Urbanization, commercialization, and industrialization, if uncontrolled, would bring whites and blacks together under conditions in which traditional or customary forms of racial dominance could no longer be expected to work. Slavery had been abolished, and deportation or absolute territorial segregation was ruled out because of the extent to which both the South and South Africa were dependent on black labor. Furthermore, segregation could be presented to the world as a more just and equitable system than straight-out white supremacy (or, to use the expressive Afrikaner word for unabashed white domination, *baaskap*). As John Cell has argued, one of the great advantages of segregationist ideology was that its claims of "separate but equal" could mystify the process of racial domination and permit the illusion of justice and fairness. Despite the major difference in form and function between Jim Crow and native segregation, there was enough similarity to inspire comparable reactions from black leaders and movements in the two societies.[19]

One might think of Jim Crow as the effort to maintain caste distinctions between racially defined groups that lived in close physical proximity, shared elements of a common American culture, and often competed for the same work. The pressure to extend and codify a preexisting pattern of customary social separation arose in the first instance from the desire of whites to draw a color line in facilities—moving and stationary—where the races would scarcely have been congregating and jostling each other if the South had not been in the throes of urban and industrial modernization. But there would have been no need for new laws if African-Americans had not been making rapid gains in education, economic efficiency, and political-legal sophistication—advances that made many of them unwilling to defer to informal or extralegal means of denying them equal access to public amenities. As southern blacks migrated to the North in increasing numbers between 1890 and 1914 and en masse after 1914, the question of whether Jim Crow would follow them across the Mason-Dixon line became a live issue; for white opposition to "social equality," while most intense in the South, was scarcely limited to that region.

Native Segregation was a scheme intended to perpetuate the state of vulnerability and exploitability in which African societies found themselves after a European military conquest that, for many of them, had only recently taken place. Outside of a tiny Westernized elite, most Africans remained rooted in traditional cultures and communities. But increasing numbers were present in the developing, white-dominated areas as migrant workers. The danger, as conceived by white supremacists, was that they would find permanent homes there, become detribalized, and imbibe dangerous Western notions of democratic rights that would lead to a new kind of threat to white rule. In other words, they would become very much like African-Americans. Native segregation can therefore be viewed as a preemptive strike against the more competitive and disorderly pattern of race relations that existed in the United States and to which the Jim Crow laws were a response. They were also, as Shula Marks has pointed out, a recognition of the continued vitality and "pulsating remains" of African kingdoms and chiefdoms. But this was more true in Natal where the Zulu monarchy persisted as a source of popular allegiance than it was in the Cape where a century of border wars, black migrations, and missionary activity had significantly weakened traditional authorities and solidarities by the end of the nineteenth century.[20]

A good way to understand how South African and southern segregation could be so close in spirit and ideology and so different in method and procedure is to view the former as a kind of macro-segregation that was meant to make the micro-segregation of the Jim Crow South unnecessary. (The coming of micro-segregation, or "petty apartheid," laws to South Africa after 1948 reflected in part the failure of the original segregation scheme to arrest or control the movement of blacks to the white urban areas and the growth of a substantial "modernized" segment of the black population that demanded inclusion in a common democratic society.) Of course the underlying fact that black South Africans would be a majority in a unified and color-blind South African polity, while African-Americans were a minority even in the southern states, gave to South African white supremacists a sense of urgency and concern about their own future dominance or even survival that North American racists, except perhaps in the "black belts" of the Deep South, could rarely match.

Between about 1910 and the American entry into World War I in 1917 some white Americans advocated what amounted to a nationalization of the southern Jim Crow system, and others proposed and attempted to legislate a form of territorial or "possessory" segregation that went beyond the social segregation laws that were now firmly in place in the southern states. The attempt to segregate the United States in a more formal and comprehensive way was not consciously modeled on the precedent of the Union of South Africa, but the Natives' Land Act as an example of macro-segregation did provide some inspiration for proposals to disengage the races in a more radical fashion through legal controls on residence and landholding. As white supremacists consciously or unknowingly called for an imitation of South African-type segregation, protest-oriented members of the black elite and their neo-abolitionist white allies mounted a campaign to prove that all forms of legalized segregation were contrary to American ideals. The result of their efforts was the successful launching of the twentieth-century civil rights movement.

African-Americans Mobilize Against Segregation

Organized black responses to the rise of official segregation in the United States and South Africa came predominantly from the same elites that had been in the forefront of protest against disfranchisement or restriction of voting rights. It is in fact somewhat artificial to separate the suffrage and segregation issues. White-imposed separatism and denials of the right to participate in a common electoral system were two aspects of a white supremacist effort to disempower blacks and keep them "in their place" at the bottom of the racial hierarchy. In a sense, however, denial of voting rights was prior to, and a precondition for, efforts at comprehensive legalized segregation. If blacks had continued to vote freely in the South after Reconstruction and if a nonracial suffrage had been extended from the Cape to the rest of South Africa after the war of 1899–1902, massive segregation might have been averted or at least have taken a form less demeaning to blacks and less damaging to their interests and aspirations. Political powerlessness meant that what blacks wanted and needed did not have to be taken into account in deciding how to implement the segregation idea, and it was therefore guaranteed that the result would be greater inequality rather than the actual realization of "separate but equal."

In his famous "Atlanta Compromise" speech of 1895, Booker T. Washington offered black acquiescence to social separation in return for economic opportunity. "In all things that are purely social," he told his predominantly white audience, "we can be separate as the fingers, yet one as the hand in all things essential to mutual progress." At this point the predominantly white crowd burst into a great cheer. But Washington did not clearly indicate that such public conveyances as railroads and streetcars were "purely social" places in which separation must be enforced. Black leaders, including those who protested vigorously against racial discrimination, had long made a habit of disavowing aspirations for "social equality," so long as the term was defined narrowly as requiring some restriction of the right to choose one's intimate friends and associates, rather than expansively as the equal treatment of strangers seeking access to the same public places and facilities. (The latter they usually referred to as "civic" rather than "social" equality.) Washington was influenced by that tradition, and he took no more kindly to being denied first-class railway accommodations than did other members of the black middle class. But at times he was tempted by the prospect of "separate but equal" facilities if they could somehow be guaranteed. Hence he vacillated during the 1890s between quietly opposing mandatory segregation and working to make it genuinely equal. The ambiguity of his position made it possible for him to be conciliatory to southern whites, whose support he needed to further his program of industrial education for southern blacks, without, for the moment, losing the good will of protest-oriented northern black leaders.[21]

In the famous critique of Washington that he included in *Souls of Black Folk* (1903), W. E. B. Du Bois did not accuse his subject of actively supporting Jim Crow laws. He merely noted that "radical" white southerners had interpreted his Atlanta speech as a "complete surrender of the demand for civil and political equality." Washington's failure to correct this impression by taking a strong public stand against segregation laws had helped, during the years since his address, to facilitate "the legal creation of a distinct status of civil inferiority for the Negro." The charge undoubtedly overestimated Washington's influence on the actions of southern state legislatures, but a growing number of educated, middle class blacks, especially in the north where such people did not have to tread so carefully to avoid being terrorized or even lynched, shared Du Bois's belief that Washington's efforts to accommodate the white South were giving the impression that blacks had no objection to being Jim-Crowed. If that impression were not countered, legal segregation might well spread from the South to the North and find new avenues of expression.[22]

Representatives of the northern black elite did not find it easy to turn against Booker T. Washington and embark on a crusade against segregation. Washington's enormous popularity among liberal and moderate whites in both the North and the South gave him access to resources and political leverage that no other black leader could come close to matching. He developed a virtual stranglehold on the white philanthropy that sustained black education, and, especially during the first administration of Theodore Roosevelt (1901–05), he controlled the patronage appointments that Republican presidents continued to make available to black supporters. He was also known for his promotion of black business enterprise; the Negro Business League (founded in 1902) was based on the assumption that increasing separation of the races offered new opportunities to entrepreneurs who would go after the black market. As ghettos began to form in northern cities, a new group of businessmen catering primarily to blacks began to replace an older economic elite that had relied on white clients or customers. These emerging black capitalists were receptive to Washington's doctrine of self-help and to his accommodationist politics. Washington also had the power to deny his critics a fair hearing in the black community. Many black newspapers depended on the subsidies that Washington gave them and could thus be relied upon to reflect his views and ignore people or movements of which he did not approve. It therefore took an extraordinary sense of moral outrage and an acute feeling of imminent danger to one's personal prospects and ambitions to bring a member of the educated black elite into direct and public conflict with Washington. But by 1905, partly out of sheer frustration at Washington's authoritarianism and ability to control the life chances of middle class African-Americans, a group of blacks, mostly northern professional men, were ready to launch a militant movement repudiating Washington's accommodationist tactics and calling for outspoken agitation against all officially sanctioned forms of racial segregation and exclusion.[23]

The Niagara Movement, founded by 29 members of the educated black elite at Fort Erie, Ontario, in 1905, was the brainchild of W. E. B. Du Bois and reflected his thinking throughout its brief existence. The paradox of the movement was that its membership was all black, but some of its patron saints were white abolitionists, and its ostensible goal was the incorporation of African-Americans into a reconstructed American democracy in which the color line had been abolished. Although Du Bois's call for "the conservation of races" and espousal of "Pan-Negroism" in the 1890s had seemed to point to some form of black separatism, his political thought and behavior at the time of his break with Booker T. Washington in 1903 was more in the spirit of Frederick Douglass than of Martin Delany or Alexander Crummell. Then, as at other times in his career, Du Bois was in fact walking a tightrope between the view that blacks were a permanently distinct people with a destiny apart from white Americans and the seemingly contrary notion that white and black should work together to create a more inclusive American nationality. As David Levering Lewis has pointed out, Du Bois viewed the "double consciousness" of the African-American that he described so eloquently in *Souls of Black Folk* ("his twoness,—an American, a Negro; two souls, two thoughts, two unreconciled strivings, two warring ideals in one dark body") as the potential source of a strong hyphenated identity rather than as a pathological condition that had to be resolved by rejecting the United States and embracing Africa. His aim, he made clear, was not to choose between being an American and being a Negro but to gain the right to be both. Washington's acquiescence to southern segregationism drew out the integrationist or Americanist side of Du Bois's racial thought, because it threatened black compliance with an invidious and damaging form of racial separation that he believed would impede the progress of the race.[24]

For Du Bois and other founders of the Niagara Movement, the politics of protest against denial of equal rights was an interracial American tradition deriving from the antislavery movement and not a suitable arena for the conspicuous display of racially specific symbols or loyalties. Of the three "friends of freedom" whose memories were especially honored at the Fort Erie meeting—William Lloyd Garrison, Frederick Douglass, and Albion W. Tourgée (the Reconstruction era "carpetbagger" who returned north to become the most conspicuous white champion of black equality during the Gilded Age), two were white. In homage to another white "friend of Freedom," the second meeting of the movement was held at Harper's Ferry, West Virginia, the scene of John Brown's raid. Since the centenary of Garrison's birth fell in 1905, Du Bois prepared a document he called "The Garrison Pledge of the Niagara Movement," which began as follows: "Bowing in memory of that great and good man, William Lloyd Garrison, I, a member of the race for whom he worked and in whom he believed, do consecrate myself to the realization of that great ideal of human liberty which ever guided and inspired him." Besides basic civil liberties and the right to vote, this neo-Garrisonian agenda included "the freedom to enjoy public conveyances and freedom to associate with those who wish to associate with me." The universalist, interracial character of the movement's official ideology was also on display in the "Niagara Address to the Nation" that Du Bois read to the 1906 meeting: "Thank God for John Brown! Thank God for Garrison and Douglass! Sumner and Phillips, Nat Turner and Robert Gould Shaw. . . . Thank

God for all those today, few though their voices may be, who have not forgotten the divine brotherhood of all men, white and black, rich and poor, fortunate, and unfortunate."[25]

William Monroe Trotter, the fiery and contentious anti-Washington radical from Boston who was second only to Du Bois as a catalyst for the Niagara Movement, must have been delighted to take the Garrison pledge, because he lived and breathed the New England abolitionist tradition and modeled his entire career on Garrison's example, to the point of emulating the great abolitionist's vituperative, often ad hominem rhetoric. Unlike Du Bois, he never evinced the slightest interest in a distinctive African-American culture, and he consistently opposed all forms of publicly sanctioned separation of the races, even those that more pragmatic black protest leaders came to view as advantageous in the short run. (Yet, as if to demonstrate how difficult it is to dichotomize black leaders as "separatists" or "integrationists," he later broke with the NAACP because of its domination by white liberals and formed a rival organization that was virtually all black in membership.)[26]

Why then did the Niagara Movement depart from the interracial abolitionist precedent by limiting its membership to blacks? One reason was that there were at that moment very few whites who would have been interested in joining forces with anti-Washington black radicals. Most whites who believed themselves to be "friends of the Negro" in the northern antislavery tradition—such as Oswald Garrison Villard, the influential New York editor and grandson of the great abolitionist—were still under Washington's spell, having been persuaded that he spoke for the great majority of African-Americans. Especially since Trotter had been jailed for his role in the "Boston riot" that silenced Washington in Boston in 1903, white racial liberals had tended to view the anti-Washington radicals as intemperate troublemakers. But the cause of black civil rights was not entirely without white supporters in 1905. John Milholland, a wealthy industrialist and unreconstructed Radical Republican, had founded the Constitution League in 1904; its program of immediate and full enforcement of the rights blacks were meant to enjoy under the Fourteenth and Fifteenth amendments was almost identical to that put forward by black radicals at Fort Erie. As it happened, a close working relationship quickly developed between the Constitution League and the Niagara Movement. The former had black members who were also Niagarans, and Du Bois at one time actually held office in both organizations. In some states the two organizations became almost indistinguishable. But no formal merger took place, and whites were at no time admitted into full membership the Niagara Movement, even though some professed an interest in joining.[27]

The most likely explanation for the movement's racial exclusiveness was that it had a second purpose in addition to agitation for equal civic and political rights that was the sole raison d'être of the Constitution League. Besides starting a new agitation for fulfillment of the Garrisonian dream of a racially egalitarian American republic, it was meant to arouse and mobilize a specific segment of the African-American community against the ascendancy of Booker T. Washington and what it represented. It was in these years that Du Bois was most insistent on the need for a "talented tenth" of "college-bred Negroes" to assume leadership of the black struggle for equality. Actually this was vastly overstating the size of the educated elite, since in 1900, as Du Bois himself had shown, there were only 2,304 black college graduates in the entire United States. At the apex of this small group were a mere 390 graduates of predominantly white colleges of whom a handful, including Du Bois and Trotter with their Harvard degrees, had matriculated at prestigious Ivy League institutions. A majority of the talented tenth were college or secondary school teachers, but there were 221 clergymen, 83 physicians, 62 lawyers, 53 civil servants, and a mere 47 businessmen.[28]

Washington's deemphasis on higher education and his concentration on industrial training was of course the original source of Du Bois's dissatisfaction with the Tuskegee philosophy. In his mind, the notion that blacks should learn to make a living as farmers, artisans, and small businessmen before they aspired to a more refined culture and political activism threatened to deprive the race of an educated vanguard that could shape its ideals and purify its aspirations. This was an unmistakably elitist philosophy. In addressing a black audience in Louisville in 1900, Du Bois has asked his auditors to acknowledge "the fact of human inequality and difference of capacity" and recognize that "there are men born to rule, born to think, born to contrive, born to persuade." As Arnold Rampersad has noted, Du Bois's belief in "salvation by aristocracy" actually displayed less confidence in the mass of black people than Washington's down-to-earth materialism: "Du Bois' notion of culture was founded on the principle that cultural fertilization descended from the top down into the masses of people."[29]

The Niagara Movement did not acquire mass support, and its leaders did not expect it to. Its membership never exceeded 400, and Du Bois indicated at one point that he thought 150 was about the right size. What characterized the Niagarans above all was that they drew their membership almost exclusively from the "college-bred" elite and especially from the most highly educated and professionally qualified among them. Most came from northern urban areas, but there were a few members or supporters on the faculty of southern black colleges, especially in the Atlanta enclave of black institutions where Du Bois himself was then working. Besides reviving the protest tradition of the abolitionists, the Niagara Movement was an effort to further the leadership aspirations of the black intelligentsia against the "self-made men" for whom Washington spoke. Washington's own lack of college training and the support for him that came from black entrepreneurs who had arisen from "the school of hard knocks" threatened the prestige and influence of college-bred and liberally educated African-Americans.[30]

Washingtonians were certainly vulnerable to the charge that their self-help, "separate-as-the-fingers" philosophy played into the hands of white supremacists and encouraged their assault on the kind of "civic rights" that the educated elite highly prized. The language of the Niagara Movement's platform statement on segregation revealed a set of concerns shared to some degree by all black people who aspired to be treated with dignity but which were felt with special acuteness by those who had the means and the cultural preferences to want the first-class treatment and elegant surrounding that normally went with personal achievement and high social status in early-twentieth-century America:

> Any discrimination based simply on race or color is barbarous. . . . Differences made on account of ignorance, immorality, or disease are legitimate methods of fighting evil, and against them we have no word of protest; but discrimination based simply and solely on physical peculiarities, place of birth, color of skin, are relics of that unreasoning human savagery of which the world is and ought to be thoroughly ashamed. We protest against the "Jim Crow" car, since its effect is and must be to make us pay first-class fare for third-class accommodation, render us open to insults and discomfort and to crucify wantonly our manhood, womanhood, and self-respect.[31]

All black professionals and intellectuals, even in the North, did not follow Du Bois's lead in challenging Washington's leadership. Some, such as Kelly Miller, the dean of Howard University, remained on the fence hoping to reconcile the competing factions of the black middle class. Others, such as the Boston lawyer William H. Lewis, a graduate of Amherst and Harvard Law School, found it to be in their professional interest to take the side of the powerful Washington against the insurgents. But it remains true that the Washington-Du Bois controversy was rooted in the conflict of two conceptions of black leadership—one based on education and intellectual or professional distinction and the other on economic success and efficiency. Du Boisians and supporters of the Niagara Movement placed a higher premium on "civic equality" and the struggle against segregation, while the Washingtonians emphasized self-help, racial solidarity, and the potential benefits of a separate black economy. There was no absolute or necessary contradiction between the two approaches; indeed they might be seen as mutually reinforcing aspects of the black struggle for power and justice in American society. It was the promotion of segregation as the white supremacist's solution to the race problem in the early twentieth century—and Washington's apparent acquiescence in it—that impelled Du Bois and other members of the "talented tenth" to risk rupturing the black middle class by adopting an uncompromising stand of opposition to white-imposed racial separatism. Paradoxically, their sense that the issue had to be thrashed out within the black community led them to pursue this goal through a racially exclusive movement.[32]

The Niagara Movement devoted most of its energies to resisting the spread of segregation (especially separate schools) to the North, and it had some success in these endeavors. But it failed in its aim to become the new voice of black America and began to decline after its first two or three years of existence. Its failure was due in part to internal factionalism and feuding, but its inability to get off the ground was due mainly to the effectiveness of Washington's implacable opposition. Since most black newspapers were subsidized or influenced by Washington and most liberal white journals viewed him with reverence, the Niagarans got little sympathetic attention from the press. Lacking the resources to mount expensive publicity or lobbying campaigns, they were unable to raise funds from white philanthropists and wealthy "friends of the Negro" because Washington again stood in the way. There was also the perennial problem that even those whites who had the strongest theoretical commitments to racial equality were reluctant to back an all-black organization over which they had no control. Such circumstances forced Du Bois and his supporters to shift their tactics, abandon the Niagara effort, and throw their lot in with a new interracial—and initially white-dominated—civil rights organization.[33]

By 1909, a few white members of the social justice wing of the American Progressive movement were at last ready to acknowledge that African-Americans were suffering from an extreme form of social injustice. This broadening of concern from the plight of the white working class, and especially the women and children who worked in factories and lived in slum conditions, to the racial discrimination that intensified the problems of blacks came in part from the experiences of white social workers and settlement-house residents with poor African-Americans in major northern cities. But the event that precipitated the founding of a new organization was the Springfield, Illinois, race riot of 1908.

William English Walling, the scion of a Kentucky slaveholding family who had become a settlement house worker, a socialist, and a leading progressive journalist, wrote in the *Independent* for September 3, 1908, that a tide of racism was flowing northward from the segregated South and had now reached the hometown of the Great Emancipator. He called for a movement to stop this flood and begin a new effort to fulfill the dreams of racial justice and equality that he identified with Lincoln and the northern cause in the Civil War. Soon afterwards, he joined with Mary White Ovington, a social worker from an abolitionist family who had taken a special interest in New York's African-American population, and Henry Moskowitz, a Hungarian-born Jewish social worker, to lay the groundwork for an interracial organization to combat violence and discrimination against blacks. Among the early recruits to the campaign was Bishop Alexander Walters of the African Methodist Zion Church, a former president of the Afro-American Council who had held aloof from the Niagara Movement but had presided over the council's rejection of Washington's hegemony in 1907. Walling also enlisted two of the leading women in the social justice movement, Lillian Wald and Florence Kelley. But his most important recruit was Oswald Garrison Villard, the grandson of William Lloyd Garrison, who was editor and publisher of the New York *Evening Post* and a wealthy philanthropist with a special interest in the plight of African-Americans. Villard was persuaded to head the effort, and he issued the call for 60 prominent whites and blacks to meet on Lincoln's birthday in 1909 to form a new national organization to advance the cause of black civil rights.[34]

The National Negro Committee, which at its second meeting in 1910 renamed itself the National Association for the Advancement of Colored People, was essentially a coalition of white interracial liberals—a large proportion of whom were descendants of abolitionists or, if old enough, had themselves been Radical Republicans during the Reconstruction era—and the Niagara Movement black militants. (Booker T. Washington was conspicuously absent at the founding meetings.) The program of the new association was virtually the same as that of the Niagara Movement. It called for enforcement of the Fourteenth and Fifteenth amendments in the southern states in order to restore free access to the ballot box and open public facilities to blacks on an unsegregated basis. It also demanded federal action against lynching and other violence against African-Americans and was vigilant in its efforts to identify and resist any tendencies to extend the southern Jim Crow system to the northern states. To accomplish these objectives, it sponsored litigation, lobbied legislative bodies, and engaged in efforts to educate the public about the extent of racial discrimination and why it was unjust and unjustifiable. Focusing almost exclusively on the civil and political rights of blacks, it paid little attention to their economic problems. Another interracial organization founded at about the same time, the National Urban League, assumed the main responsibility for dealing with the employment, housing, and social welfare of the hundreds of thousands of African-Americans who were migrating from the southern countryside to the cities during the second decade of the twentieth century. The Urban League was a relatively conservative and accommodationist social agency that concentrated on adjusting the needs of black migrants to the requirements of American capitalism. The reformist NAACP's lack of an economic program meant that the mainstream movement for black advancement in the United States would develop with few ties to the labor movement and little affinity for socialist perspectives on the condition of blacks and other underprivileged groups in American society.[35]

Most of the national officers of the NAACP in its earliest years were prominent white progressives, but from the beginning the rank-and-file membership, as well as the leadership of the local branches, was composed mainly of middle-class blacks. What gave many African-Americans confidence in the new organization, beyond their approval of its forthright commitment to civic and political equality, was Du Bois's conspicuous role in its founding and leadership. Some of the Niagara militants were uncomfortable with the dominant position assumed by whites. Trotter detected too much moderation at the 1909 meeting and spoke from the floor in favor of more radical positions on the enforcement of civil rights. Presumably because of his vociferousness and reputation as a hothead, he was left off the committee of 40 (including a dozen blacks) that was appointed to carry on the work of the association. Other black radicals excluded from the governing body included J. Milton Waldron, a Niagara Movement stalwart, and Ida Wells-Barnett, the celebrated anti-lynching crusader. (Wells-Barnett was later added, but she never forgave the original slight; as a result, the most courageous and charismatic of black female activists never got deeply involved in the work of the association.) Trotter remained a member until 1913 when he broke with the NAACP entirely to form a more militant, black-led organization. Du Bois, however, enjoyed the respect of the founding whites; he was one of two black members of the association's first seven-person board of directors (the other was the relatively moderate Bishop Walters).[36]

As an officer of the NAACP, Du Bois was responsible for research and publicity, and he performed this function mainly through his editorship of the association's official magazine *The Crisis*. Du Bois was primarily an intellectual and publicist; for the most part he was content to let others—in the early years mainly whites—conduct the NAACP's legal and lobbying efforts and oversee the activities of local branches. It soon became apparent that *The Crisis* was the voice of Du Bois himself rather than of the association's national leadership. He directed it primarily at black readers, and not merely at African-Americans; from the start the journal went beyond the association's focus on equal American citizenship, devoting space to expressions of black cultural nationalism and providing a platform for Pan-Africanist ideas that were of little interest to the NAACP's white officers.

Friction between Du Bois and Oswald Garrison Villard, the first chairman of the board, quickly developed over how independent *The Crisis* could be and still be considered the official organ of the NAACP. Du Bois demanded total autonomy and held his ground until Villard resigned over the issue in 1914 and was replaced as chairman by the Jewish scholar and philanthropist Joel Spingarn. Du Bois developed a better working relationship with Spingarn and remained on relatively good terms with Mary White Ovington, the secretary of the association. Spingarn and Ovington respected Du Bois's powerful intellect and understood that the credibility of the NAACP among blacks depended on Du Bois's identification with it. But tension persisted over his adamant insistence that the NAACP be in effect a kind of dual organization with two coequal branches, one headed by a white and the other by himself. The white section was primarily concerned with undoing the wrongs that whites had done to African-Americans, while the black side also sought to instill in African-Americans a positive sense of identity and a thirst for cultural revitalization. As *The Crisis* grew in circulation and international recognition (its readership would reach 100,000 by 1920), Du Bois became the most influential black intellectual in the world. His efforts to harmonize blacks' aspirations for constitutional rights with an equally insistent claim to membership in a great African race with a proud history and a distinctive cultural heritage resonated well beyond the boundaries of the United States.[37]

Among those who read *The Crisis* with great interest were members of the educated elite of black South Africans, who were simultaneously engaged in efforts to organize and protest against their own system of segregation. Despite the obvious differences between Du Bois's situation and theirs, they found the African-American debate on how to respond to Jim Crow relevant to their own discussions of how best to confront the challenge of "native segregation."

The National Congress in Comparative Perspective

In February 1909, the national convention to draw up a constitution for the Union of South Africa released to the press the product of its deliberations—a draft that denied blacks voting rights in three of the four provinces and restricted membership in parliament to Europeans. In response, a small number of highly educated Africans—the equivalent of Du Bois's "talented tenth"—took the lead in efforts to unite Africans across tribal and provincial lines to resist the imposition of a constitution that provided whites with a monopoly of substantive political rights and representation in the governing bodies of the new state. Earlier African political activity had, for the most part, been confined within the boundaries of the four colonies that were now being unified and had usually been intra-tribal as well. As we have seen, the Ethiopian movement in the churches had asserted a new ideal of spiritual unity across ethnic divisions, but religious Pan-Africanism had produced no clear political direction, and its fragmentation into several competing sects had precluded the kind of solidarity that was required to meet the challenge of white unification. What Africans desperately needed in 1909 was a common European policy to counter the common "native policy" that was a principal aim of the whites who were framing a constitution for the Union of South Africa.[38]

To become law, the South Africa Act had to be endorsed by the four colonial assemblies and then approved by the British parliament. African opinion was virtually unanimous against the act, and organized protest quickly developed. The obvious response was to call a convention of Africans from all the colonies of South Africa to register this opposition. The South African Native Convention, hosted by the Orange River Colony Native Congress, met in late March 1909 in Bloemfontein. Among the delegates were the leading members of the educated elite, some of whom would play prominent roles in the formation of the South African Native National Congress three years later. According to historian André Odendaal, the meeting of the convention "was a seminal event in the history of African political activity in South Africa. It was the first occasion in which African political leaders and fledgling political associations co-operated formally. The meeting was a major step toward formation of a permanent national African political association."[39]

On the basis of a creed strikingly similar to the one being enunciated in the same year by the National Negro Committee (later the NAACP) in the United States, the convention called for a clause in the constitution that would in effect replicate the Fourteenth and Fifteenth amendments to the American Constitution. It would provide for all inhabitants of the Union "full and equal rights and privileges, subject only to the conditions and limitations established by law and applicable alike to all citizens without distinction of colour or creed." Elected president and secretary of the convention, respectively, were those stalwart champions of the Cape color-blind franchise Walter Rubusana and A. K. Soga. The vice president was John Dube of the Natal Native Congress, who would later be first president of the South African Native National Congress.[40]

Despite African protests, the colonial parliaments approved the color bar provisions of the proposed constitution, and a revised draft was sent to the British parliament for ratification. Believing that the British tradition of fair-minded liberalism was their last line of defense, African and Colored leaders rallied behind a white champion of color-blind constitutionalism, W. P. Schreiner, who proposed to lead a delegation to Britain to plead the cause of South African blacks. Schreiner, brother of the novelist Olive Schreiner and a true believer in the letter of the Cape liberal tradition, had led the white opposition to including racially restrictive clauses in the proposed constitution. He was that rare phenomenon among early twentieth-century white South Africans, a true racial liberal comparable to the neo-abolitionists who helped found the NAACP.

The African elite professed great admiration for Schreiner, and one African newspaper even dubbed him "our South African Abe Lincoln." Included in the delegation to Britain that Schreiner headed were Dr. Abdul Abdurahman of the (predominantly Colored) African Political Organization; John Tengo Jabavu, the moderate Cape leader who had held aloof from the efforts of his more radical Cape rivals to unify Africans across colonial boundaries but who shared their opposition to the South Africa Act; and three representatives of the South African Native Convention: Rubusana and Daniel Dwanya from the Cape and Thomas Mapikela of the Orange River Colony. John L. Dube of Natal, who was ostensibly in Britain for other purposes, served as a de facto member of the delegation. Despite the fact that the delegation represented the entire spectrum of educated black opinion and notwithstanding its strenuous efforts to mobilize British public and parliamentary opinion against the South Africa act, it failed in its mission. The government was too strongly committed to devolving power to whites in what it deemed to be colonies of European settlement to hear pleas for the rights of "natives"—even when, as in the exceptional case of South Africa, they made up a substantial majority of the population.[41]

After parliament approved the South Africa Act and the white-dominated Union became a fait accompli, African protest politics lost some of the momentum that had sent the delegation to London. The Native Convention failed to assemble again or assume a permanent form; its leaders were left clinging to the slender hope that the white minority would rule benevolently and consent to the gradual extension of political rights to "civilized" blacks throughout South Africa. A qualified suffrage for Africans and Coloreds remained in force for the Cape Province and was given special protection in the constitution—it could only be abrogated by a special two-thirds vote of both houses of parliament sitting together. Since their right to the suffrage was based in part on access to landholding, a court decision would later hold that the Natives' Land Act of 1913 did not apply to Cape Africans. Their own voting and landowning rights apparently safeguarded, many of the Eastern Cape Xhosa who had been in the forefront of the suffrage agitation in the period between the signing of the Treaty of Vereininging in 1903 and the seating of the first Union government in 1910 retreated into political passivity or concentrated on local issues. Preparing to assume their place in the vanguard of the movement for African rights were educated Africans from areas where no modicum of African rights was entrenched, especially Zulus from Natal and Tswana and Sotho from the Free State and the Transvaal.[42]

The first two years of Union saw the beginnings of agitation for the systematic "segregation" of whites and Africans. The government, led by Louis Botha and his South African Party, sought to harmonize the interests of Afrikaners and English-speaking South Africans at the expense of Africans, who became pawns to be moved about and sacrificed as suited their white overlords. In 1911 African contract workers were denied the right to strike and an act regulating mines and factories authorized the exclusion of blacks from skilled, responsible, and high-paying jobs. By 1912 legislation to implement the South African Native Affairs Commission's recommendation for a division of the country into black and white areas was being considered by parliament. Threatened with this juggernaut of segregationism, a group of African professional men, based mainly in Johannesburg, seized the initiative and called for another national conference. Their clear intention this time was to form a permanent organization that could unify Africans in defense of their rights and interests against a white government that had revealed its discriminatory intentions.[43]

The prime movers in the effort to form a native congress were four African lawyers, educated abroad, who had recently begun to practice in Johannesburg—Alfred Mangena, Richard Msimang, Pixley ka Izaka Seme, and George Montsioa. Among the first black South Africans to practice law, they all came from Christian enclaves and had received a mission education before being sent abroad for advanced training. Mangena, Msimang, and Seme were Zulus from the Kholwa (African Christian) community of Natal, and Montsoia was a Barolong (Tswana) from Mafeking. Their legal training had taken place in England, and they had been called to the bar or articled there (Mangena and Seme were barristers, Msimang and Montsoia solicitors). Returning to South Africa at the time of Union, these four young lawyers were living exemplars of the mission ideal of elevating Africans to "civilized" status; furthermore they had been thoroughly immersed in British traditions of personal rights and liberties. What they saw happening in South Africa was a flagrant violation of everything they had been taught to revere, and they were prepared to take vigorous action on behalf of their less fortunate and enlightened compatriots.[44]

Although Mangena, the first black barrister to practice in South Africa, was the senior member of this quartet, it was Pixley Seme who took the lead in summoning a new congress. Before going to England to study law at Jesus College, Oxford, Seme had been a student in the United States, first at Mt. Hermon School in Massachusetts and then at Columbia University, where he received a B.A. in 1906. Little is known about the specific influences brought to bear on him during his American sojourn, but undoubtedly he was exposed to African-American thought during the height of the Washington-Du Bois controversy. While at Columbia, he gave an oration on "The Regeneration of Africa" that sounds remarkably like Du Bois, especially the Du Bois who advocated "Pan-Negroism" in his 1897 address on "The Conservation of Races." Seme spoke of the unique "genius" possessed by each race of mankind; argued that Africans were no exception, as demonstrated by the pyramids of Egypt, "all the glory" of which "belongs to Africa and her people"; noted that the long dormant African genius, stimulated by contact with a more "advanced" civilization, was reviving as blacks developed a healthy "race consciousness"; and predicted that "a new and unique civilization is soon to be added to the world." The great obstacle to the regenerative process was fragmentation and tribal rivalries, but this divisiveness was being overcome, partly through the agency of Africans educated abroad who "return to their country like arrows, to drive darkness from their land." The young Seme was obviously an heir of the great tradition of nineteenth-century black nationalism, which had come to fruition in the early thought of Du Bois. In harmony with this tradition, he combined a belief in the "ancestral greatness" and innate capabilities of Africans with a belief that "the essence of efficient progress and civilization" was "the influence of contact and intercourse, the backward with the advanced."[45]

What Seme had in mind when he took the lead in calling for a national congress of Africans was set forth in an article he wrote for an African newspaper in 1911:

> There is to-day among all races and men a general desire for progress and co-operation, because co-operation will facilitate and secure that progress. . . . It is natural therefore, that there should arise among us this striving, this self-conscious movement, and sighing for Union. We are the last among all the nations of the earth to discover the precious jewels of cooperation, and for this reason the great gifts of civilization are least known among us today. . . . The South African Native Congress is the voice in the wilderness bidding all

> the dark races of this sub-continent to come together once or twice a year in order to review the past and reject therein all those things which have retarded our progress, the things which poison the springs of our national life and virtue; to label and distinguish the sins of civilization, and as members of one house-hold to talk and think loudly on our home problems and the solution of them.[46]

According to this formulation, Seme was proposing an agency for African self-improvement and cooperation across ethnic and regional divisions that would not necessarily be a protest organization. He was not advocating "de-tribalization" or the merging of white and black South Africa into a single, nonracial nation; for he stressed the fact that chiefs and traditional leaders were being invited to the forthcoming assembly and would be given a prominent place in the organization. He was himself a Zulu and proud of it; later he would marry the daughter of the Zulu paramount chief, Dininzulu, whom he had previously persuaded to serve as an honorary vice president of the Native Congress. Like the other lawyers involved in founding the South African Native National Congress (SANNC) Seme retained a strong ethnic allegiance and often had chiefs as clients. His governing idea was not the obliteration of tribal loyalties and distinctions among Africans but rather the subordination of traditional allegiances to a broader conception of black nationality. This of course required exorcising "the demon of [inter-tribal] racialism." As his distinction between "the gifts of civilization" and "the sins of civilization" might suggest, he thought it essential that Africans respond selectively to modernizing influences, absorbing those that were beneficial and compatible with their best traditions and rejecting those that were not.[47]

When the congress assembled in Bloemfontein in January 1912, Seme responded to the Union government's assault on black rights by making protest an explicit function of the organization. In his keynote address, he complained that "in the land of our birth, Africans are treated as hewers of wood and drawers of water." In the new Union formed by the white people, "we have no voice in the making of laws and no part in their administration." Consequently, it was necessary to form an association for the dual purpose of "creating national unity and defending our rights and privileges." In Seme's conception, the new congress combined the self-help and racial-solidarity themes of Booker T. Washington's followers with the protest orientation of Du Bois and the NAACP. While a law student in England, Seme had sought Washington's endorsement of his efforts to establish an organization of African students. Some correspondence ensued and they actually met on one occasion. For reasons that are somewhat unclear, the meeting was unsatisfactory—possibly, according to Louis Harlan, because Washington refused to "take a more active interest in African causes." It is also possible that Washington perceived in Seme some of the militancy and aggressive political activism that he associated with the Du Bois camp in the United States.[48]

If the SANNC at its founding contemplated NAACP-type constitutional protest as one of its aims, it is equally clear that Booker T. Washington's example and philosophy of self-help as the key to group advancement had a strong hold on the new organization's leadership. This influence was particularly evident in the career and thought of John L. Dube, who was elected first president of the congress despite his absence from the convention. Dube was so honored because he was the most prominent black educator in South Africa and the leader of Natal's African Christian or Kholwa community. He was also a man who had been so influenced by Washington that he had modeled his own career on that of Tuskegee's principal. Like Seme, Dube was a protege of American Congregationalist missionaries in Natal who had been sent to the United States to be educated, although several years earlier than Seme. After studying at Oberlin, he returned to South Africa as a teacher and missionary in 1892. In 1897, on a subsequent visit to America, he visited Tuskegee and fell under Washington's spell. In 1901, he founded Ohlange Institute, a school for Africans under their own direction, that was a self-conscious imitation of Tuskegee.[49]

In the acceptance speech that he sent to the congress, Dube made explicit his debt to Washington, calling him "my patron saint" and "my guiding star." He had chosen to emulate Washington "because he is the most famous and best living example of Africa's sons; and . . . because like him I, too, have my heart centered mainly in the education of my race." But he believed with Seme and the other founders of the SANNC that current circumstances required more than education and self-help: "While I believe that in education my race will find its greatest earthly blessing, I am forced to avow that at the present juncture of the South African Commonwealth, it has a still more pressing need—the need for political vigilance and guidance, of political emancipation and rights."[50]

The first secretary of the congress, the Barolong journalist Sol Plaatje, was also a keen admirer of Washington. His Tswana-language newspaper, *Koranta ea Becoana*, did much to publicize the Tuskegee Institute and Washington's philosophy among black South Africans. In the early 1920s, several years after Washington's death, Plaatje visited Tuskegee and acquired some film footage of it that he later showed to African audiences who, besides being fascinated by what was likely to be their first exposure to motion pictures, were shown what Plaatje regarded as a dazzling example of black achievement. But his lifelong devotion to Washington did not prevent Plaatje from admiring Du Bois and the NAACP. He read *The Crisis* and quoted Du Bois frequently in his writing; while visiting the United States in 1921, he met Du Bois, accepted his invitation to attend the NAACP convention in Detroit, and ended up addressing the delegates on conditions in South Africa.[51]

Taking sides in the Washington-Du Bois controversy, or having a similar conflict of their own about the competing claims of political action and economic self-help, was a luxury that the founders of the National Congress did not believe they could afford. Washington was admired, not so much for his ability to accommodate whites, as for his success in building up a school of his own that had impressed the world. As Louis Harlan has recognized, "the separatist concept that was Tuskegee conservatism in the American context became radical nationalism in the African." African visitors to Tuskegee were generally more impressed with the fact that a black man was in charge than that the form of education being offered was calibrated to mollify white fears of "over-educated" African-Americans. When Dube applied Washington's concept of industrial education to Africans in his Ohlange Institute, he gave whites the impression that he was training a docile work force, when in reality he was educating much more broadly than this concept suggested. His real aim, according to Manning Marable, was to "create a class of independent and aggressive entrepreneurs" and "moderate political activists," thus laying the foundations for "an independent, aggressive intellectual class." A Washingtonian emphasis on building character and the acquisition of skills and capital was being fused with Du Bois's conception of a "talented tenth" committed to leading their race or nation to freedom and equality.[52]

Furthermore, the policies of the South African government made political action necessary before the process of accumulation and individual enterprise that Washington recommended could plausibly begin. Du Bois had charged that the South's denial of civic and political equality to blacks fatally hobbled them in the economic race of life, but Washington could at least reply that the rights to purchase land and take one's labor to a better market were not formally abrogated. In South Africa, however, blacks worked under special contracts which made leaving work a penal offense, and in 1912 the parliament was preparing legislation that would not allow Africans to purchase or rent land in more than 90 percent of the country. It was therefore more glaringly obvious than in the American case that agitation for equal rights under the law was an essential precondition for the economic and entrepreneurial development of the African population. After the Land Act was passed, Congress protested that its consequence for Africans was "To limit all opportunities for their economic improvement and Independence: To lessen their chances as a people of competing freely and fairly in all commercial enterprises." In South Africa, therefore, the "petty-bourgeois" goals of Booker T. Washington could not be pursued with the remotest hope of success if political activity and protest were eschewed; a DuBoisian agitation for constitutional rights was patently necessary to give Africans any chance at all in a competitive, capitalist economy.[53]

If the protest orientation of the National Congress in its early years resembled that of the NAACP, the fact that it was composed exclusively of blacks and lacked white liberal sponsorship and involvement made it more analogous to the Niagara Movement or to the Afro-American Council. What was more, the implication that it was the "congress" of a colonized people addressing an imperial overlord expressed a straightforward national consciousness that no African-American movement could duplicate. (The prototype was the Indian National Congress founded in 1885.) As the original inhabitants and the majority of the population in South Africa, Africans could imagine themselves as the South African nation in ways that were not open to the African-American minority when they thought about the future of the United States. One reason—beyond their belief that the Union constitution protected their established rights—that kept some African leaders in the Eastern Cape from joining the congress or playing a prominent role in it was the special tradition of interracialism that survived, if only in attenuated form, in that province. The moderate John Tengo Jabavu refused to participate in the founding of the SANNC because he did not believe in racially exclusive movements; instead he founded a rival organization, the South African Races Congress, which was theoretically open to members of all racial groups.[54]

Seme's Pan-African rhetoric and the all-African membership of the Congress did not commit the SANNC to the Ethiopianist goal of "Africa for the African." Its official objective was a common democratic society in which Africans, whites, coloreds, and Indians would be equal; its separatist character reflected the weakness of white racial liberalism in the Union of South Africa more than it did the strength of Africanist ideology. There was no abolitionist tradition that predisposed an influential group of whites to champion the public equality of blacks. The closest equivalent was "the Cape liberal tradition," and most of the heirs of this tradition had supported the South Africa Act. Only W. P. Shreiner had emerged as a prominent and consistent defender of black political rights throughout the Union. In the northern provinces there were virtually no whites who dissented from the white supremacist consensus. Consequently the go-it-alone posture of the National Congress could be regarded as a pragmatic necessity for the pursuit of equal rights rather than as the model for an ethnically defined African nation. The subsequent history of the congress would be filled with tension between cosmopolitan and ethnocentric perspectives but would see the inclusive and incorporationist perspective—called at different times "multi-racialism" or "nonracialism"—prevail again and again.[55]

Besides being constituted as a national rather than an interracial movement, the SANNC differed from the early NAACP in another obvious respect. The Western-educated elite, which had monopolized representation at the National Convention of 1909, moved to enlist the support of traditionalists by inviting the chiefs of the various tribal groups to the founding meeting, giving them special recognition, and making them honorary vice presidents of the SANNC. The organization's constitution, which was not actually approved until 1919, provided for an upper house of chiefs in addition to the assembly of elected delegates. It was a clear reflection of the basic differences between the two black communities and their histories that no equivalent source of traditional authority existed among African-Americans. But this difference does not vitiate comparison between the early American civil rights movement and early African nationalism in South Africa. Chiefs did not in fact have much influence in running the congress and making its policy. Their role was indeed "honorary," and the real source of authority was an educated elite with a worldview strikingly similar to that of the African-American "talented tenth" who constituted the membership of the Niagara Movement and the black component in the NAACP (which would assume executive leadership of the association by 1920).[56]

The first executive committee of the SANNC was composed of four clergymen, three lawyers, two teachers, an editor (Plaatje), and a building contractor—an occupational profile strikingly similar to that of the early black board members of the NAACP, four-fifths of whom were professionals with ministers and lawyers again predominating. The African-American professional middle class—small as it was relative to that of white Americans—was much larger and more firmly established than the black South African elite; as we have seen, the development and progress of the former served to some extent as a model for the latter. As the careers of leaders like Seme and Dube demonstrate, some of the ethos of the African-American middle class had been absorbed and brought back to South Africa by early nationalist leaders who had received part of their education in the United States. What led to the creation of new national protest organizations at about the same time was not, however, a process of emulation. There is no evidence that the founders of the National Congress were influenced or inspired in any way by the recent establishment of the NAACP in the United States. What drove them to comparable action was a similar sense of how segregation threatened the aspirations of educated, middle-class blacks.[57]

The black educated elites of the two countries had acquired professional status and competence on the expectation that their qualifications would lead to "careers open to talent" in the white-dominated society. The educations they received from mission institutions and in liberal, ostensibly color-blind universities in the northern United States or Great Britain had inspired hopes that by becoming "civilized," they would have expanded opportunities. As we have seen, optimism about the prospects of incorporation into a progressive, liberal-capitalist society seemed firmly rooted in the "color-blind" or theoretically nonracial liberalism of the mid-Victorian era, the ideological current that sanctioned slave emancipation in the United States and was the basis for the impartial franchise and equality under the law that distinguished the "liberal" Cape Colony from the other, more blatantly discriminatory South African colonies or republics to the North. A belief that Reconstruction-era constitutional reforms opened the way for a steady advancement to full equality within American society had raised the spirits of African-American achievers. Similarly, a conviction that British-inspired Cape liberalism with its promise of "equal rights for every civilized man" was the wave of the future for the entire area south of the Zambezi had given hope to educated Africans that fair treatment under the benevolent umbrella of the British empire was only a matter of time. These expectations were shattered around the turn of the century by an upsurge of white racism and the beginnings of segregationism as a deliberate public policy with applications far beyond what middle class blacks had previously acknowledged to be legitimate provisions for social differentiation.[58]

The impulse to organize in defense of individual rights, rather than merely striving to succeed as individuals, was a defensive reaction by black elites against white aggression. The Jim Crow laws and practices of the United States, and their tendency to spread from the South to the North, heaped the same indignities on black professionals as they did on servants and field hands. The new South African constitution denied educated blacks the political opportunities and governmental responsibilities for which they felt themselves fully qualified, and the discriminatory, segregationist laws of the Union parliament put a low ceiling on black aspirations for social and economic advancement. Calling attention to the particular class and status concerns of the elite is not meant to expose a lack of commitment to the interests and rights of black peasants and workers. Recognizing that no black meritocracy was possible under white supremacy forced the conclusion that individual talent and virtue could be rewarded only by gaining legal and political equality for all members of the oppressed and subordinated racial groups. If the special economic problems of the lower classes received less attention than socialist or labor-oriented radicals thought they deserved, it was because of the elite's sincere belief that an open, nonracial capitalist society was in the best interests of all classes.[59]

The ethos shared by elite black activists in the two societies during the second decade of the twentieth century is well represented in a notable book published in 1920—*The Bantu Past and Present* by Silas M. Molema. Of a slightly younger generation that the founders of the SANNC, Molema came from a prominent Baralong family with chiefly connections. After studying at Lovedale, he went abroad in 1914 to study medicine at the University of Glasgow and was still in Scotland, having recently received his degree, when his book was published. He returned to South Africa in 1921 and began a long career as a prominent physician and sometime leader of the African National Congress. *The Bantu Past and Present* was mostly a somewhat rambling compilation of historical and ethnographic information about black South Africans, but it concluded with a discussion of contemporary race relations and made some explicit comparisons between the American and South African situations.

What the southern United States and South Africa had in common were enforced patterns of "complete separation" in "all walks of life." Acknowledging the obvious demographic and historical differences, Molema found a fundamental similarity in the extent to which blacks were exposed to "a strong feeling of dislike and contempt." In the United States, the Negro's "educational and industrial progress have [*sic*] fanned the flames of colour prejudice, and his political aspiration have fanned them more, till, between black and white, there is drawn by the latter a rigid line in all matters social, religious, civil, and political."

The South African situation was not identical because the educated or "civilized" element was a much smaller proportion of the whole black population; "for the Bantu are educationally and socially far behind American Negroes." But the tiny African elite was experiencing the same painful rejection as its more numerous African-American counterpart. It was, he argued, the "civilized" and not the "uncivilized" Bantu who experienced the most prejudice. Their pain was acute, because "the more civilized a man is, the more keenly is he apt to feel the stigma of the prejudice he encounters, the disabilities which he is placed under solely by reason of his colour, and the determination of the ruling class to ignore his intellectual attainments, forcing him down to the level of his rudest brethren."

Molema noted the contempt with which black teachers and ministers were treated by white educational and religious leaders, the expulsion of Africans from civil service jobs as clerks and interpreters so that whites could hold these positions, and the exclusion of black lawyers and physicians from professional facilities and opportunities—"from everything that is white." He concluded that "a civilized black man . . . keenly feels the stigma which is eternally placed upon him; he bitterly resents the social ostracism to which he is subjected." He then quoted some passages from the first chapter of Du Bois's *Souls of Black Folk* on the despair of cultivated African-Americans whose strivings had run up against a similar wall of white rejection.[60]

Echoing Du Bois's call for leadership by a "talented tenth," Molema stressed the necessity of a self-conscious elite to uplift and represent the black masses. University educated blacks "are the interpreters, the demonstrators and exponents of the new and higher civilization to their struggling brethren. They are also the mouthpieces and representatives of their fellow-countrymen to the advancing white race." No more than Du Bois, however, was Molema calling for a slavish imitation of everything white and a categorical repudiation of black culture and the African past. The educated elite was to play a complex mediating role: "They should study, uphold, and propagate their national customs and institutions, only modifying and abolishing such as are pernicious, and seem calculated to clash with the best in civilization and to arrest progress." For Molema, as for the young Du Bois, public equality and progressive development did not require cultural uniformity based on European models or a self-betraying renunciation of Africa.[61]

By the time Molema's book was published, Du Bois and to some extent the other blacks who were giving direction to the NAACP had moved beyond the unabashed elitism of the years of the Niagara Movement and the beginnings of the twentieth-century civil rights struggle. Black workers were by then joining the association in substantial numbers and social-democratic perspectives were having some influence on black staff members such as Du Bois and James Weldon Johnson (the first African-American executive secretary). In the SANNC, soon to be renamed the African National Congress, the principal founders—whose basic attitudes Molema expressed—were being challenged by spokesmen for an emerging African workers' movement. But the belief that the professional middle class were the natural leaders of the race and that its concerns and aims should predominate in the framing of a black or African political agenda had taken deep roots and would survive the upheavals of the 1920s in both organizations.[62]

Besides being similar in the social-psychological profiles of their leadership, the NAACP and the National Congress were remarkably alike for at least three decades in the reformist assumptions that they held and in the means that they employed for the advancement of blacks. For the most part, they sought liberation from prejudice and discrimination through nonconfrontational means: the publicizing of black grievances, the lobbying or petitioning of public authorities and legislative bodies, and the support of legal challenges to discriminatory laws and policies. They assumed that the whites in power could be influenced by black opinion because of shared liberal-democratic values that whites were knowingly violating when they discriminated against blacks. Protest based on the expectation of significant reform could therefore take place within the existing constitutional framework. Violent or revolutionary action was ruled out from the beginning, but the congress, impressed by Gandhi's passive resistance campaigns on behalf of South African Indians, listed nonviolent action as a possible means of protest. Although Congress members and supporters participated in nonviolent actions in 1913 and again in 1919, mass civil disobedience did not become a central tactic of the organization until the 1950s. Generally, as in the NAACP, there was a fear that mass action could get out of control and impede the campaign to persuade whites of the justice of the black cause by alienating those who might otherwise be sympathetic.[63]

The NAACP's legalistic reformism had a justification that the congress's lacked: it had the letter of the United States Constitution as a basis for its egalitarian claims. There was no way the South African constitution could be construed as a charter of rights for Africans. For a time, the hope persisted that the British parliament, which had the right to disapprove South African legislation until 1930, could be persuaded to disallow the most flagrantly discriminatory laws on the grounds that they violated the rights of British subjects. But by the 1920s it was clear that this was a vain hope. The decline and atrophy of the congress between the two world wars, at a time when the NAACP was relatively vigorous and enjoying some limited successes, especially in the courts, can be attributed partly to the inescapable fact that a reformist campaign for civil rights had an inherently better chance to yield fruit in a legal and constitutional context that could be interpreted as favorable to black rights than in one that was patently unfavorable. In 1930, Dr. Albert B. Xuma, a young African physician and future president of the ANC, applied the knowledge that he had gained during thirteen years studying and interning in the United States to make the contrast explicit. After noting the violation of American constitutional principles that could be found "in certain sections of the United States," he affirmed that "liberty and justice is the foundation stone in the law of the land. It gives hope and citizenship rights to all alike. . . . In test cases of serious consequence the Supreme Court of the United States has sometimes, if not always, upheld the spirit and letter of the Constitution." He then proceeded to tell the contrasting story of proliferating color-bar legislation in South Africa from the Natives' Land Act of 1913 to the Native Administration Act of 1927. Although he may have romanticized somewhat the American situation in 1930, there was substance to Xuma's view that the American standard was equality under the law, whereas the official South African policy was differential and discriminatory treatment.[64]

Resisting the High Tide of Segregation, 1913–1919

The NAACP and SANNC had barely come into existence when the segregation movement intensified and expanded, finding new and more radical applications for its principles. The year 1913 saw major new segregationist initiatives in both societies, and black attention was focused more sharply than ever before on the meaning and consequences of government-enforced racial separation. During the war that broke out in Europe in 1914 the mettle and resolve of both organizations were tested under conditions that were fraught with ambiguity. In some respects the war period provided new opportunities to challenge discriminatory racial policies, but it also tempted black leaders to accommodate to segregation in the hopes of receiving a better deal within its parameters. Black thought and action on the segregation issue between 1913 and 1919 took somewhat parallel forms in the two societies, but efforts to affect government policy had differing degrees of success. These varying outcomes helped to put race policy in the two societies on divergent paths, one leading to gradual improvement in the status of blacks and the other to gradual deterioration.

In the United States, the election of Woodrow Wilson in 1912 as the first president since the Civil War born south of the Mason-Dixon line threatened to nationalize the southern Jim Crow policy. The armed forces had been segregated since the enrollment of substantial numbers of African-Americans during the War for the Union, but in the civil service blacks worked alongside whites. In 1913, the 24,500 blacks who worked for the federal government constituted slightly more than 5 percent of all government employees. If they were somewhat underrepresented on the basis of population, it was nonetheless evident that the federal civil service provided far and away the best opportunity that blacks possessed for secure, white collar employment. In Washington, D.C., a concentration of black officeholders gave a firm foundation to a distinctive black elite that not only was locally influential but also played a conspicuous role in national organizational efforts involving the black middle class.[65]

It was therefore of great consequence to the black elite that the Wilson administration had no sooner taken office than it began to segregate the federal civil service. The separation of government employees by race seemed to threaten the dignity and equality of blacks in the last place where the access to government opened up during the Reconstruction era was still a tangible reality. (The last black congressman from the southern states had left office in 1901.) If the way Jim Crow functioned in the South was any indication, separation would surely lead to a decrease in possibilities for promotion on the basis of merit, if not in the chances of getting a government job in the first place. The NAACP and William Monroe Trotter's National Independent Political League (which had supported Wilson in 1912) protested vigorously against the plan. NAACP president Oswald Garrison Villard proposed to Wilson that he call a national conference to review race policies, and Trotter carried the campaign into the President's office; during the interview that followed tempers flared, and Wilson ended up showing Trotter the door because of his allegedly disrespectful demeanor. The failure of all efforts to arrest segregation of the federal civil service was a profound blow to the hopes of the black elite that Jim Crow would remain a matter of southern state legislation and not become nationalized.

Du Bois consequently gave up on the strategy that he had previously endorsed of blacks using their votes to maneuver between the major parties in hopes of a better deal; like Trotter, Du Bois had supported Wilson in the election out of disappointment at the failure of the preceding Taft administration to follow Republican traditions of appointing blacks to office and out of wishful thinking based on a vague pledge Wilson made during his campaign to be fair to African-Americans. Republicans' indifference to their black supporters and the Democrats' endorsement of segregationism seemed to leave African-Americans bereft of political leverage no matter how they voted.[66]

If elected politicians were unwilling to stem the tide of segregation, perhaps the federal courts would respond to black contentions that at least some forms of legalized separation entailed the kind of racial discrimination explicitly outlawed by the Fourteenth Amendment. Between 1913 and 1917 the NAACP was preoccupied with the growth of a kind of segregation that went beyond differential access to public accommodations that had been sustained in *Plessy v. Ferguson* and attempted to separate the races on the basis of residency. The most extreme proposal that received a public hearing was the plan of Clarence Poe, the editor of an influential North Carolina farm journal, to segregate the rural population of the South by prohibiting whites and blacks from purchasing land in areas to be designated by law as reserved for the occupancy of one race or the other. Poe was directly influenced by South African segregationist thought and precedent. He was in fact attempting to do for the American South what the Natives' Land Act would do for South Africa. Although it provoked considerable anxiety among black leaders, Poe's plan did not gain significant support among white supremacists or result in any state legislation, principally because white political leaders feared that it would give too much independence to blacks and put many of them geographically and economically beyond the reach of white employers who wanted to hire them at low wages or take them on as disadvantaged sharecroppers.[67]

A more threatening and effective effort to implement the principle of residential or spatial segregation was the movement for mandatory neighborhood segregation in urban areas. In 1910 a Baltimore city ordinance prohibited members of either race from moving into blocks where the other race predominated. In 1912, the Virginia assembly voted to authorize the state's municipalities to designate white and black neighborhoods and restrict residence and the purchase of property in order to encourage racial separation. By 1913 several cities in the southern and border states had some kind of residential segregation ordinance.[68]

The movement to insure that blacks and whites did not live cheek-to-jowl had implications that went beyond differential access to public accommodations for the purpose of preventing "social equality." It had an obvious economic significance: it threatened to block the path of self-improvement and entrepreneurship that, according to Booker T. Washington, blacks could follow even if they were socially segregated. By limiting the right of blacks to buy and sell property in response to market opportunities, it put a ceiling on their ability to accumulate wealth and improve their physical environment. In a pamphlet on "The Ultimate Effects of Segregation" published in 1915, black educator William Pickens—a future high official of the NAACP—focused on the residential segregation ordinances and the damage they would do to the practical, worldly aspirations of blacks. Official ghettos, he argued, would deprive municipal authorities of any incentives to improve conditions in neighborhoods designated for blacks; these ghettos would inevitably deteriorate, and African-Americans who aspired to a respectable middle-class way of life would be unable to escape from slum conditions by moving to a better neighborhood. The claim of the segregationists that these ordinances placed equal burdens on both races and thus conformed to the Fourteenth Amendment seemed absurd to Pickens: "The segregation law *in effect* means that those who have no homes shall not acquire homes of those who have homes; and aspires to constitutionality by adding that those who have homes shall also not acquire homes of those who have them not."[69]

Booker T. Washington realized that his basic program was jeopardized by the residential separation laws, and their enactment impelled him to make his first public protest against segregationist legislation. His famous

New Republic article entitled "My View of the Segregation Laws," which was not published until shortly after his death in 1915, was specifically directed at the residential ordinances. Like Pickens he noted that the vice district was always in the area designated for black occupancy; the result was that the whorehouse might be next door to the schoolhouse and that "when a Negro seeks to buy a house in a respectable street he does it not only to get police protection, but to remove his children to a locality in which vice is not paraded." As a promoter of black business enterprise, Washington complained that the segregation laws never prohibited whites from owning stores and businesses in black neighborhoods and thus were inherently inequitable since, as a practical matter, there were no equivalent opportunities for blacks in white neighborhoods. Washington was thus impelled, posthumously as it turned out, to adopt the protest orientation that he had previously eschewed. Segregation was clearly beginning to threaten the prospects of those for whom he spoke—the black strivers who had been willing to tolerate racial separatism so long as it did not preclude the chance to make a decent living, accumulate property, and establish a middle-class home.[70]

When the NAACP went to court to challenge the residential segregation ordinances it spoke for a vital concern of all blacks who aspired to middle-class status, whether they happened to consider themselves followers of Washington or Du Bois. For the Washingtonians, it was an extension of segregation beyond what could be allowed to occur without public protest, while for the Du Boisians it represented a further development of the kind of practices and policies that they had always found intolerable. The NAACP won a notable victory when it challenged the Louisville Ordinance in the federal courts and had it declared unconstitutional in 1917 on the ground that it denied the Fourteenth Amendment right not to be deprived of property without due process of law. The toleration of segregation by federal courts was shown to have limits: separate railway cars and separate seating on streetcars were, for the time being, constitutional, but legally mandated separate neighborhoods were not.[71]

The importance of this victory has not been fully appreciated because it obviously did not prevent black ghettoization by restrictive covenants, zoning laws, and other devices. But black ghettos have at least been allowed to grow, and neighborhoods in major American cities have gone through brief "integrated" periods before becoming all-black. What would have happened, one wonders, if the core idea of the residential segregation laws had been effectively implemented; if, in other words, blacks had been forbidden by law—as well as being discouraged by prejudiced realtors and the threat of extra-legal violence—from moving into white neighborhoods. More living space for blacks would have to be provided through formal decisions by predominantly white political authorities. Since it would have been politically difficult if not impossible to displace white residents of inner-city neighborhoods, politicians and planners would have had to establish separate black suburbs in the less desirable segments of the urban periphery. In short, something like the black townships and "group areas" of South Africa might have resulted.

The South African Native National Congress had been formed in anticipation of new legislation requiring land segregation, and its fears were realized when the Natives' Land Act was passed in 1913. This law, which inspired Clarence Poe's abortive plan for rural segregation in the American South, became the object of a sustained protest campaign that served as the principal focus of the congress's activity for several years. Full implementation of the act was delayed pending the report of a commission that would study the map of South Africa and draw lines on it to designate precisely which areas would be reserved for Africans, but the law's immediate abolition of "squatting" or sharecropping on white-owned land by Africans who did not work part of the year directly for white farmers had the immediate effect, especially in the Orange Free State (where Africans had never had the right to own land) of causing mass evictions. By 1914, homeless blacks with herds of livestock that no longer had a place to graze were clogging the roads and tracks of the Free State.[72]

The congress agreed that the act as written had to be opposed, but there was division in its ranks about whether to accept the principle of spatial segregation and argue for a fair division or reject the idea out of hand. John L. Dube, the president of the SANNC, like his "patron saint" Booker T. Washington, was more comfortable as an accommodationist than as a protester, and his tendency to seek a better deal for Africans within a segregationist framework would weaken and dilute the campaign against the Land Act. In a letter to the American supporters of his industrial school written at the time of his election to Congress's presidency in 1912, Dube confessed that he had undertaken the responsibility out of a sense of duty "to keep in check, the red-hot republicanism that characterize [*sic*] some of our leaders, and is calculated to injure rather than help our cause." In 1914, he wrote to Prime Minister Botha and represented the Congress as making "no protest against the principle of segregation so far as it can fairly and practically be carried out."[73]

There was no question in the minds of Congress's leadership that the Land Act itself was the prelude to a patently inequitable distribution of territory, and when the government failed to respond to pleas and petitions, the SANNC dispatched a delegation to Great Britain in an effort to get parliament to overrule the act. As in the campaign against the South Africa Act in 1909, African leaders put their faith in the British traditions of liberalism and legal equality that had significantly shaped their own view of the world. The delegates—Dube, Plaatje, Walter Rubusana, Saul Msane, and Thomas Mapikela—carried with them a petition objecting to the act because of the limitations it placed on the right to acquire property, enter freely into contracts, or move about at will in search of economic opportunity. They appealed in particular to the humanitarian lobby that claimed descent from the antislavery movement and professed to be vigilant against its revival in a new form. "Is not this a form of slavery which will gradually lead to conditions rendered insupportable for self-respecting human beings?" the petitioners asked. To allow the Land Act to stand would violate the cherished ideal that "every coloured subject shall remain free beneath the Union Jack."[74]

But faith in British humanitarianism was misplaced. When the delegation sought the aid and sponsorship of John Harris and the Anti-Slavery and Aborigines' Protection Society, they put themselves in the hands of a pressure group that viewed land segregation as "aborigine's protection" rather than as a form of oppression. Their attitudes shaped by experience with "native" minorities in countries such as New Zealand, paternalistic humanitarians in Britain tended to think first of preventing settler encroachment on aboriginal lands. As the price of his support, Harris persuaded the delegation to sign a document objecting to some details of the Land Act but accepting the idea that there should be some kind of statutory division between African and European areas. After the rest of the delegation returned to South Africa shortly after war broke out in Europe, Sol Plaatje, a particularly devoted advocate of the Cape liberal tradition of "equal rights for every civilized man" in a common South African society, was left to carry on by himself. Cooperation between Harris and the resident voice of the SANNC in Britain quickly broke down. In 1916 Plaatje published his *Native Life in South Africa*, an eloquent polemic that vividly portrayed the sufferings of South African blacks under the act. Although he did not debate the theory of segregation in the book, Plaatje made it clear that the act was totally unacceptable, a position that put him at odds with the Anti-Slavery and Aborigines' Protection Society.[75]

In 1916, shortly after the first edition of Plaatje's book had appeared, the Beaumont commission reported on how the land should be divided under the act and assigned Africans a mere 7.4 percent of the area of the Union with vague promises of future augmentation. All hopes of an equitable division dashed, the congress intensified its campaign to have the Land Act disallowed by the British parliament. But the issue of whether the Congress was opposed to segregation in principle or only to the unfair application of it in the Land Act was not immediately resolved. In early 1917, Richard Selope Thema, the relatively conservative general secretary of the SANNC, wrote to Travers Buxton, Secretary of the Anti-Slavery Society, conceding the principle of segregation but denying that it could be effected fairly by an all-white South African government: "While the Bantu people will gladly welcome the policy of territorial separation of the races if carried out on fair and equitable lines, they cannot bind themselves to support a government that cannot carry out the principle with justice. In fact no democratic government can do justice to the unrepresented in Parliament." Buxton seized on the first part of this quotation to make the claim that the congress accepted the principle of the Land Act and that Plaatje's militant opposition could be disregarded by those responsive to African opinion. When word of Thema's concession of the principle got back to South Africa, a militant anti-segregationist faction, led by Seme and Saul Msane, forced the resignation of Thema and Dube for sanctioning the theory of territorial separation and strongly endorsed Plaatje's position. As Thema wrote to Buxton after the accommodationists had been ousted, "the Congress is divided into two camps with regards to the policy of segregation." Plaatje, who had emerged as the spokesman for the dominant faction, was offered the presidency; after he declined it for personal reasons, S. M. Makgatho, the vigorous and relatively radical leader of the congress in the Transvaal, was chosen. Makgatho and the new executive secretary, Saul Msane, stiffened up the SANNC's opposition to government policies, especially to the Native Administration bill of 1917, which proposed that Africans be represented in electoral assemblies with limited powers within their own territories but not in the central government.[76]

The dispute in the congress over land segregation bears a rough resemblance to the earlier Washington-Du Bois controversy in the United States, with Dube and Thema hoping to establish a basis for black self-help within the framework established by white segregationists, and Seme, Plaatje, Makgatho, and Msane resisting any concessions to white authority that would compromise the ideal of equal citizenship. But divisions on the segregation issue were not as clear-cut in either case as this abstract formulation would suggest, and the image of a clash between separatist accommodationists and integrationist protesters is an oversimplified view of the debates. As we have seen, Washington was ambivalent about the segregation of public transport and at the end of his life took a public stand against residential and territorial segregation. As the full text of his letter to Buxton reveals, Thema (who spoke for Dube as well) would be willing to contemplate dividing the land on racial lines only if blacks were represented in the government and had a voice in deciding the lines of demarcation.

If the putative accommodationists were not really accepting a white-imposed plan of segregation, neither were those who adopted an apparently more radical stance repudiating all forms of racial separatism on principle. Du Bois was strongly committed to the encouragement of separate black institutions, a Pan-African ethnic consciousness, and a distinctive African-American culture. His opposition to legally imposed segregation did not mean that he favored the total assimilation of blacks into Euro-American society. Equal political and civic rights were for him a necessary foundation for autonomous group development. Plaatje and Seme also had Pan-Africanist views, as well as strong attachments to their particular ethnic communities. Sounding more like Washington than Du Bois, Plaatje and other protest-oriented Congress leaders often disavowed any aim of "social equality with the white man," and they did not in fact object to the general pattern of separate facilities, institutions, and associations for white and black in South Africa so long as separation was voluntary or consensual and not merely a front for white monopolization of power and privilege. Indeed the congress itself was a manifestation of separatism despite its platform of political inclusion. In neither case can opposition to the laws and policies that whites used to make blacks "separate and unequal" be construed as advocacy of social and cultural assimilation.[77]

The willingness of the Dube-Thema faction to accommodate to government segregationist policies if they could influence their implementation and derive some practical advantage from them was analogous to the response of Du Bois and the NAACP to the mobilization of African-Americans for participation in the First World War. When the United States took up arms against Germany in 1917, the NAACP supported the war effort and called for equal opportunities for blacks in all its phases, including the armed forces. Traditionally, blacks in the armed forces had been enlisted in separate units under white officers, precisely the way colonial powers utilized "native troops" in their own wars. When it became clear that there was no chance of fully integrating the First World War army, Joel Spingarn, the white president of the NAACP, backed by Du Bois and *The Crisis*, decided to settle for the attainable objective of a separate officer training camp for black officers in an effort to shift the line of military segregation from the horizontal to the vertical. If blacks had to fight in separate units, Spingarn and Du Bois argued, they should at least have their own black officers.

This compromise proposal proved acceptable to a government that was willing to pay a limited price for black support of the war effort, but it caused a major controversy among the leaders of the black protest movement. William Monroe Trotter and his National Equal Rights League attacked it as an ignoble concession to the principle of segregation, and within the NAACP itself there was sufficient opposition to prevent the organization from officially endorsing the camp. But it was difficult for a pragmatic, reformist movement for black advancement to avoid compromising with segregation when the choice was between some movement toward the equalization of facilities and the status quo of separate-and-flagrantly-unequal. On future occasions, the NAACP would favor upgrading separate black schools and equalizing teachers' pay, despite the fact that such a stand conceded the legitimacy, or at least the inevitability, of educational segregation. Du Bois's endorsement of a separate training camp was the product of a logic that would sometimes prove compelling, the realization that segregation per se was not the issue; it was rather how to increase the resources and opportunities available to the black community.[78]

If we look in a broader way at elite black responses to the First World War, we find remarkable parallels between the basic positions of the NAACP and the SANNC. The association's patriotic disposition was carried to its logical conclusion in Du Bois's famous editorial in

The Crisis of July 1918, calling on blacks to "close ranks" and give wholehearted support to President Wilson's crusade to make the world safe for democracy. "Let us, while this war lasts," wrote Du Bois, "forget our special grievances and close our ranks shoulder to shoulder with our white fellow citizens and the allied nations that are fighting for democracy."[79]

In 1914, the congress took a similar stand and suspended agitation to the point of calling back the delegation sent to Britain to protest the Natives' Land Act. Like the NAACP, the congress believed that its hopes for justice and equality for blacks would be enhanced by wartime loyalty and contributions to the national cause. The SANNC even encouraged blacks to enlist in the armed forces, despite the fact that the only way they could do so was to volunteer for labor contingents that were not only kept out of combat in France but were kept in closed compounds so that the inmates could have no contact with local whites or even with black soldiers from other allied countries. The total lack of any reward for black patriotism and the refusal of the government to employ black manpower except under the most humiliating circumstances made it difficult to maintain the no-protest policy. In 1916, the grievances of black military laborers, the report of the land commission, and the prospect of the new discriminatory actions by parliament forced Congress to renew political agitation in the midst of the war. Unlike African-Americans, who appeared to be getting something for their loyalty—in addition to the officer's camp, the Wilson administration appointed blacks to positions as advisers on African-American mobilization—black South Africans faced renewed humiliations and no suspension of government assaults on their rights and interests. For the moment at least, the parallel strategies of seeking a better deal by offering to stand fast with whites at a time of national peril seemed more productive in the American case than in the South African.[80]

In retrospect, it is clear that the period 1913–18 represented a critical stage in the history of white supremacy in both the United States and South Africa. In both cases the segregation movement surged forward and made clear its intentions to make legal separation of the races as thorough and comprehensive as the interests of whites required or permitted. But in the United States blacks and their white liberal allies managed to hold segregation in check, not rolling it back but at least limiting its growth; in South Africa, the congress failed to erect any effective barriers to a continuous and accelerating increase in the kind of segregationist policies that at a further stage of development would be called apartheid.

The achievement of the NAACP during Wilson's presidency has for the most part been overlooked by historians, who have tended to stress the failure to block segregation of the civil service. It is undeniable that this action was a serious setback for the cause of civil rights. Intrusion of the southern caste system into the federal government, Professor Kelly Miller of Howard University warned in 1914, "would thereby sanction all of the discriminatory legislation on the books of the several states and justify all such legislation in the future." But southerners in Congress attempted to go much further in making Jim Crow federal policy; they introduced about two dozen bills during Wilson's first administration that included mandating segregation in the civil service and in interstate transportation, repealing the Fifteenth Amendment, and outlawing intermarriage by federal statute. Partly as a result of the lobbying of the NAACP and its allies, none of these proposals became law, and only the anti-miscegenation bill even got out of committee. Since the segregation of federal employees remained executive policy rather than a legal requirement, federal law remained relatively free of overt racial discrimination.[81]

Besides helping to hold the line against federal Jim Crow laws, lawyers representing the NAACP won two notable court victories during this period. In 1913, the Supreme Court held that "grandfather clauses"—the blatant form of voting restriction which limited the suffrage to those whose ancestors could vote before the enfranchisement of blacks during the Reconstruction era—violated the Fifteenth Amendment. And, as we have seen, the Court held in 1917 that residential segregation laws could not be reconciled with the Fourteenth Amendment's prohibition on depriving citizens of life, liberty, and property without due process of law. At the time the United States joined the war in Europe, it was clear that, within the South, the segregation of public facilities and a general pattern of de facto disfranchisement were established facts of life for African-Americans and would face no serious challenge in the immediate future, but it was equally evident that the legalized caste system would not spread North and that constitutionally permissible segregation would not include the requirement that the races reside in separate areas or neighborhoods. James Weldon Johnson, who served as the NAACP's first black executive secretary during the 1920s, accurately assessed the contribution of the early NACCP in 1934: "When the N.A.A.C.P. was founded the great danger facing us was that we should lose the vestiges of our rights by default. The organization checked that danger. It acted as watchman on the wall, sounding the alarms that called us to defense. Its work would be of value if only for the reason that without it our status would be worse than it is."[82]

The NAACP's accommodation to military segregation during the war was not disastrous, partly because it was accompanied by new opportunities for educated blacks to become officers, but, more important, because it was rewarded by a new willingness on the part of the Wilson administration to take black interests and opinions into account in making public policy. It is true enough that most of the gains blacks made during the war did not survive the armistice. But at least a precedent was established that in time of war or national emergency, blacks should have some say about the terms on which their cooperation might be assured. Wilson's appointment of black advisers on African-American mobilization for war foreshadowed the New Deal's similar efforts to consult blacks on their stake in measures to deal with the Great Depression.[83]

If African-Americans came out of the war period with something to show for it that would not completely disappear in the generally reactionary decade that followed, South African blacks gained nothing from their support of the war and in fact saw their rights increasingly assaulted. Seeing no need for black cooperation in what was regarded as strictly a white man's war, the South African government pushed new measures for territorial and political segregation while the conflict was still going on. The Native Administration bill, which mandated a separate and subordinate set of political institutions for Africans based on the territorial divisions projected by the Land Act, did not actually become law during the war, but it was clearly on the agenda for the future, and the SANNC had no weapons in its arsenal that might avert it. Another delegation was dispatched to Britain in 1919, but there was no reason to expect that it would have any greater success than earlier deputations. The frustration and disillusionment of the period between the passage of the Land Act and the end of the war would force the South African Native National Congress (soon to be renamed the African National Congress) to explore new and more radical methods of protest and resistance in the immediate postwar years.[84]

The contrast between the limited success of the NAACP in combating the spread of segregation in the United States and the total failure of the SANNC to do likewise in South Africa was clear evidence of a major difference in the long-term prospects of the two movements. NAACP would face a long uphill struggle, and its elitist character and commitment to legalistic reformism would prevent it from being fully equal to the task of ending government-enforced segregation in the South. But at least the uphill direction of its struggle was established. The congress, on the other hand, was on a downward slope as the 1920s began. Lacking the constitutional legitimacy and influential white allies that made the NAACP's task easier, the SANNC faced a grim future of futile protests against new white supremacist initiatives. Not until it abandoned its efforts to reform an inherently racist constitutional system and reached beyond the elite to draw the masses into the struggle would it have the capacity to shake the foundations of segregation.

· 4 ·

"Africa for the Africans": Pan-Africanism and Black Populism, 1918–1930

Working-Class Protest and Middle-Class Organizations, 1918–1921

World War I altered the course of world history in ways that strongly affected black protesters in the United States and South Africa. It halted the imperial expansion of the West (although it would take another world war to reverse it), shook the faith of Western nations in the inevitability of peaceful progress under the auspices of liberal capitalism, spawned in Soviet Communism a competing program for modernizing the world, and encouraged a new upsurge of ethnic nationalism under the banner of "self-determination." For colonized and oppressed groups whose domination had taken the form of racial oppression, it inspired dreams of liberation from the tyranny of the West or the white race but did not provide the means to realize those dreams or settle the question of whether the struggle should be based on the Marxist vision of an uprising of the working classes throughout the world or on the Zionist model of national self-realization in a territory of one's own.

The Marxist-Leninist path to liberation won the active adherence or fellow-traveling sympathies of a few black intellectuals in the United States and South Africa during the 1920s, but the impulse to substitute class for race as the keynote of the struggle would not energize significant numbers of Africans or people of African descent until the 1930s or '40s. In the years immediately following the Great War, a racially defined nationalism—or, more specifically, Pan-Africanism of one kind or another—was the principal rival of the incorporationist liberalism that had dominated the thinking of prewar elites and uneasily survived into the postwar era as the official—but not uncontested—doctrine of the National Association for the Advancement of Colored People and the South African Native National Congress (renamed the African National Congress in 1923).[1]

The main demographic trend of the period in both countries was the massive movement of black population from the countryside to the cities, which began with the homefront labor shortage of the war years but continued unabated up to the time of the Great Depression. It was among the newly urbanized black and brown people of Harlem, Chicago, Johannesburg, and Cape Town that the new racial consciousness first took root. Its most dramatic and potent manifestation was Marcus Garvey's Universal Negro Improvement Association (UNIA), which spread from its base in Harlem throughout the African diaspora as well as to parts of Africa itself. Black South Africa showed great interest in Garvey's vision of "Africa for the Africans," despite having only a limited opportunity or inclination to join his association and participate directly in its activities. The new South African organization that expressed the mood of the 1920s and demonstrated the kind of mass appeal that characterized the Garvey movement in the United States was the Industrial and Commercial Workers' Union of Africa (ICU), under the charismatic leadership of Clements Kadalie. As its name suggests, the ICU had a Pan-African perspective; its leader was not even South African in origin but in fact an immigrant from Nyasaland (now Malawi) who never learned to speak a South African native language. At times the ICU espoused a black nationalist ideology similar to Garvey's and even shared officers with the relatively small South African branch of the UNIA. The ICU followed a somewhat different path and ended up at a different point on the spectrum of black ideologies. But such a divergence may tell us something important about the contrasting circumstances that conditioned efforts to create mass movements among blacks in the two countries during the 1920s.

Both the UNIA and the ICU have been characterized as "populist" movements. One of the most elusive and problematic terms in the lexicon of historians and political scientists, populism is more easily defined in terms of what it is not than of what it is. But with the help of the political theorist Margaret Carnavon and the historian Lawrence Goodwyn, it may be possible to arrive at a concept of populism that applies to both the UNIA and the ICU and serves to distinguish them from rival black movements of the liberal or Marxist type. The essential and most obvious ingredient of populism is opposition to the exercise of authority by established or self-appointed elites who profess to act for the good of the people, in favor of a leadership that presents itself as the actual voice of the people. Populists normally encourage forms of popular participation in decision-making that at least give the appearance that the people are ruling directly. The concept of an enlightened vanguard, a meritocracy, or a "talented tenth" called upon to elevate the benighted or undeveloped masses is almost as antithetical to populist thinking as an aristocracy based on birth. A *black* populism would be opposed not only to white rule but also to the claims of educated or "advanced" black elites —such as those traditionally in charge of the National Congress and the NAACP—to represent black or African communities and direct the main movements of protest and resistance.[2]

The second distinguishing feature of populism is that its concept of "the people" is broader than the Marxian conception of the working classes and includes, often at the core of the movement, social groups that orthodox Marxists would describe as "petit bourgeois." The American agrarian movement of the 1890s, for example, contained both landowning farmers and sharecroppers, with the former providing the leadership. In the South African domestic protest movement of the 1980s, tension developed between "populists," who believed in a united front that included middle-class blacks as part of the oppressed group, and "workerists," who argued that the struggle against apartheid ought to be a strictly proletarian affair. This inter-class character of populism normally leads to a program that is neither purely socialist nor consistently liberal-capitalist and may be viewed as kind of "third way." Formally democratic economic institutions—cooperatives, employee-owned businesses, or corporations that limit the amount of stock any individual may own—are likely to play a key role in the populist conception of a "cooperative commonwealth" or reformed market economy. Unlike radical socialism, populist economic doctrine respects private property so long as it is widely distributed; unlike liberal capitalism it opposes large concentrations of wealth and the kind of economic power that comes from the top down rather than the bottom up.[3]

Although it is not essential to a definition of populism that it have an ethnic or racial character (there have been movements of a populist type in ethnically homogeneous societies such as Denmark), it is often the case that "the people" struggling for self-determination are members of a distinctive cultural community as well as being non-elite citizens or subjects of a particular government. The interwar Eastern European populisms described by Margaret Carnavon were often rooted in the consciousness of ethnic groups seeking to become nations. Obviously a *black* populism would have to involve racial or ethnic consciousness, although the precise meaning of black nationality or peoplehood would not, in either the United States or South Africa, be self-evident; it would have to be constructed in opposition to more parochial loyalties or against the grain of more inclusive, potentially interracial conceptions of nationality.[4]

If the populism of the UNIA and the ICU filled a vacuum left by the organizations of the black elite, it was not before the NAACP and the SANNC had made serious efforts to get in touch with urban working-class blacks in the hope of forming a broader movement. Between 1918 and 1920 there was an upsurge of black working-class mobilization and protest in both countries, and part of the leadership of the older "talented tenth" organizations moved to embrace the new grass-roots activism. The ultimate failure of their efforts to create a mass movement by synthesizing the economic needs and concerns of the laboring class with their own traditional emphasis on civil and political rights provided an opening for the new black populist movements.

The story of working-class protest and its effect on the older organizations in the two or three years after the war is so similar in the two societies that its broad outlines can be conveyed by a single narrative. During the years between the outbreak of the war in Europe in 1914 and the armistice of 1918, an increase in manufacturing to meet wartime demands and a shortage of labor (due in the United States before 1917 to the curtailment of European immigration) helped to draw massive numbers of blacks from the countryside to the cities, where many of them found employment in industrial or service occupations. Social historians of both the United States and South Africa tend to see the war years as crucial in the emergence of an urban industrial working class among blacks. At the end of the war, there was a sharp increase in prices that cut deeply into real wages and the threat of widespread urban unemployment as the war-fueled demand for goods wound down. These developments affected the better-placed white workers as well the newly proletarianized blacks, and a wave of strikes by organized white labor followed on the heels of the armistice in Europe. Blacks, who were suffering even more from the postwar economic dislocations, began to view the strike as a weapon they might also use. Between 1918 and 1920 there was an unprecedented upsurge of strikes by blacks or involving blacks and efforts to form unions among them. In the United States this labor militancy spread even to the Deep South, where attempts were made to organize black timber and paper mill workers in Louisiana and sharecroppers in Arkansas. In South Africa, the main locus of militant African labor activity was on the Witwatersrand among mineworkers and the municipal employees of Johannesburg.[5]

The middle-class protest organizations—the NAACP and SANNC—responded to this pressure from below by supporting some strikes and insisting more strongly than before on the right of blacks to join or form unions. A huge increase in NAACP membership in 1918–19 was apparently due in large part to the recruitment of black workers, and the organization undertook to represent the interests of this new constituency. James Weldon Johnson, the first black field secretary of the NAACP, affirmed late in 1919 that the "mightiest weapon colored people have in their hand is the strike" and recommended that it be used to demand "abolition of Jim Crow conditions," as well as to improve the economic position of black workers. Levi Mvabaza, one of the founders of the SANNC, was personally involved in the 1918 strike of Johannesburg night soil workers that signaled the onset of black labor militancy in South Africa; for their direct support of the strikers, he and four other congress leaders were arrested and brought to trial for inciting to violence. At a meeting of the Transvaal Native Congress to protest the arrest of the striking "bucket boys," Mvabaza gave strong expression to the left-wing sentiments that were percolating in the organization: "The capitalists and workers are at war everywhere in every country. . . . The white workers do not write to the Governor-General when they want more pay. They strike and get what they should."[6]

In 1919, after the strikes by municipal workers had been forcibly repressed, the Transvaal Congress moved to direct action on a broader front and attempted the kind of mass resistance to legalized white supremacy that James Weldon Johnson was merely contemplating. The resulting campaign of civil disobedience against the hated pass laws was not the first such action by Africans. In 1913 and 1914 black women in the Orange Free State had successfully resisted new regulations extending the pass laws to females. But the 1919 actions on the Witwatersrand—which took the form of collecting passes from thousands of black workers and turning them in to the authorities—was the first time the congress or one of its branches had involved itself in mass civil disobedience against racist laws. The anti-pass campaign focused on a form of labor control that effectively prevented African workers from exercising the basic rights that all white workers took for granted. (It was the fact that they carried passes binding them to white employers that made strikes by Africans illegal.) Since pass laws also inconvenienced and humiliated middle-class blacks, agitation against them was the perfect vehicle for inter-class solidarity among Africans.[7]

The National Congress's interest in organizing black labor peaked in 1920 when several SANNC leaders were involved in attempts to form a single national union for African and Colored workers. Thereafter, however, efforts within the congress to redirect the black protest movement along economic or class lines quickly ran out of steam. Similarly, the NAACP in the United States retreated from its efforts to encompass the cause of black labor and reverted by 1921 to its earlier emphasis on civil rights and the status concerns that resonated most strongly with the middle class. Neither the congress nor the NAACP proved capable during the early 1920s of broadening its base much beyond the educated elite and thereby preempting some of the terrain on which their populist rivals would flourish.[8]

To explain these comparable failures to create a cross-class coalition to fight for the most basic needs of the laboring majority, one needs to look at the United States and South Africa separately. Of critical importance in the American case was the crushing defeat of militant labor by the forces of organized capital in the early 1920s. The destruction of radical industrial unions that might have opened their ranks to blacks left behind a labor movement composed mainly of conservative craft unions that were likely to maintain rigid barriers against black membership. Outside the South, blacks did not predominate in any particular line of work, except in such special niches of the railroad industry as pullman car porters and redcaps; consequently there was no real future for a separate black labor movement. In the mid-1920s, the socialist A. Philip Randolph began his long struggle to organize the sleeping-car porters, but from the start his objective was to use this all-black union as the entering wedge for black participation in an interracial labor movement.[9]

In the black belt of the Deep South, where a racially based labor movement was conceivable, efforts of blacks to unionize met with vicious repression. The effort to form a black sharecropper's union in rural Arkansas in 1919 led to a riot and subsequent reign of terror that may have claimed the lives of over 200 blacks, making it one of the most murderous racial pogroms in American history. The NAACP defended the 122 blacks charged with murdering the three whites killed in the disorders and with other alleged offenses arising from the "deliberately planned insurrection of negroes" conjured up by the white supremacist imagination. In 1923, the association won a landmark victory when the Supreme Court set aside the convictions on the grounds that fair trials had been impossible given the atmosphere of mob rule in which they took place. But the decision did not uphold the right of sharecroppers to organize in defense of their economic interests. Under the circumstances existing in the South during the 1920s, the NAACP could well argue that the priority it gave to establishing the right to a fair trial—a goal that it pursued not only through litigation but also by lobbying for federal anti-lynching legislation—was justified as a necessary first step to gaining economic rights. NAACP assistant secretary Walter White, who spearheaded the campaign against lynching, viewed the threat of "rope and faggot" for "uppity" blacks as the cornerstone of an oppressive labor system.[10]

In South Africa, the effort of the left wing of the ANC to make workers' struggles the congress's main concern was also the casualty of a pattern of repression that prevented an African labor movement from getting organized to the point where it had real bargaining power. The agitation of 1919 did not lead to abolition of the pass laws, although some exemptions were later introduced that eased the burden of the educated elite and thus made confronting the authorities on the issue seem less urgent. The deflation and unemployment of 1921 cut the ground out from under working-class agitators and movements and decreased the pressure from below on the ANC. In 1923 the Urban Areas Act tightened the controls on black migrants to urban areas and made organizing black workers more difficult than ever. In these circumstances, the more conservative element of the ANC, composed mainly of professionals and businessmen who sought to distinguish themselves from workers by seeking special recognition as "civilized" Africans, reasserted its dominance over the organization.[11]

White repression was thus deterring both the NAACP and the ANC from putting an economic or class-based struggle at the top of their agendas. In the American South, the association found that it had to deal with extralegal terror—lynching and pogroms—before it could get to the point of defending unionization and strikes. In South Africa, official repression of black labor through pass laws and influx controls made a trade union orientation seem increasingly futile or dangerous. But ideology was also a factor in both cases. As has been suggested, the educated elite that dominated the ANC was divided between strong adherents of liberal capitalism and a minority that was receptive to socialist ideas. If anything, the NAACP was even less susceptible to the kind of economic radicalism that would sustain a militant working-class struggle. This stance was partly the result of the continued dominance of the association's board of directors by liberal white philanthropists, but most of the black officers and active members were also proponents of liberal capitalism; they normally voted Republican because the party of big business was also the party that had freed the slaves and passed the Fourteenth and Fifteenth amendments. (The Democratic party, although somewhat more sensitive to the economic concerns of working people, tolerated the segregationism of its southern wing.) W. E. B. Du Bois, by this time a democratic socialist, was beginning to find this economic liberalism constraining, but most of his colleagues in the association saw their cause as the elimination of racial caste in an otherwise satisfactory capitalist democracy.[12]

That these elite protest organizations did not become mass movements of a populist type in the 1920s was due to more than their inability to reconcile the interests of middle-class and working-class blacks. Individual rights—political, civil, or economic—were of vital concern to all Africans or African-Americans, but focusing on them to the exclusion of cultural issues could not satisfy the need of people who had been humiliated by white supremacy for sources of group pride and a positive sense of identity. Du Bois understood this, and in the pages of *The Crisis*, he called for black cultural revitalization as well as for the right to vote or sit anywhere in the streetcar. It was also understood by Levi Mvabaza, who accompanied his 1919 advocacy of the rights of black workers with an attack on white missionaries and an assertion of the collective rights of his people to the land they had once possessed: "The white people teach you about heaven and tell us after death you will go to a beautiful land in heaven. They don't teach you about this earth on which we live . . . if we cannot get land on this earth neither shall we get it in heaven. . . . The God of our chiefs, Chaka, Moshoeshoe, Rile, Sandile, Sobuza, Lentsoe, etc., gave us this part of the world we possess."[13]

As organizations, neither the NAACP nor the ANC was able during the 1920s to give adequate expression to the powerful feeling among blacks that they had been wronged not just as individuals or members of social classes but as a people or nation. The affirmation of black or African identity was a task that would have to be undertaken by other movements.

Elite Pan-Africanism

In the United States and South Africa blacks could become conscious of being a people or a nation only through accepting a concept of "race" originally constructed by white supremacists and then turning it to positive uses. Ethnicity in a purely cultural sense was problematic in the American case and dangerously divisive in the South African. African-

Americans were descended from people violently extracted from a variety of African cultures. The experience of servitude had not obliterated African cultural traits, but it had attenuated and homogenized those that survived, syncretized them with Euro-American beliefs and customs, and thereby given rise to the "slave culture" that Sterling Stuckey has found to be the cultural root of African-American nationalism. Developing alongside the slave culture, however, was the "middle-class" culture of free blacks and elite slaves who were more deeply influenced by Euro-American norms of thought and behavior. Blacks from these relatively favored backgrounds or those that assimilated their values provided most of the visible leadership for the black community in the post-Civil War period and into the twentieth century. A large proportion of them were of mixed racial origins, and if white American society had not imposed its peculiar "one drop" rule for determining who was a Negro, this status group might have evolved into a distinct racial category like the "mulatto" or "colored" stratum that developed in the West Indies. Although nineteenth-century nationalists called for the solidarity of all people of African descent, the culture to which they appealed as a basis of unity was heavily influenced by the Western conceptions of "progress" and "civilization" with which the educated middle class identified. The most creative African-American thinking of the early twentieth century, preeminently the cultural thought of W. E. B. Du Bois, was devoted to the search for a higher synthesis of African roots, slave culture, and middle-class progressivism. Du Bois's underlying assumption was that blacks, as defined by the American racial order, had a common history linking them politically and culturally with people of African descent throughout the world.[14]

Constructing a black or African identity in South Africa was a different kind of enterprise; rather than fashioning a common history and culture out of fragments of tradition and experience, it required forging formerly antagonistic ethnic communities into a united front against white oppression. But as in the American case, a broader racial or Pan-Africanist consciousness of one kind or another was an essential ingredient in the solution. The very act of founding a national congress of Africans in 1912 had projected a rudimentary sense of race and nationality. On another level of consciousness, identification with African-Americans contributed significantly to the fashioning of a cosmopolitan black identity. And this was not simply a matter of Western-educated "progressive" Africans being inspired by Frederick Douglass, Booker T. Washington, T. Thomas Fortune, and W. E. B. Du Bois. In what may have been the first book written by a Zulu in his native language, Magema M. Fuze, a Christian convert who remained loyal to the Zulu monarchy, wrote shortly after the turn of the century about "our people now in America" in a way that conveyed the essence of what black America could mean to Africans who were not part of the progressive elite. After being transported to America and made slaves of the whites, he wrote, American blacks, "through the labor and great effort of their grandfathers in early times, were eventually able to obtain their release from slavery, and by the time they had children of their own, they had been freed and were important men in their own right. They have multiplied exceeding, being numbered by hundreds and thousands in that country to which their grandfathers were taken and sold."[15]

As in the United States, however, the problem of class or status differences among blacks complicated the process of providing specific content for an all-encompassing black or African identity. Would the unifying ideology be the liberal progressivism of educated elites, which would have to be imposed from the top down, or a new, more radical Africanism based on common elements in the experience of black workers and peasants? And to what extent would solidarity from the bottom up be based on a synthesized or reconstructed traditionalism and to what extent on new identities created by migration and industrialization?

During the decade or so following World War I, Pan-Africanism was an insurgent force in political thought and expression throughout the black world. But there were three distinct varieties of the impulse to establish links between blacks in the New World and in Africa that competed for the allegiance of black intellectuals and activists in the United States and South Africa. There was a conservative Pan-Africanism associated with the legacy of Booker T. Washington and represented in the 192Os by J. E. K. Aggrey; a liberal reformist Pan-Africanism led by Du Bois and manifested in the four Pan-African conferences that took place between 1919 and 1927; and the populist Pan-Africanism of Marcus Garvey and the Universal Negro Improvement Association.[16]

The origins of the conservative or accommodationist strain in postwar Pan-Africanism can be found in Booker T. Washington's intermittent involvement with Africa between 1900 and his death in 1915. Washington's attention was initially drawn to Africa because colonial officials and educated Africans solicited his advice on how to train "natives" for useful roles in Europe's African colonies. Believing that the problems of American Negroes and indigenous Africans were similar, Washington enthusiastically promulgated the Tuskegee model of industrial education. His commitment, however, was not great enough to induce him to travel to Africa and provide advice in person. In 1903, he turned down invitations to pay officially sponsored visits to South Africa and Rhodesia, explaining that he had too much to do at home. But his contacts with educated Africans, either by exchange of letters or when they turned up at Tuskegee, led him to contemplate a form of Pan-African cooperation. "I myself, at various times," he wrote to one of his African correspondents in 1906, "have advocated an understanding between the intelligent Negroes of the world, looking to the developement [*sic*] of the masses." In 1912 he presided over an International Conference on the Negro at Tuskegee, but most of the attendees were white missionaries rather than "intelligent Negroes"; only one black African is known to have been present.[17]

Washington's conception of Pan-African activity was of educated blacks and benevolent whites joining forces to assault the ignorance and immorality of the masses of black people throughout the world. But he inadvertently helped to inspire a nationalistic conception of black solidarity. According to the historian Kenneth King, "with many Africans it was not so much the white money behind Tuskegee as the black president and all-black staff that gave it much of its appeal"; for the young Marcus Garvey, who thought he was following in Washington's footsteps when he founded the Universal Negro Improvement Association, Tuskegee was (in King's words) "an island of black pride and racial solidarity."[18]

During the 1920s, the leading proponent of black-white cooperation for the redemption of Africa—the accommodationist side of Washington's message—was James Emman Kwegyir Aggrey, a native of the Gold Coast (now Ghana) who spent twenty-two years in the United States, mostly as a student and teacher at Livingstone College in North Carolina, before returning to Africa in 1920 as an emissary of the Phelps-Stokes Fund. The original target of this fund's philanthropy had been black education in the American South. It had strongly supported Booker T. Washington's program of practical, industrial training, and for this reason its 1913 report on Negro education had incurred the hostility of W. E. B. Du Bois. It was partly to head off a similar reaction from some segments of black opinion that the fund appointed Aggrey to its commission to survey the education of colonial Africans and make recommendations for its improvement. Aggrey strongly disapproved of Marcus Garvey's militant version of Pan-Africanism, and his principal mission during his tour of Africa was to counter Garveyite propaganda and try to halt the spread of its influence.[19]

Aggrey should not be written off as a stooge of white imperialists who deserves no place in the history of Pan-Africanist thought. He was a man of strong convictions and the authentic heir of a major tradition of black nationalist expression. Like Washington, he was an advocate of racial accommodation, but he arrived at this position by a different route. His inspiration was essentially religious rather than down-to-earth and pragmatic; in fact he was a latter-day exponent of the Ethiopian myth that had played such a conspicuous role in nineteenth-century black nationalism. As a child in the Gold Coast, Aggrey had been converted to Christianity and educated by white Methodist missionaries. It is not known precisely how or why he emigrated to the United States at the age of 22 to become a student at Livingstone, the main seat of learning of the African Methodist Episcopal Zion Church. AMEZ missionaries were active in the Gold Coast, and it is likely that he had some contact with them before coming to America. In any case, he was ordained an elder in the AMEZ Church in 1903, just a year after he had earned a B.A. from Livingstone. As a Christian who took the Golden Rule seriously, Aggrey could not condone racial animosities or conflicts, but this did not mean that he lacked race pride or favored an amalgamation of the races. His favorite metaphor for improved black-white relations was reminiscent of Washington's fingers-and-hand analogy except that it was explicitly color-coded: a piano with the black and white keys remaining quite distinct while playing in harmony.[20]

As a minister in one of the two great African Methodist denominations, Aggrey was heir to the Ethiopian myth and the romantic racialist tradition of viewing blacks as a people with a special aptitude for Christianity. In a speech before the Foreign Missions Conference of North America in 1922, he used the story of the Magi to illustrate the differing gifts that the three great races brought to Christianity. The whites brought the gold of material success, Asians the "frankincense of ceremony," and blacks the "myrrh of child-like faith," which is of course the essential ingredient: "We look for a Christ who loves all men, Who came to die for the whole world; we believe in God as a child believes. If you take our childlikeness, our love for God, our belief in humanity, our belief in God, and our love for you, whether you hate us or not, then the gifts will be complete."[21]

Aggrey was often heckled by blacks in South Africa, both in 1921 and during the second visit of a Phelps-Stokes Commission in 1924, because many in his audience believed that white Christians would take advantage of any turning-of-the-other-cheek and use it to perpetrate further injustices against unprotesting blacks. History would later show that the Sermon on the Mount could serve the cause of black liberation only when it was accompanied by acts of nonviolent resistance against unchristian racism.[22]

Aggrey's visits to South Africa were associated with efforts to export a white-sponsored scheme for interracial cooperation originating in the American South. In 1919, with backing from the Phelps-Stokes Fund, the Commission on Interracial Co-operation was founded in Atlanta. Its purpose was to improve race relations by bringing the "best" representatives of both races together for discussions and problem-solving. The framework of legalized segregation was taken for granted, and the object was to improve facilities for blacks in order to make "separate but equal" more than a legal fiction. In South Africa, the fund subsidized the Joint Councils of Europeans and Natives that brought together white liberals and educated Africans. But the ANC kept the Joint Councils at arm's length and discouraged the involvement of its officers. Like the Commission on Interracial Cooperation, the Joint Councils endorsed some forms of segregation—they acceded to the general pattern of reserved territories and separate political institutions—but protested vigorously against economic color bars that advantaged white workers and discouraged black industry and initiative. As historian Paul B. Rich has noted, South African white liberals were influenced in the 1920s by currents of American thought that rejected assimilationist models of group relations in favor of some form of racial or ethnic pluralism.[23]

The most prominent black leaders to participate in the racial cooperation movements were Robert R. Moton, Booker T. Washington's successor at Tuskegee, and Davidson Don Tengo Jabavu, son of John Tengo Jabavu and a professor at Fort Hare Native College. Moton, who carried on in the

Washingtonian accommodationist tradition, showed less interest in Africa than his predecessor, but at times he gave rhetorical support to a conservative black nationalism. In his installation address in 1916, he noted with approval the very fact about Tuskegee that impressed African nationalists: "While [Washington] always sought the advice, criticism, and help of the white race, he drew the 'color line' when it came to the actual work of the institution." Jabavu, who had visited Tuskegee while Washington was still alive, made precisely the same point when he noted in 1920 that "the absence of white men, the successful administration of a purely negro facility, . . . is not among the least significant phenomena of Tuskegee."[24]

Jabavu, who was more of an intellectual than Moton—indeed, he was the most prominent black intellectual in South Africa during the 1920s—gave full and eloquent expression to the moderate, accommodationist Pan-Africanism that Washington had inspired. In 1913, while a university student in England, Jabavu traveled to the United States and spent several weeks at Tuskegee, where he fell under the spell of Washington and his philosophy. In two volumes of essays and addresses on the race question, *The Black Problem* (1920) and *The Segregation Fallacy* (1928), he called for black South Africans to follow the example, not simply of Washington himself, but of African-Americans generally. In the earlier work, he dismissed the Washington-Du Bois controversy as merely a matter of tactical responses to the differing circumstances of southern and northern blacks and held up the overall record of black American achievement as a model for native South Africans. During his visit to the United States, he had marveled at thriving black educational institutions, churches, and businesses and at the emergence of a middle class of professionals and entrepreneurs. Although most of his program was derived from Washington's gospel of agricultural and commercial self-help, he included Du Bois among his heroes. "If you meekly belong to what was a down trodden [*sic*] people," he told one of his audiences, "then you may take courage and contemplate the pitiful negro slaves of 1850 in the United States who today lead all black races in culture—(see their wonderful magazine the 'Crisis' of New York)—and behold how extensive your opportunity is to make good."[25]

In his 1928 book of essays, Jabavu expressed his dismay at the radical segregationist policies of Prime Minister J. B. M. Hertzog and his Nationalist-Labor "pact" government. Because Hertzog had enacted the economic color-bar legislation that inspired the opposition of the Joint Councils and went on to propose elimination of the Cape African franchise, Jabavu departed from the self-help moderation that had characterized his earlier writings to engage in vigorous protest. He compared Hertzogian segregation to "the Jim Crow system of the southern states of America, where it has resulted in lynching and terrible race conflicts." As an alternative, he endorsed the Cape liberal tradition of a qualified, impartial suffrage and equality before the law, which he described as a combination of public equality and voluntary social separation. Such a compromise, he argued, was precisely what Washington had advocated; paraphrasing the principal of Tuskegee, Jabavu called for black and white South Africans to be "as separate as the fingers in matters social but strong as the hand in national solidarity."

Because he attacked the Garvey movement and the ICU, kept aloof from the ANC, praised Aggrey and the Joint Councils, and conceded that it might take centuries for Africans to become fully civilized and thus able to constitute a political majority in South Africa, Jabavu appears in retrospect to have been a conservative spokesman for the interests of an African elite. But, like Washington at the end of his life, he took a strong stand against the extension of segregation beyond what he had earlier decided he could tolerate, a position that would make him a leader of efforts to create an African united front during the 1930s. As a preeminent representative of the black intelligentsia and the direct heir of the Cape liberal native tradition that his father had helped to establish, he was seeking a middle ground between blind acquiescence to white domination and the kind of radical, anti-white populism he saw in the Garvey movement and the ICU. What South Africa blacks needed to do, he believed, was "to substitute for untrained leaders a number of intellectual spokesmen of the type of Booker T. Washington and J. E. K. Aggrey, who will call attention in a vigorous but constitutional manner to some of the glaring examples of injustice and who will at the same time furnish constructive schemes for the amelioration of the conditions of the people, on lines of cooperation with the friendly section of whites."[26]

The reformist Pan-Africanism associated with Du Bois can be distinguished from the accommodationist variety espoused by Washington, Aggrey, and Jabavu because it explicitly anticipated the eventual liberation of Africans from European rule. But the international Pan-African movement never came to a clear decision on whether the model of African self-determination applied in a literal way to South Africa with its substantial and permanent white minority or whether it had any political—as opposed to cultural or inspirational—relevance to the situation of the African-American minority in the United States. Hence the differences between the accommodationist and reformist varieties of Pan-Africanism were not as sharp or as substantial as might first appear to be the case.

The international Pan-African movement was formally established in 1900, when a small number of prominent blacks from Africa, the West Indies, and the United States gathered in London, under the leadership of the Trinidadian lawyer Henry Sylvester Williams and Bishop Alexander Walters of the AME Zion Church of the United States, to call for the just and humane treatment of people of African descent throughout the world. Its resolutions were moderate in tone, and the permanence of European rule over its African posssessions was not directly challenged. Du Bois attended, but he did not dominate the proceedings or in fact even attach great importance to them at the time. It appears that no South Africans were present, but the conference sent a memorial to Queen Victoria protesting the treatment of native peoples in South Africa. The

Pan-African Association, which was founded as a result of the conference, lasted for only about a year, and hopes for an international black movement went into abeyance until Du Bois revived them during the war years.[27]

In his important book *The Negro*, published in 1915, Du Bois set forth the conception of Pan-Africanism that would guide the deliberations of the postwar congresses. Much of the work was a history of the black race that stressed its past achievements and future potentialities. When he came to defining Pan-Africanism, Du Bois was careful to avoid putting it on a narrow, chauvinistic foundation. Increasingly influenced by democratic socialist ideas, he saw no conflict between "a unity of working classes everywhere" and "a unity of the colored races"; both were steps toward "a new unity of man." According to the historian William Toll, he "fused race consciousness with, rather than opposing it to, class analysis" and thus gave new meaning to the African-American struggle for equal rights, making it part of a worldwide struggle against the oppression of classes, races, and nationalities which stemmed from an unholy alliance of white racial arrogance with capitalist imperialism. Consciousness of race did not lead to black separatism and parochialism but to participation in a great international movement against the color line. "There is slowly arising not only a curiously strong brotherhood of Negro blood throughout the world," he wrote, "but the common cause of the darker races against the intolerable assumptions and insults of Europeans has already found expression. Most men in the world are colored. A belief in humanity means a belief in colored men."[28]

But Du Bois found a special role for blacks in the larger struggle of "darker races" against Western imperialism. With a touch of the romantic racialism that inspired Aggrey, he asked if the liberation struggle of colored peoples would require the kind of violence that was then raging in Europe, or "will Reason and Good prevail? That such may be the case, the character of the Negro race is the best and greatest hope; for in its normal tradition it is, at once, the strongest and gentlest of the races of men." Du Bois's Pan-Africanism of 1915 was not a call for colonial revolution but rather an appeal to the decency and rationality that he believed was within the capacity of all human beings and might be rekindled in Europeans and Euro-Americans by the black example of moderation and nonviolence.[29]

Du Bois's memorandum for the Paris Pan-African Congress of 1919 reflected this nonviolent reformism; it internationalized D. D. T. Jabavu's Cape liberalism more than it anticipated an ideology of anti-colonial revolution. In it, Du Bois stressed such objectives as "political rights for the civilized" and "development of autonomous governments along the lines of native customs with the object of inaugurating gradually an Africa for the Africans." The resolutions that came out of the congress were in the same cautious spirit. Africans "should have the right to participate in the government as fast as their development permits," or in other words when "they are civilized and able to meet the tests of surrounding culture." Carrying on in the great tradition of nineteenth-century pan-Negroism, Du Bois and the other representatives of the international black elite that met in Paris continued to view Western civilization as embodying a universal standard of human progress and enlightenment to which people of color were expected to conform. What was wrong with whites, essentially, was that they had failed to live up to their own best values.[30]

The Pan-African congresses—they met again in 1921, 1923, and 1927—did not actually call for the complete independence of African colonies from European hegemony, in part because of the role played in their deliberation by Francophone Africans like Blaise Diagne, who represented Senegal in the French national assembly and was a minister of the French government. Du Bois was obliged to concede that whether a particular colony became part of a sovereign African state in the future or was assimilated into a Euro-African democratic nation depended on the disposition of the colonial power. But in his discussion of Pan-Africanism in *Darkwater*, a book of essays published in 1920, he set forth his personal vision of a "new African world state" that would achieve the objective of "Africa for the Africans guided by organized civilization." He recommended specifically that the former German colonies, and perhaps those of Belgium and Portugal as well, be placed under the direct control of the League of Nations and gradually brought up to a standard of civilization that would entitle this new African state to full independence.[31]

Leading black Americans, he argued, should be put on the international commission set up to guide this emerging nation toward self-government, but their participation should not be viewed as encouraging New World blacks to emigrate to Africa: "The Negroes in the United States and the other Americas have earned the right to fight out their problems where they are, but they could easily furnish from time to time technical experts, leaders of thought, and missionaries of culture for their backward brethren in the new Africa." Here again Du Bois drew on the tradition of nineteenth-century Pan-Negroism, which had seen a special role for African-Americans as the vanguard of the race. Years later, Du Bois provided his clearest statement of the African-American stake in the independence of Africa. "I do not believe," he wrote in the *Pittsburgh Courier* in 1936, "that it is possible to settle the Negro problem in America until the color problems of the world are well on the way toward settlement. I do not believe that the descendants of Africans are going to be received as American citizens so long as the people of Africa are kept by white civilization in semi-slavery, serfdom, and economic exploitation."[32]

In practice, Du Bois's Pan-Africanist enterprise was quite congruent with the "talented tenth" reformism of the NAACP. It was essentially a matter of the international black elite joining forces with friendly and progressive whites in the major Western nations to publicize black grievances and petition their governments to recognize the basic human rights of Africans and people of African descent. Unike the rival Garvey movement, this version of Pan-Africanism did not seek mass support or eschew cooperation with white humanitarians. At the meeting in Belgium in 1921 whites actually outnumbered blacks in the audience, and the guest list for the London meeting in 1923 included several prominent whites. Just as the capitalist liberalism of most of the white philanthropists and middle-class blacks who directed the NAACP prevented Du Bois from incorporating his socialist ideas into the association's program, so the need to accommodate moderate to conservative elements of the international black elite, especially the Francophone African officialdom, kept his larger vision of a struggle against capitalist imperialism from finding expression in the congresses' resolutions.[33]

Nevertheless, Du Bois failed to gain the sustained backing from the NAACP for his international efforts. In the flush of the immediate postwar interest in the "self-determination" of subject peoples around the world, the association highlighted the question of Africa's future at its 1919 convention but in fact gave only limited and grudging support to the Paris congress. In the 1920s, the association made clear its position that the destiny of Africa was tangential or irrelevent to the basic aim of securing equal citizenship for African-Americans. Lacking significant support from the "talented tenth" in the United States and making no strong impression on the masses of black people anywhere—if inclined to Pan-Africanism at all they were likely to prefer Garvey's populist version—Du Bois's initiative had little historical impact except to help provide a genealogy for the more vital and influential Pan-African movement associated with the decolonization of Africa that emerged after World War II. (And this connection would not have been made so strongly if a more openly radical Du Bois had not been around to draw attention to his earlier efforts.)

In South Africa, the Pan-African congresses of the 1920s had little impact, except briefly on the fading, old-guard leadership of the ANC. Sol Plaatje, who was in contact with Du Bois, displayed a keen interest in the 1919 and 1921 congresses but was unable to attend either. John L. Dube did represent the ANC at the 1921 meeting, but he neither played a conspicuous role in the proceedings nor maintained an interest in the movement after he returned to South Africa. In general, Du Boisian Pan-Africanism seemed to have had no greater attraction for the ANC than it did for most of the leadership of the NAACP.[34]

Populist Pan-Africanism: The Garvey Movement

The third major strain of Pan-Africanism—the populist variety associated with Marcus Garvey—struck a more responsive chord in the ANC, as it did in the hearts and minds of black people from an astonishing variety of social and cultural backgrounds throughout the world. No one can be certain how many members Garvey's Universal Negro Improvement Association actually had at its high point—Garvey's own figures of four million in 1921 and six million in 1923 were undoubtedly inflated—but there can be no doubt that its numbers far exceeded those of any other black organization in history, both within the United States and internationally. (One historian's estimate of a milllion members in the United States and an equal number elsewhere at the movement's height is not beyond the realm of possibility.) Its appeal to race pride and call for the development of Africa into a great independent nation or empire were clearly what millions of blacks throughout the world had been waiting to hear.[35]

Garvey was born in Jamaica in 1887 and began his career as a printer and trade unionist. His roots were in the skilled and educated upper strata of the black working class of the island, and he resented the predominantly "colored" or mulatto group that constituted the middle class of Jamaica's three-tiered race/class system even more than he did the tiny white upper class that maintained British colonial hegemony. In 1912, he traveled to Great Britain, where he encountered more prejudice against blacks than he had expected and ended up associating with a small group of Africans who were protesting against conditions in their home colonies. According to his autobiography, the great inspiration for his subsequent career as a "race leader" was the reading in London of Booker T. Washington's *Up from Slavery*, which showed what a black man could do and also what needed to be done: "I asked, 'Where is the black man's Government?' 'Where is his King and his Kingdom?' 'Where is his President, his country and his ambassador, his army, his navy, his men of big affairs?' I could not find them, and then I declared, 'I will help to make them.' " On the boat returning to Jamaica in 1914, he fell into conversation with a West Indian who was returning from Basutoland (now Lesotho) in southern Africa with his Basuto wife, and learned of "the horrors of native life in Africa." Only days after he arrived back in Jamaica, he founded the Universal Negro Improvement Association and African Communities League in an effort to deal with the common problems of New World and African blacks.[36]

The UNIA began with more modest and narrowly Washingtonian objectives than Garvey's later recollections might suggest. Its initial aim was to establish "a Tuskegee in Jamaica." To gain advice on how to set up his projected "Industrial Farm and Institute," Garvey wrote to Washington and, after receiving an encouraging response, resolved to travel to the United States to visit Tuskegee and confer with its principal. But by the time he arrived in New York in 1916, Washington had died, and Garvey's objectives had broadened. He decided to shift his base of operations from Jamaica to the United States and turn the UNIA into a multi-faceted organization for black uplift and solidarity. He seems to have thought briefly about joining forces with the NAACP, but a visit to its offices gave him the impression that it was staffed by a mulatto elite similar to the one that had snubbed him in Jamaica. In 1917, Garvey and a handful of initial recruits established the New York branch of the UNIA.[37]

The meteoric rise of the UNIA from these beginnings to the great convention of 1920, attended by thousands of members from branches all over the United States and around the world, reveals that Garvey was a truly remarkable propagandist and organizer. Word about the movement was spread through its newspaper, the *Negro World*, and the single activity that attracted the most attention and support was the Black Star Steamship Line launched in 1919, which proposed to establish communications between the black diaspora in the New World and Africa through a fleet of vessels owned by the UNIA and tens of thousands of black shareholders. Other enterprises—including factories, retail stores, laundries, and restaurants—were also initiated, presaging a separate black economy far beyond the imaginations of Booker T. Washington and the Negro Business League. By 1920 the organization had developed all the trappings and auxiliaries of an African empire-in-exile, with its own uniformed paramilitary units, titles of nobility, and patriotic rituals.[38]

But Garvey's personal charisma, boundless energy, and organizing skills would not have been enough to launch a mass movement if circumstances had been less favorable. World War I inspired movements for independence among a great variety of oppressed or thwarted nationalities and ethnic groups. The imperial powers were physically weakened as the result of the war they had fought among themselves, and the principle of self-determination that became associated with the allied cause not only brought the actual liberation of most of Ireland from British rule and various Eastern European societies from Austro-Hungarian domination but also stimulated ongoing efforts to create an independent India and a Jewish homeland in Palestine. Although Garvey's enterprise was in some ways more similar to Zionism than to the other nationalist movements given stimulus by the war, his own favorite model for an African liberation movement was the anti-British struggle in Ireland.[39]

The time was also ripe for the African-American community, which would provide the majority of UNIA's membership and most of its financial support, to embrace more radical forms of racial separatism and self-help. The war had a paradoxical effect on black Americans: it gave them a new sense of their potential economic strength, but at the same time it inspired pessimism about the prospects for political and civil equality in the United States. The cut-off of European immigration and the war-stimulated boom in manufacturing had opened new employment opportunities, leaving many blacks temporarily more prosperous than they had been before the conflict. But the fact that blacks had some money to invest does not explain why they invested it in Garvey's Africa-oriented black separatist enterprises. What turned them in that direction was disillusionment with the fact that patriotic black participation in the war effort had not brought progress toward full citizenship, a lowering of the barriers of segregation, or even protection from racial violence. One of Garvey's first major addresses as head of the UNIA in the United States was a vigorous protest against the East St. Louis race riot of July 1917. The subsequent riots of 1919 served to make blacks receptive to Garvey's message that blacks had no future of equality or self-determination in a strictly American context but should look to an independent Africa as the instrument of their liberation. As early as February 1919, Garvey announced that "Negroes have got to win their freedom just as the Russians and the Japanese have done—by revolution and bloody fighting. Negroes in the United States cannot do this. They would be hopelessly outnumbered and it would be foolish to attempt it. But in Africa, where there are over four hundred million Negroes, we can make the white man eat his salt."[40]

These remarks reveal that Garvey's priority was African liberation, but he did not at this time regard battling for equal rights in the American context as futile provided that it was linked to the larger struggle. In a speech of April 19, 1919, devoted mainly to attacking Du Bois's gradualist approach to African self-determination, Garvey made explicit the connection between African-American and African liberation. "Be ready for the day when Africa shall declare for her independence," he advised a convention of 3000 UNIA delegates. "And why do I say Africa when you are living in the West Indies and America? Because in those places you will never be safe until you launch your protection internally and externally. The Japanese in this country are not lynched because of fear of retaliation. Behind these men are standing armies and navies to protect them."[41]

But after 1921 Garvey came close to severing the link he had earlier made between protesting injustice in America and overthrowing colonialism in Africa. By 1922 he was seeking an accommodation with the Ku Klux Klan and other extreme white supremacist groups in an effort to generate support for government-subsidized emigration to Africa leading to a "total separation" of the races. Black Americans who believed in the potential efficacy of protesting against lynching, disfranchisement, and segregation now concluded that Garvey was an enemy of their cause. Although it would oversimplify and distort the post-1921 Garvey movement as a whole to describe it simply as a "back to Africa" movement, Garvey expressed himself at times in precisely those terms, at least when he was contemplating the long-term future of the black race. His "True Solution of the Negro Problem" (1922) was unambiguous in its advocacy of emigration as the only permanently viable alternative to racial oppression:

> If the Negro were to live in this Western Hemisphere for another five hundred years he would still be outnumbered by other races who are prejudiced against him. He cannot resort to the government for protection for government will be in the hands of the majority of the people who are prejudiced against him, hence for the Negro to depend on the ballot and his industrial progress alone, will be hopeless as it does not help him when he is lynched, burned, jim-crowed and segregated. The future of the Negro, therefore, outside of Africa, spells ruin and disaster.[42]

But Garvey did not hew to a rigid line on this issue, and in 1923 he backtracked from his dismissal of the domestic struggle. "To fight for African redemption," he wrote, "does not mean that we must give up our domestic fights for political justice and industrial rights. . . . We can be as loyal American citizens or British subjects as the Irishman or the Jew, and yet fight for the redemption of Africa, a complete emancipation of the race." Garvey rarely expressed himself in this bicultural vein; had he done so consistently there would have been little difference between his Pan-Africanism and that of Du Bois.[43]

To understand the role of ideology in the Garvey movement, it may be helpful to invoke George Rude's distinction between "derived" and "intrinsic" ideologies. Enunciated by Garvey and official representatives of the UNIA was a "structured" set of beliefs that has a specific intellectual history. This ideology obviously had some grass-roots appeal or it would not have aroused the support of black communities in the United States and elsewhere. But it could be roughly adhered to without being embraced in all of its aspects and implications; popular or "intrinsic" beliefs would intervene to create a mix that could differ in significant ways from official movement doctrine. Garvey, in his most characteristic rhetorical vein, seemed to have given up on America, but many of his followers clearly had not. Local Garveyites in Cleveland or Los Angeles might in fact view race pride and solidarity as a means of getting ahead in a liberal capitalist America rather than as a disengagement from American realities and prospects. In South Africa, Garveyism would be filtered through a different set of lenses and associated with other "intrinsic" beliefs.[44]

Considered as a "derived" or "structured" ideology, Garveyism was based in part on the social Darwinist racial theories that were popular in the Western world around the turn of the century, with the significant difference that the concept of inherent black incapacity that was normally central to the white supremacist version was emphatically rejected in favor of the view that all races, including blacks, were potential winners in the struggle for existence. Garvey vividly described the evolutionary challenge facing blacks in his *Philosphy and Opinions*: "We are either on the way to a higher racial existence or racial extermination. This much is known and realized by every thoughtful race and nation; hence we have the death struggle of the different races of Europe and Asia in the scramble for the survival of the fittest race." William H. Ferris, a leading Garveyite intellectual, made the debt to social Darwinism even more explicit in a 1920 editorial in *The Negro World*: "The Negro has recently learned what the Teutons learned in the forests of Germany two thousand years ago, namely that the prizes of life go to the man strong enough to take and hold them. And he, the Negro, now realizes that the Darwinian and Spencerian doctrines of the struggle for Existence and the Survival of the Fittest are the hard facts, which he must face in his struggle upwards."[45]

Garvey and the official ideologists of his movement believed that the races as defined by the physical anthropologists of the nineteenth and early twentieth centuries were real entities that had always existed and always would exist. (In contemporary parlance they had an "essentialist" view of race.) Second, they held that racial prejudice and antagonism were natural and inevitable when two races came into contact. In one of his earliest writings after arriving in the United States, Garvey attributed the "progress" made by African-Americans since emancipation to "the honest prejudice of the South"; it gave them "a real start—the start with a race consciousness." Garvey saw no reason to condemn white supremacists for their "honest prejudice"; they were simply acting out of their God-given race consciousness, and blacks—instead of harboring illusions about a human brotherhood—should learn to do the same. In an Easter Sunday sermon on "The Resurrection of the Negro" in 1923, he called on blacks to replicate "the attitude of sovereignty" shown by "the great white race" and "the great yellow race." On another occasion in 1923, he wrote that the UNIA acknowledged that the white man could do what he wanted in his own country: ". . . that is why we believe in not making any trouble when he says that 'America is a whiteman's country,' because in the same breath and with the same determination we are going to make Africa a black man's country." The great races, he argued on several occasions, were engaged in a struggle leading to "the survival of the fittest group." The only way to avoid such a conflict was for each race to concentrate within its own natural "habitat." "The time has really come," he announced in his *Philosophy and Opinions*, "for the Asiatics to govern themselves in Asia, as the Europeans are in Europe and the Western world, so also is it wise for the Africans to govern themselves at home. . . ."[46]

A third element of Garvey's racialism was a conception of race purity, which led him—in a way reminiscent of Edward Blyden—to oppose intermarriage and distrust mulattos, especially when they assumed positions of race leadership. But he never actually attempted to exclude mulattos from his movement. Accepting the white American definition of the races—the rule that "one drop of Negro blood makes a man a Negro"—he called for "100 per cent. Negroes and even 1 per cent. Negroes" to "stand together as one mightly whole to strike a universal blow for liberty and recognition in Africa." His call for race purity and race integrity did not mean drawing a color line within communities of predominantly African descent; it was aimed rather at encouraging pride in African ancestry and discouraging future intermixture with whites. A number of relatively light-skinned African-Americans were prominent in the UNIA in the United States, and in South Africa the movement attracted substantial support among the Coloreds of Cape Town. His attack on Du Bois and the leaders of the NAACP was not on their light skins per se, but rather on what he took to be their color prejudice and amalgamationist goals—"They believe that in time, through miscegenation, the American race will be of their type." He also charged that the NAACP represented a privileged and snobbish "colored" elite like the one that he had found so objectionable in the West Indies.[47]

Garvey based this view of the African-American leadership partly on the impression he had received when he first arrived in the United States that he was being snubbed by the NAACP and that the association had ignored his claim to race leadership because he was so black. Some later comments of Du Bois and other members of the African-American elite on Garvey's physical appearance lent credence to his belief that he was the victim of color prejudice. But the real source of friction was ideological rather than phenotypical. The NAACP did not really aim at amalgamation of the races, but rather at their public equality in the United States. Since Garvey either repudiated this goal as unrealizable or, at best, gave it a secondary and subordinate role in his crusade for African liberation, there was no way that the NAACP—or even Du Bois with his elite, reformist version of Pan-Africanism—could come to terms with the mature Garvey movement. The unfortunate way that the dispute enflamed class or race prejudices on both sides helped to obscure the real issues.[48]

The secret of Garvey's popular success was not an affirmation of genetic blackness, but rather an appeal to existential blackness, the condition of having been "discriminated against throughout the world" because of African ancestry; his cause was the transmutation of that identity from a source of shame and discouragement into a basis for pride and achievement. His followers did not have to agree with, or even understand, his social Darwinist ethnology to get the point that they were all in the same boat and needed to unite against a common oppressor that had defined itself as "the white race." Considered as a form of populism, the Garvey movement drew on a sense of "the people" that had to be racially defined because no other basis for solidarity existed or was conceivable in the United States of the 1920s.

It has often been noted that Garvey was not so much a black cultural nationalist as a political one. This was inevitable given his international frame of reference and the state of his knowledge. There were distinctive black cultures in various parts of Africa and the New World, and anthropologists and historians would eventually discover some vital interconnections among them, but there was only one common black culture that Garvey was aware of that he could draw on—black religious nationalism or Ethiopianism. As we have already seen, Garvey made use of the Ethiopian prophecy to provide an aura of religiosity to his movement. But the ethical content of Ethiopianism did not jibe well with Garvey's social Darwinism. In fact he implicitly repudiated the romantic racialist celebration of a distinctive black character that conformed to the meekness and pacifism prescribed in the Sermon on the Mount. Garvey's ambition was to endow blacks with the same kind of evolutionary "fitness" that had allegedly served whites so well. The Ethiopianist hope or expectation that an African nation or empire would manifest a higher morality and not merely replicate the lust for power and wealth that had allegedly characterized the expansion of Europe was at odds with the basic thrust of Garvey's Darwinian nationalism. His vision of an independent African state or empire conformed more closely to the conceptions of national and racial fitness that had justified Western imperialism than to any strong sense of the moral and cutural distinctiveness of black people.[49]

On the unifying theme of "race first," Garvey built a movement among black people that was populist in several senses. Considered in terms of class, it was neither a working-class nor a middle-class movement but rather an inter-class coalition. Judith Stein has recently demonstrated the "bourgeois" character of the UNIA's leadership and many of its activities, but she has not disproved the older view that the rank-and-file membership of the movement was predominantly working-class. Garveyism divided the black community along a different axis than a class-conscious bourgeois or proletarian movement would have done; it in effect distinguished between the "people"—who were workers, clerks, and struggling small-business owners—and the educated elite of professionals and the relatively successful entrepreneurs—precisely the kind of African-Americans who were influential in the NAACP and the National Urban League.[50]

Garveyism was also distinctively populist in its economic philosophy—to the extent that it developed one. Often conceived of as a form of black capitalism, the Garvey movement envisioned an economic order that was closer to the "cooperative commonwealth" ideal projected by white populist movements of the late nineteenth and early twentieth century than it was to the American businessmen's gospel of unfettered private enterprise. UNIA enterprises such as the Black Star Steamship Line and the Negro Factories Corporation were organized in a quasi-cooperative fashion with strict limitations on the number of shares any individual could own and with dividends paid to the UNIA treasury rather than to shareholders. Garvey's opposition to African-Americans joining labor unions, which is often portrayed as evidence of his economic conservatism, derived from the fact that in the United States these unions were white-dominated and apparently incapable of treating blacks as equals. In his native Jamaica, both before and after his North American period, he promoted unions as a desirable form of race organization.

In one of the clearest statements of his economic ideology, Garvey praised capitalism for its contribution "to the progress of the world," but contended in populist anti-monopoly fashion that "there should be a limit to the individual or corporate use or control of it." Specifically, he recommended that no person should be allowed to have more than a million dollars and no corporation the control of more than five million. Admittedly, this contradicted Garvey's call on other occasions for the emergence of black Carnegies and Rockefellers, but there is no evidence that he attempted to make himself a wealthy man, despite the opportunites that his enterprises might have offered to an acquisitive entrepreneur. Cooperation for mutual benefit of the race was the main theme of his economic initiatives, just as it was of his political program. Garvey devoted relatively little attention to economic matters and never fully articulated an alternative to capitalist and socialist forms of production, but his basic instinct was to seek a middle way that respected private property, small business, and a free market but resisted concentrated wealth and favored cooperative enterprises over large corporations owned by the rich.[51]

The career of the Garvey movement in the United States was meteoric; it gained strength until 1923 when Garvey was convicted of mail fraud and began to decline after he was sent to prison in 1925. When he was pardoned in 1927 and promptly deported to Jamaica, he left behind no more than a shadow of the mass movement of the early 1920s. There can be no doubt that the personal charisma and magnetism of Garvey was an essential element in the movement's success and that his removal from the scene left a vacuum that no one else could fill. His personal downfall stemmed from the failure of his economic enterprises, especially the Black Star Line and from the successful efforts of the African-American elite, including leaders of the NAACP, to persuade an already hostile United States government to act against him—a campaign that was motivated more by revulsion to his dealings with the Ku Klux Klan than to the alleged financial improprieties that provided the actual basis for prosecution and deportation. Historians have acquitted Garvey of criminality, finding that it was corrupt associates and his own mismanagement or poor judgment that made him vulnerable to persecution.[52]

The American UNIA, however, was more than Garvey writ large. While it would be wrong to ignore his personal significance—charismatic individuals do sometimes affect the course of history, and the African-American community has shown a tendency over the years to personify its aspirations in the form of a single, heroic race leader—it would be equally mistaken to assume that Garvey created something out of nothing and had perfect control over his own creation. What is known about local chapters of the UNIA suggests a greater continuity with the Washingtonian gospel of economic self-help in a segregated environment than Garvey's Africa-first rhetoric might suggest. The most active Garveyites, it appears, were likely to be inhabitants of black urban ghettos in both the North and the South who were neither part of the educated elite nor members of the lower class of poorly paid, unskilled, and often unemployed laborers and servants. Workers with some skills and some property or savings and proprietors of marginal businesses with a ghetto clientele formed the nucleus of UNIA chapters. These "strivers" were not people who had given up on the American dream of material success. But, in the tradition of Booker T. Washington, their aspirations had narrowed to a purely economic sufficiency within a capitalist context. They were attracted to Garvey's message of racial pride and solidarity because, unlike either the liberal elite who backed the NAACP or the black radicals of the 1920s who hoped for an interracial socialism, they viewed the American color line as a permanent fact of life and sought whatever advantages might come to blacks through collective effort within the context of segregation. Their celebration of Africa was more a symbol to inspire hopes for the success or viability of social and economic separatism in the United States than a commitment to become politically involved in the struggle for African independence. The possibility of actually emigrating to Africa seems to have entered the consciousness of relatively few American Garveyites.[53]

As a genuinely popular—and populist—form of black nationalism, the Garvey movement played a critical role in the development of African-American thought. It created a mass movement by synthesizing two ideologies with deep roots in the black American experience—nineteenth-century pan-Negroism, including some elements of the Ethiopianist religious tradition, and the Washingtonian philosophy of economic self-help and group solidarity. It was no accident that Garvey's patron saints were Washington and Edward Blyden. But Garvey limited his appeal and compromised his legacy when he moved to an accommodation with white racists after 1921, which included not only an endorsement of segregation laws but also a reluctance to protest, as he had done earlier, against lynching and racial pogroms. Such an orientation did not reflect race pride and assertiveness, but quite the contrary. African-American nationalists of later years would draw much inspiration from Garvey, but they would rarely repeat his mistake of trying to do business with the worst enemies of their race.[54]

Black Populism in South Africa: Garveyism and the ICU

Historians have recently discovered that the Garvey movement had a greater impact in South Africa than anywhere else outside the United States and the British West Indies. But the precise extent and character of that influence and how it interacted with indigenous patterns of thought and action remain unclear. The most intriguing question concerns the relationship between Garveyism and the main expression of black populism in South Africa during the 1920s—Clement Kadalie's Industrial and Commercial Workers Union of Africa (ICU).[55]

The Garveyite slogan "Africa for the Africans" already had a history in South Africa as the motto of Ethiopianist religious separatists around the turn of the century. But Garveyism gave it a straightforward political significance that clearly went beyond the question of who should control mission churches. In the minds of anxious whites, its central ideas seemed to be that African-Americans would return to Africa and that Europeans in Africa would be "driven into the sea." Taken literally and pushed to their logical outcome, the doctrines of the UNIA constituted a much more radical challenge to the rule of a white minority in South Africa than they did to the dominance of a white majority in the United States. Here there could be no accommodation with white racists short of partitioning the country in a way that reversed the division of the Land Act of 1913 by giving most of South Africa back to the Africans. In practice, however, most of those influenced by Garveyism in South Africa did not push the doctrine of "Africa for the Africans" to its logical conclusion. Except for some millennialist movements in the Transkei that incorporated rumors about the coming of Garvey and black Americans into their expectations of a sudden and miraculous elimination of their white rulers, Africans who fell under the spell of the Provisional President of Africa did not normally announce that they intended to expel or subordinate the white minority. Often they invoked race pride and racial solidarity in the course of demanding basic civil and political rights in a multi-racial society.

The UNIA never became a mass organization in South Africa. Although seven branches were formed by 1926—more than in any other African country—five of them were located in the greater Cape Town area (the others were in East London and Pretoria), and it is doubtful that total membership ever exceeded a few hundred. The association's success in Cape Town was due partly to the stimulus provided by a lively group of West Indian and African-American expatriates, who in South Africa were classified as "Coloreds" rather than as Africans. It seems likely that a majority of Cape Town Garveyites were from the Colored community. If so, access to a positive "Negro" identity would have relieved the discomfort of being caught in the middle between Europeans and "natives."[56]

The leading exponent of Garvey's message in South Africa was based in the Western Cape and was well suited to bring the message of "Negro" unification to a population with a high degree of phenotypical and cultural diversity. James Thaele, the son of a Basuto chief and a Colored mother, began his education at Lovedale and then spent ten years as a student in the United States, earning two degrees at all-black Lincoln University in Pennsylvania. An enthusiastic disciple of Garvey while still in America, he returned to South Africa in 1923 and became active in several organizations in Cape Town. His principal achievement was the creation, for the first time, of a strong ANC branch in the Western Cape. Thaele made this breakthrough by overcoming the image of the congress as a strictly "native" organization and recruiting a substantial Colored membership. He was also for a time one of the principal leaders of the ICU, was active in the local UNIA, and in 1925 founded a newspaper, obviously modeled on Garvey's *Negro World*, called *The African World*. This journal, which functioned as the official organ of the Western Cape ANC, carried on its masthead the slogan "Africa for the Africans and Europe for the Europeans." The Garvey movement had already attracted sympathetic attention in the ANC before Thaele returned to South Africa, but his presence as an important congress leader helped to maintain interest in the UNIA.[57]

As a source of rhetoric and symbolism, if not of specific plans and projects, the Garvey movement for a time had a pronounced influence on the South African National Congress. In 1920, the congress newspaper *Abantu Batho* gave extensive and sympathetic coverage to the great UNIA convention of that year. In 1921, the president of the Cape Province African Congress (and the future president-general of the national ANC), the Reverend Z. R. Mahabane, gave an address reviewing the past year which gave special attention to the UNIA convention—"an event of far-reaching importance and very great interest to the African." As he assessed the problems of blacks in South Africa, Mahabane expressed himself in unmistakably Garveyite language:

> The European came to Africa, robbed the African of his God-given land and then deprived the African of all rights of citizenship in a country originally intended by Providence to be his home. Why, did not the Almighty in His wisdom and prescience divide the earth into four continents—Europe, Africa, Asia, and America? The man whom He created was planted on this earthly planet. He made them white, black, and yellow. To the white man He gave Europe to be his abode, Africa He gave to the black man, and Asia he allocated to the Yellow man. America God seems to have intended as the land of the surplus population of each of the three great divisions of mankind named above. The amazing thing today is that the white man claims Africa as the white man's country, and by his legislative action he has practically excluded the black man. This is injustice.[58]

Mahabane, however, was not proposing that Africans take back their birthright by force. He was suggesting rather that they "launch a big constitutional fight" for "the divine right of peoples" to "self-determination." He went on to conclude that a just and peaceful future for South Africa would result from "the full and free cooperation of all white and black races of the land." Soon to be a supporter of the Joint Councils movement, Mahabane showed how it was possible to combine Garveyite rhetoric with a relatively moderate, constitutionalist approach to the black struggle in South Africa.[59]

The example of Garveyism may have helped inspire the congress to change its name in 1923 from the South African Native National Congress to the more Pan-African-sounding African National Congress. In 1927, the congress added the UNI motto "One God One Aim One Destiny" to its letterhead. But it is doubtful whether there was a "wholesale embrace of Garveyism" that signaled the "radicalization" of the ANC after 1925, as Robert A. Hill and Gregory A. Pirio have argued. The eclecticism that Mahabane demonstrated in 1921—a combination of the Garveyite rhetoric of black self-determination with a moderate, even accommodationist, attitude toward the white presence in South Africa—was characteristic of several ANC leaders who appeared to be embracing Garveyism. When Sol Plaatje visited the United States in 1921–22, he participated eagerly in UNIA activities, speaking to several branches, but he also addressed the NAACP convention and was dazzled by a visit to Tuskegee. Plaatje returned from the United States with a heightened conception of the value of black pride and self-help, but this did not alter his stance as a moderate within the ANC, a spokesman for collaboration with large white-owned corporations against the more militant and confrontational posture of ANC president-general S. M. Makgatho. One of Garvey's most devoted disciples among the ANC leaderhip was T. D.

Mweli Skota, secretary general from 1923 into the 1930s. Instrumental in the change of name in 1923 from "native" to "African," Skota combined his devotion to Garvey with a dedication to enhancing the status of the black elite. His *African Yearly Register*, published in 1930, was a Who's Who of eminent Africans who had succeeded in distinguishing themselves from the masses either by the "progressive" attainments valued by Western civilization or by traditionalist criteria. Maneuvering adroitly between moderate and radical factions of the ANC, Skota was able to synthesize his Garveyism with either liberalism or Marxism depending on the current disposition of the top leadership of the ANC.[60]

Even James Thaele, probably the most thoroughgoing Garveyite within the ANC, did not interpret "Africa for the Africans" to mean that blacks should literally take back South Africa from the white settlers. He went beyond Mahabane's "big constitutional fight" by advocating (but not actually practicing) Gandhian nonviolent resistance. He also called upon Africans to withdraw from white mission churches and for the black separatist denominations to unite in one great African church. But he did not follow the Garveyite logic of "Africa for the Africans" to the point of calling for a black takeover of South Africa. According to historian Peter Walshe, "Garvey was . . . a symbol of Negro success in a world dominated by whites, but Thaele's admiration of his hero certainly stopped short of any rigorous defense of his African projects." A temporary radicalization of the ANC did take place after 1927 when the ANC elected J. T. Gumede president-general to succeed the more conservative Z. R. Mahabane and at the same time called on the South African government to use its good offices with the American government to obtain Garvey's release from prison. Like his predecessor, Gumede employed Garveyite rhetoric on occasion, but the radicalism that led to his ouster from the top post of the ANC in 1930 was his embrace of Marxism and the Soviet Union, not his Africanism. Many in the ANC were skeptical about Garveyism or indifferent to it, and there was no apparent correlation between attraction to the world-view of the UNIA and militancy, or lack of it, in the battle against white supremacy in South Africa. As in the case of grass-roots Garveyism in the United States, the emphasis on race pride and solidarity was more often invoked for the purpose of improving the position of blacks within a segregated society than for the revolutionary goal of overthrowing European domination.[61]

The impact of Garveyism in South Africa went well beyond the interest expressed by some members of the ANC leadership. Rumors and distorted notions about Garvey and his projects fueled millennialist movements in rural areas. The expectation of these movements was that black Americans would come in force to liberate their South African cousins from white rule. Such hopes also contributed to the grass-roots support for the great mass movement of the 1920s, the Industrial and Commercial Union of Africa (ICU)—an organization that, unlike the ANC but like the UNIA in the United States, deserves to be characterized as an expression of black populism. In the late 1920s, the ICU organized in rural areas and interacted with the local African movements that prophesied the coming of the *Ameliki* (Americans).

Conditions in the countryside and in the "native reserves" made rural Africans receptive to prophecies of miraculous redemption. In basic economic terms, the problem was that there were too many people on too little land. White expropriation of most of the best farmland, which was made official policy in the Natives' Land Act of 1913, had left Africans without the means to produce an agricultural surplus or even to maintain self-sufficiency. Because of overuse and erosion, once-fertile land deteriorated in quality, making it increasingly difficult to scratch out a living in the areas assigned to Africans. To maintain their families, increasing numbers of men had to leave their *kraals* to work for extended periods in the mines or as unskilled laborers in the rapidly growing urban areas. The class of relatively prosperous market farmers that had emerged in the late nineteenth century and provided the economic base for the "progressive" African elite was in serious decline by the 1920s. This reversal of fortunes, and the blatantly white supremacist government policies that hastened it, deprived the liberal incorporationist ideology of "equal rights for every civilized man regardless of race or color" of any material basis in the rural life of the early twentieth century. At the same time, exposure to Christian beliefs and to the cosmopolitanism of the mines and cities had weakened the hold of tribal customs and values as well as respect for chiefs and other traditional authorities. The fact that traditional leaders and forms of governance were often coopted or manipulated by white officials further reduced their prestige and credibility among Africans who were looking for relief from worsening economic and social conditions and a way of viewing the world that offered some hope for the future.[62]

When both progressive and traditionalist ideologies failed to provide solutions to the desperate economic and cultural plight of rural Africans during the 1920s, many turned to the separatist Christian churches. In 1921 only about 50,000 blacks out of a total of 1.3 million rural black Christians belonged to Ethiopian churches rather than white-dominated mission churches. By 1936, there were over a million. "The appeal of these independent churches," according to the historian Helen Bradford, "lay in their proto-nationalism, and in their pledges to lead blacks literally and figuratively to the promised land." Some of these churches were of American origin, especially the African Methodist Episcopal Church and the black Baptists. The South African AME hierarchy in the 1920s was notably conservative and accommodationist; in its quest for respectability and official toleration, it opposed Garveyism or anything like it. But on the local level, AME members and even some clergymen participated in millennialist agitation.[63]

A number of smaller African-American sects also had followers in South Africa. Among these was the African Orthodox Church, which was directly allied with the Garvey movement. Under the leadership of Archbishop Daniel William Alexander, a Colored former Anglican priest from Cape Town, it began in 1924 to recruit black clergy and parishioners from Church of England missions. Another was the Church of God and Saints of Christ, which in 1912 gained the allegiance of a former Methodist lay preacher, Enoch Mgijima. Five years earlier Mgijima had seen a vision telling him that the end of the world was fast approaching. When he indicated to his American co-religionists that Armageddon would take the form of a war between the black and white races, he was excommunicated and thereafter carried on independently as the head of his own Israelite sect. In 1920, a number of signs, including apparently the foretaste of race war that he perceived in the Garvey movement, led Mgijima to assemble his followers in a permanent encampment near Bulhoek in the Eastern Cape to await the end of the world. When the Israelites refused orders to vacate the government land that they occupied, they were attacked by military forces on May 24, 1921. In the ensuing massacre, at least 183 of the approximately 1200 unresisting Israelites were killed.[64]

The prophet Mgijima was in the long line of quietistic millenarians who have gathered at various times in the history of Christianity to await the end of the world. His misfortune was that he massed Africans in defiance of the authority of a white South Africa that lived in fear of a black insurrection. The doctrines associated with Garveyism came more directly into play when rumors spread among the African population that black Americans, perhaps with Garvey himself at their head, were literally on their way to free black South Africans from white rule. News about the Black Star Line was transmuted into the image of a black American task force bound for South Africa, and Garvey's projected Black Eagle Flying Corps became an African-American air force preparing to rain bombs on the citadels of white power in South Africa. Some of those who responded to such rumors believed that the United States was a black nation and that the Versailles conference had awarded South Africa to the Americans, a decision that the white authorities in South Africa were disregarding and trying to conceal. (It is likely that such rumors were based not only on misinformation about the Garvey movement but also on a creative interpretation of Du Bois's proposals for African-American involvement in international control of the former German colonies.) Such fantasies were rife in the Transkei especially in the early 1920s. When J. E. K. Aggrey visited the principal town of Umtata in 1921, many in the enormous crowd that came to hear him did so under the impression that he was a black American in the Garvey mold who had come to proclaim the end of white domination. They were puzzled and disappointed by his actual message.[65]

Five years after Aggrey's visit such notions were still circulating, and an organized movement based on the myth of the coming of the Americans took shape in the Transkei under the leadership of a man who called himself Dr. Wellington and professed to be an American Negro representing Garvey. He was actually a Zulu born in Natal and baptized Elias Wellington Butelezi. Butelezi began his career as a confidence man by practicing Western medicine without a license in Natal and Basutoland (now Lesotho); but in 1925 he met Ernest Wallace, a West Indian representative of the UNIA in Basutoland and became a convert to Garveyism. Moving to the Transkei in 1926, he passed himself off as a black American born in Chicago who had arrived in Africa on one of Garvey's ships to proselytize for the UNIA. What he actually preached in the Transkei was not simply the orthodox Garveyite message of black pride, solidarity, and commitment to the ultimate liberation of Africa from European colonial rule; he also latched on to the local fantasies about the impending arrival of black American liberators and gave them added credence. He traveled throughout the Transkei and some of the districts of the Eastern Cape Province, winning converts to his doctrine of miraculous redemption. Updating the postwar rumors of an American claim to South Africa, Wellington contended that King George had awarded South Africa to the United States for aiding Britain in the war and that Prime Minister Jan Christiaan Smuts had been ready to comply, but that the new prime minister J. B. M. Hertzog (elected in 1924) had reneged and was about to go to war with a mythical black-dominated America in order to preserve white rule in South Africa. Wellington regaled his followers with prophecies of a millennium achieved through the exercise of black American power in the form of ships, troops, and bombing planes. It was the aircraft that caught the fancy of his audiences, and they were inspired to act by Butelezi's warning that destruction would rain from the air, not only on Europeans, but also on Africans who did not belong to the Wellington movement. Besides joining the movement and paying for its identification cards, adherents were advised to paint their houses black, kill their white pigs, and destroy possessions that were white or derived from pigs. Many did so, and when the Americans failed to arrive as predicted in 1927, Wellington held on to much of his following by claiming that liberation had to be postponed because not all the pigs had been killed. But in that same year, the authorities became alarmed, arrested Wellington, and banished him from the Transkeian territories.[66]

The historical significance of the Wellington movement is difficult to determine. Wellington himself was probably an imaginative charlatan who knew how to profit from popular hopes and fears. The large sums of money he collected for memberships in the UNIA apparently went into his pockets rather than into the coffers of the organization. Also, his activities and message had a tenuous or ambiguous relationship to the history of African resistance to European domination. He did not ask his followers to challenge or confront white power; just as Mgijima had called on the Israelites to wait quietly for the coming of God and the end of the world, Wellington called on his followers to await the coming of the Americans. According to Robert Edgar, the historian of the Wellington phenomenon: "A fundamental passivity characterized his movement. Although his message indicted European rule and raised expectations of imminent change, he stopped short of demanding a frontal assault on European power."[67]

But the Wellington movement was not merely a case of mass delusion. The synthesis of Garvey's message of racial redemption with localized Xhosa beliefs about ritual purification through killing swine was a creative reconstruction of the world that made sense to people who were torn between traditional cultures that no longer worked and an imposed European culture whose rewards they were denied. Their mythic view of Garvey and black American liberators did not promote effective political action, but it did keep alive the sense that white rule was unjust and projected a vision of reality that conflicted with the one that white supremacists were trying to impose. The fact that aspects of that vision were demonstrably false meant that those who took it literally were condemned to failure and frustration. But the symbolic import of Wellington's gospel was true to the perception of many Africans that they had been forcibly divested of their birthright and that divine justice—as well as millions of black people throughout the world—were on their side as they struggled against their condition. Millennialism is often a cry against injustice, and the Wellingtonian version was a cry against a form of racial injustice. As such, it was a manifestation of the same thirst for liberation that animated the separatist church movements with which it was closely allied. Besides claiming to be a representative of the Garvey movement, Wellington also professed to be a minister of the AME Church, and separatist congregations appear to have been especially receptive to his message. Helen Bradford has contended that Wellingtonians were mostly "disaffected school people" (those who had attended mission schools but had become alienated from the paternal direction of white missioniaries) or "half-educated Christians"—in other words, the same kind of people who joined separatist churches. Like Ethiopianist religion, the folk Garveyism that expressed itself in the Wellington movement was based on a redemptionist vision combining modern and traditional elements that made the salvation of the world dependent on the liberation of blacks from bondage to white overlords.[68]

Garveyism of a more orthodox and less fanciful kind played for a time an important role in the labor organization cum mass movement known as the ICU. The Industrial and Commercial Workers' Union of Africa was founded in Cape Town in 1919 under the leadership of Clements Kadalie, an immigrant from Nyasaland (now Malawi). In association with white socialists and trade unionists, Kadalie helped to instigate a strike of Colored and African dock workers that was well supported and impressive, although ultimately unsuccessful. Broadening his efforts, he then set out to establish "one big union" among the subordinated races of South Africa. Because Kadalie was a black man who did not belong to any of the African ethnic groups of the Union, spoke none of their languages, and had a Cape Colored wife, he personified aspirations to transcend tribal and color divisions in the name of a kind of Pan-African or Pan-Negro unity. Sometimes he was mistaken for an African-American, and one reporter thought he feigned a black American accent. His personal ambition, he revealed privately in 1920, was to become "the great African Marcus Garvey," and his flamboyant style of leadership and rousing oratory were clearly reminiscent of the founder of the UNIA. In 1921, he developed a close association with leaders of the Cape Town UNIA, although he never joined their organization, and for a time his union and the local Garveyites worked closely together. Kadalie remained in the driver's seat as national secretary of the ICU, but J. H. Grumbs, the West Indian UNIA activist, served for several years as president, and James Thaele, the most fervent Garveyite in the ANC, was also influential in the union.[69]

The ICU's first newspaper, *The Black Man*, founded in 1920, was so permeated with Garveyism that Garvey himself believed that it was a UNIA publication. Its successor, *The Workers' Herald*, which began to appear in 1923 and was edited by Kadalie himself, also had a Garveyite look in its first two years. It was a self-proclaimed "Race Journal"—"Are You a Race Man?" asked one 1925 headline. But the same number revealed ideological tensions in the ICU because it featured an article quoting with approval an editorial by Du Bois in *The Crisis* denouncing "back to Africa" movements as segregationist and calling for a "larger unity" that would lead to "INTERRACIAL PEACE" rather than polarization. In 1925, Kadalie was trying to maintain a united front of Garveyites, Communists, and social-democratic unionists in an organization that was still mainly concentrated in the Western Cape. Personally, if his autobiography is to be believed, he had never believed in "the slogan of 'Africa for the Africans' " and from the first resisted the efforts of Cape Town "Negroes" to make the ICU "an auxiliary of the Universal Negro Improvement Association." His essential commitment, he maintained in *My Life and the ICU,* was always to working-class solidarity regardless of race.[70]

From its inception the ICU was theoretically interracial, accepting whites as members and seeking alliances with nonracist segments of the white labor movement—although it did have a bar against white officeholders until 1927. But Kadalie's recollections do not tell the whole story of his changing attitude toward Garveyism. Before 1925 he was friendly to the UNIA and tolerant of his doctrines. An article that Kadalie wrote in late 1923 and published in the African-American socialist magazine *The Messenger* in July 1924 sheds considerable light on his attitudes toward race and class in the early 1920s. Taking note of the *Messenger*'s advocacy of racially mixed unions, he argued that South Africa would have to be an exception to "the philosophical theory of the labor movement" that prescribed interracial solidarity based on class. After describing the exclusion of blacks from recognized unions, the legal disadvantages from which they suffered because of their race, and the unrelenting hostility of white workers to their efforts to organize and strike, he concluded that an all-black movement was a practical necessity and, what was more, might gain strength from the "race consciousness" behind it. His ultimate hope, however, was that black labor would develop sufficient strength and solidarity to force white labor to accept its claims. As a short-term separatist and long-term integrationist, Kadalie differed profoundly from Garvey in basic philosophy, but at times he was willing to appeal to Garveyite "race consciousness" in order to gain support for his union. When he concluded that the Garveyites were becoming too influential in his movement and were threatening to take it over, he turned against them.[71]

In May 1925, the ICU was a relatively small organization concentrated mainly in the Cape, but it was on the verge of a great expansion that would turn it into a genuine mass movement. Kadalie proclaimed his intentions in *The Workers' Herald*:

> We are aiming at the building up in Africa of a National Labour Organization of the aboriginals, through which we shall break the wills of white autocracy. We must prevent the exploitation of our people in the mines and on the farms and obtain increased wages for them. We shall not rest there. We will open the gates of the Houses of legislature, now under the control of the white oligarchy, and from this step we shall claim equality with the white workers of the world to overthrow the capitalist system of government and usher in a co-operative Commonwealth, a system of Government which is not foreign to the aboriginals of Africa.[72]

It is evident that this was not a Garveyite statement, for it clearly privileged class over race as the ultimate basis for solidarity. But neither was it an endorsement of Marxist revolutionism, for it implied that victory over "white autocracy" would come through gaining access to the ballot box and parliamentary representation. Its remarkable invocation of the ideal of "a co-operative Commonwealth"—the watchword of the American Populist movement—suggests that Kadalie was essentially a social-democratic populist, who was organizing on the basis of race because blacks happened to be the disfranchised working class of South Africa. They could act jointly with whites in a political struggle against capitalism only after they had won their battle for civil and political equality.

By 1926 Kadalie's efforts to hold Garveyite and Communist factions of the ICU in line behind what he considered a "middle" position had clearly failed. From his perspective, neither side was willing to allow deviation from its own orthodoxy to achieve a broad-based movement. In the eyes of his opponents, Kadalie was too jealous of his personal power to tolerate dissent. In any case, Kadalie marginalized the Garveyites of Cape Town by moving the headquarters of the ICU to Johannesburg and expelled the Communists by a narrow vote of the national council. These actions, which historians friendly to either black nationalism or revolutionary Marxism have condemned as a turning away from radicalism or militancy, were actually a prelude to the most successful phase of ICU organization and agitation, as the union moved into rural areas or native reserves and galvanized a variety of local struggles.[73]

Between 1926 and 1929, the ICU penetrated the rural areas of Natal, the Orange Free State, and the Transvaal, established over a hundred new branches, and swelled in total membership from less than 30,000 to somewhere between 150,000 and 250,000, making it the first genuine mass movement among Africans. The secret of its success was that it was able to provide support and leadership for protest meetings, strikes, and boycotts in response to a range of grievances that had been building up in African communities. It mobilized labor tenants against landlords, workers against employers, squatters against efforts to drive them from European-owned land, and even women who brewed beer at home against municipal beer halls. It also mounted legal challenges to discriminatory public policies, and attempted to establish producer and consumer cooperatives. In these local struggles, it was relatively unencumbered by ideology, even the doctrines that Kadalie himself was proclaiming at the time. It was even prepared to cooperate with millennialist movements of the Wellington type. Wherever Africans were protesting injustice or exploitation the ICU seemed to be involved. Because of its populist character and ideological flexibility, it became the receptacle of ideas and feelings that emerged spontaneously in particular African communities, even very traditional ones, rather than following the example of the ANC or the Communist party by attempting to modernize the consciousness of the people it was trying to help. The kind of folk Africanism of the Ethiopian churches and the millennial movement readily found expression in the activities of ICU branches.[74]

On the national level, however, Kadalie was drifting into an alliance with white liberals and moderate socialists that would prove disastrous. Once he broke with the Communists, Kadalie was courted by a coterie of liberal "friends of the natives" who, with the best of intentions, contributed to the demise of the ICU as a national organization. Under the influence of his new white patrons, Kadalie sought help from the British Labor party, and in 1928 an organizer named William Ballinger was sent from England to help Kadalie turn the ICU into a proper trade union on the British model. Ballinger found much that was wrong, and he pressed hard for reforms of finance and structure in what in fact was a loose and inefficent organization bedeviled by the corruption of some of its officers and in the process of disintegrating at the center at the very time that it was attracting the most support on the local level. Ballinger failed to understand, however, that the ICU gained more mass support when it was serving as a vehicle or facilitator of popular protest than when it was trying to school African workers in trade unionism. He also moved too readily from an advisor's role to that of de facto boss of the ICU. Kadalie, who contributed to his own undoing by imagining for a time that he could replicate the British Labor party among disfranchised African workers, soon found that he had little power in his own organization, and in 1929 he broke with Ballinger to found a new Independent ICU (or IICU) to replace the one that had been captured by philanthropic whites.[75]

The IICU failed in its efforts to become a national organization, but it did function for several years as a vigorous local movement in the Eastern Cape. Shifting his base of operations to East London, Kadalie inherited a dynamic ICU branch that had based its appeal on a combination of reformist trade unionism and a forthright black nationalism that drew on separatist Christianity and paid tribute to Garvey. According to police reports of its meetings in 1928, its most articulate leader, Alfred Mnika, expressed militant Africanist views without hesitation. "This black nation of ours has to be recognized as a nation," he said, "and the ICU will build you into a nation if you will only stick to it." In a Pan-African vein, he predicted that "the ICU intends to reign all over Africa not in one little part." "We must be like Garvey," he urged his audience, noting that his hero "still worked hard to better the conditions of his fellow black men" even when he was in prison. After Kadalie arrived and the East London branch affiliated with the new IICU, this radical Africanism became even more pronounced and was accompanied by a call for black self-sufficiency, which meant, among other things, blacks establishing their own shops rather than patronizing white traders. With this emphasis, the organization expanded in 1929 and 1930, establishing links with the inhabitants of surrounding rural areas who were no in position to become trade unionists but responded to calls for black autonomy and self-determination. Kadalie, despite his personal reservations about Garvey and black nationalism, did not disassociate himself from the rhetoric that was providing him with a new power base.[76]

The example of the IICU in East London, as well as the experience of the ICU in earlier local struggles, suggests that the success of the movement for one big African union that Kadalie represented depended more than its titular leader liked to acknowledge on an appeal to race rather than class. Considered as populist movements, the ICU and IICU resembled the UNIA in the United States in speaking for constituencies that had been excluded, for the most part, from the industrial working class or at least from the legally sanctioned labor movement—groups that suffered in the first instance from racial or ethnic disabilities. Racial oppression, not class disadvantage, was the existential experience on which a mass movement of blacks had to be based, at least in the 1920s when the forces of white supremacy and segregation were in the ascendancy in both countries and alliances with a white working class that clung to its caste advantages were not in the cards. Kadalie, unlike Garvey but in basic agreement with Du Bois in the United States, could take a longer view and see the eventual linkage of black liberationist struggles with the battle of an international working class against capitalist domination. But even for Kadalie and Du Bois, some form of black nationalism was a necessary precursor to the wider struggle for the salvation of humanity.

Two Black Populisms: Comparing the UNIA and the ICU

To compare the UNIA in the United States and the ICU in South Africa one must do more than show similarities between two versions of populist Pan-Africanism. The ICU arose out of special South African conditions; although its origins made it receptive to Garveyite influences, it also had distinctive features that have to be treated on their own terms. Nevertheless, the parallels between the two movements are striking. Both were populist in similar ways. Defining the people to be mobilized as either essentially or for all practical purposes an oppressed racial group, the UNIA and the ICU sought to empower the people, not only against their white oppressors, but also against elites of their own race. The UNIA attempted to wrest leadership of the black struggle from the "talented tenth" entrenched in the NAACP. Although it did engage the services of a few black intellectuals such as William Ferris and John Edward Bruce, the majority of "college-bred" blacks remained aloof and were the objects of UNIA contempt as "white men's Negroes" or assimilation-minded "colored people" who were ashamed of their African ancestry. The ICU attempted at times to work with or through the ANC, but its rhetoric was often directed against "the good boys," who joined the Joint Councils, welcomed Aggrey, and constituted the moderate to conservative element in the Congress. Except for James Thaele, no prominent members of the educated elite played a significant role in the ICU. Both movements adapted themselves to the needs and opinions of local black communities in ways that more intellectually self-conscious and ideologically precise organizations would have been unable to do. The eclectic and ambiguous quality of populist ideologies is a source of strength when it comes to organizing relatively inarticulate people who know that something is terribly wrong but resist abstract and rationalized interpretations of what it is. In their economic thinking the two movements reflected the "cooperative commonwealth" idea often associated with populist movments. Both at times showed a strong interest in cooperatives and the building up of economic enterprises that were—at least in theory—of, by, and for the people.

The middle leadership of the two movements—the activists, organizers, and local officeholders—had remarkably similar social profiles. They were neither squarely in the working class nor securely in the middle or upper class of their racial or ethnic communities. The term "petite bourgeoisie" springs to mind, but it should be employed with caution given the negative connotations and aroma of reductionist Marxism that the term carries. The study of one UNIA chapter in the United States has found middle-class leadership but of " 'second level' community leaders" rather than of the wealthier, more highly educated element that gravitated to the NAACP. These activists included clergymen of the less prestigious churches, white-collar workers, skilled artisans, railroad employees, and steadily employed factory workers. Few were unskilled laborers or domestics.[77]

A study of ICU organizers in South Africa found them to be predominantly people with some education and middle-class pretensions who were threatened with downward mobility by the growth of segregation and economic discrimination. Many were younger men from middle-class Christian families who were on the verge of falling into the laboring masses despite enough education in mission schools to give them higher ambitions. Although such fears of status decline do not figure so prominently in the seemingly more hopeful expectations of American Garveyites, a toehold in the middle class requiring a struggle to move upward or avoid slipping downward would provide an adequate description of the social position of those most inclined to be activists in both movements. Being on the boundary between the middle and lower classes gave these leaders empathy with those below them whose ranks they had recently escaped from or might soon join and also some of the self-discipline and competence of the established middle class. They were clearly in a better position than the elite to give direction to popular struggles.[78]

Both movements capitalized on the betrayal of liberal-reformist hopes in the post-World War I era. In the United States, the integrationist neo-abolitionism of the NAACP survived and was not entirely without influence in the 1920s, but the reality for most African-Americans in urban ghettos was the struggle for economic survival in a separate and unequal world. The 1919 race riots and the rise of the Ku Klux Klan throughout the nation encouraged the Garveyite view that simply agitating for equal rights in the American context was futile and that a race-conscious Pan-African movement was the only route to power and self-esteem. In South Africa, the failure of a 1919 mission to Britain dashed the longstanding hope of the black elite in the National Congress for the intervention of British liberals and humanitarians to insure "equal rights for every civilized man" within the empire, and the efforts of the government after 1924 to establish new color bars and a more comprehensive segregation program further weakened expectations of a gradual lowering of racial barriers. Even the Anglophone white liberals of South Africa, on whom the congress had relied to influence white opinion on behalf of racial justice, tended to embrace an idealized version of government-imposed segregation in the 1920s, leaving many Africans with the sense that the only salvation was through separatist or independent action. If help was coming from outside, it seemed more likely to come from African-Americans or from a Pan-African movement than from white Christendom. The only question was whether solidarity should be based exclusively or primarily on race and the expectation of African nationhood or on the class foundation that was theoretically possible in a country where the subordinated racial group was also the overwhelming majority of the working class, broadly defined.[79]

The reasons why the American UNIA and the ICU failed to sustain themselves as mass movements into the 1930s were also analogous. Government repression played a role in both cases—in the United States,

Garvey's imprisonment and deportation seriously weakened the UNIA; in South Africa, a provision in the Native Administration Act of 1927 that made it a crime to foment hostility between the races and additional legislation of 1929 that provided for the internal exile of agitators gave the government powerful tools for repressing the ICU. At the same time repression helped induce Kadalie himself to moderate his positions and seek white allies, thereby putting him, at least temporarily, out of phase with the militant Africanism of many of his grass-roots supporters.[80]

But there were also seeds of failure within the movements themselves, especially as a result of the paradoxical fact that these were mass movements of a populist type that had charismatic leaders with authoritarian tendencies. Neither movement achieved the full breadth and cohesion of which it might have been capable because of vagaries at the top. Both Garvey and Kadalie shifted direction or emphasis without consulting their constituencies and purged associates who would have been valuable supporters of a more inclusive and democratically organized movement for black liberation. In 1921 Garvey narrowed his potential base by turning decisively against social radicalism, trade unions, and strikes as instruments of struggle in the United States. The following year he carried his racialism and separatism beyond what the majority of African-Americans were willing to accept when he sought to cooperate with the Ku Klux Klan in a misguided pursuit of government assistance for black emigration. Kadalie's break with both Communists and Garveyites in 1926 and his subsequent embrace of European-style social democracy was clearly not the way to develop an enduring movement with strong, grass-roots support. It is perhaps only because Kadalie was so preoccupied between 1926 and 1928 seeking the recognition and help of European social democrats that he permitted his local organizers to involve themselves in a variety of grass-roots movements.[81]

But the tendencies toward divisive sectarianism at the top of the UNIA and the ICU pointed in opposite directions. Garvey got into difficulty with African-Americans by stressing race too insistently and exclusively, while Kadalie got out of touch with his base when he put too much stress on European conceptions of class action or politics and played down the racial nationalism that was the source of much of the ICU's popular appeal, especially in rural areas. These differences stemmed in part from the fact that the two organizations were differently conceived at the outset and consequently followed distinctive trajectories. The UNIA started as a racial self-help organization inspired by Booker T. Washington and evolved into a Pan-African liberation movement that shunned alliances with the international Communist and socialist left. The ICU began as a trade union devoted to improving the pay and working conditions of what was, relatively speaking, a privileged segment of the black working population—permanent urban residents who worked for wages and were not constrained by the coercive contracts that prevented oscillating migrant workers in the mines and some other industries from organizing or striking. It broadened during the 1920s to embrace political protests against white supremacist legislation and the struggles of rural African labor tenants and squatters, interacting sympathetically at the local level with Ethiopianism and folk Garveyism—although the national leadership at the time was rejecting the orthodox Garveyism of the UNIA. The UNIA of course never admitted whites to its ranks, while the ICU not only did so, but in the end fell into a kind of white social-democratic receivership. Both movements were accused of being anti-white, and Kadalie at times emulated Garvey by denouncing whites or Europeans in ways that seemed to admit no exceptions. "The Europeans are rascals," he announced categorically at an ANC convention in 1926. But in the end Kadalie looked beyond race to an ideal of working-class solidarity and political action that Garvey would never have countenanced. Taken as a whole, the movement that Kadalie led can be described as pursuing a synthesis of class and race perspectives on South African oppression that it never fully formulated, while the Garvey movement in general carried the emphasis on race and racialism to a logical extreme and refused to give any significant weight to nonracial class perspectives on the black predicament in the United States and elsewhere.[82]

The ideological divergences between the UNIA and ICU were not accidental; they reveal that the contexts for the construction of black populism in the two societies were quite different. A race emphasis was favored in the United States because it was virtually inconceivable that an interracial folk movement could develop. Some white populists in the southern states during the 1890s had contemplated an alliance with black farmers and sharecroppers, but surging white racism and the disfranchisement of most blacks had made such efforts difficult to recall or even imagine by the 1920s. The white industrial working class had a long tradition of anti-black prejudice and most trade unions persisted in drawing the color line. Consequently, a realistic conception of "the people" for African-Americans seeking to create a mass movement that included those farthest down in the American social hierarchy was the racial group, especially the non-elite members of the black community. For those who tried to imagine how this oppressed American minority group would ultimately gain control of its own destiny, Garvey's call to participate in the creation of an independent African state offered a compelling answer. Short of achieving Garvey's African utopia, the best that could be expected was the emergence of a relatively self-sufficient black social and economic entity within the United States—a kind of "nation within a nation."[83]

In South Africa an alliance with lower- or working-class whites to form an interracial populism was even less likely to develop given the legal and economic advantages of white workers and small farmers and the importance that the white lower strata attached to the "color bars" that prevented blacks from competing with them. But a contention that black grievances were based on class as well as race had a logic to it that was lacking in the American context. Since blacks made up the overwhelming majority of the total population in South Africa and an even greater proportion of the laboring classes (broadly defined), a campaign for black liberation could be defined simultaneously as a struggle for the emancipation of the real working people of South Africa. An all-black trade union in the United States, such as the one A. Philip Randolph was trying to create among the sleeping-car porters, could serve only as an opening wedge for African-American incorporation into a predominantly white labor movement; it could not be the nucleus for a powerful racial movement, because blacks predominated in few other industries. In South Africa, on the other hand, one big black union, such as the one Kadalie worked to create, might conceivably represent the working class of South Africa as well as the national aspirations of Africans.

Kadalie failed to achieve his objective of creating a powerful union not because his increasing emphasis on class and labor was out of tune with South African possibilities, but rather because it was historically premature and did not reckon adequately with the coercive power of the white state. With only a few exceptions, the black working class of the 1920s was so repressed, underprivileged, and caught up in a disorienting transition between traditionalism and modernity that it could not be organized along trade union lines. The government's segregation policies aimed to perpetuate that condition. The ICU was most successful when it did not insist upon "structured" laborite or social-democratic ideologies and adapted its actions to the "inherent" ideologies that grew out of local circumstances and popular beliefs. But when it did so it betrayed its modernizing spirit and allied itself with forces that were likely to be irrational and regressive. To the extent that Africans actually became an industrial proletariat, and to the extent that a repressive government permitted them to organize, Kadalie's class message was logical and persuasive. He sensed the direction in which South Africa had to go, but he underestimated the political barriers that for the next half-century would limit the development of working-class consciousness and organization among Africans.

Both Garvey and Kadalie were guilty of what appears in hindsight to have been political naïveté. Garvey apparently expected that the United States government would subsidize black emigration, failing to acknowledge the continued dependence, especially in the southern states, on inexpensive, coerceable black labor. He also failed to anticipate the extent to which colonial governments and the settler regime in South Africa would use a combination of repression and cooptation to make life difficult for the UNIA and limit its ability to establish branches and make converts. Kadalie clearly overestimated the possibilities for reform in South Africa. Never a revolutionary, he believed that a combination of agitation and conventional trade union action would bring about a free and democratic South Africa. It is clear in retrospect that such an approach had no chance of success in the 1920s—or for a long time thereafter. The increasingly evident inability of his reformist, social-democratic message to shake the determination of white South Africa to maintain a monopoly on political, economic, and social power opened the way to programs for black liberation that were more radical and revolutionary. Similarly, the failure of the Garvey movement to engender sustained and effective anti-colonialist movements in the interwar period led to a more militant and less ethnocentric form of anti-imperialism. If the 1920s were the heyday of Pan-Africanists and black populists, the '30s and '40s would feature the efforts of revolutionary Marxists to put themselves at the head of black liberation struggles.

· 5 ·

"Self-Determination for Negroes": Communists and Black Freedom Struggles, 1928–1948

Reds and Blacks: Introduction and Overview

Between the late 1920s and the late '40s, the Communist parties of the United States and South Africa made strenuous efforts to place themselves in the forefront of black protest against white supremacy. Although they never achieved the mass support that the Garvey movement or the ICU had attracted during the '20s, Communists did manage to play a significant role in the struggle against white racism. The orthodox Leninist conception of the party did not require it to be a mass movement like the ICU or the UNIA; it was meant to be an elite or "vanguard," composed of carefully selected and highly disciplined revolutionaries who would give direction to the masses at appropriate times but needed to be wary of admitting to actual membership people who had not fully assimilated the Marxist-Leninist ideology. If such people were willing to accept Communists as allies in the struggle against racial oppression, anti–imperialist or anti-racist "fronts" of one kind or another could be established, within which Communists could hope to exert an influence far beyond their actual numbers.

The parties generally enforced ideological and tactical consensus in their own ranks through the practice of "democratic centralism," which meant that once a decision was made no one could question it. This policy was a strength in the sense that it gave the party cohesion and ideological coherence, especially in comparison with populist-type movements, but it could also be an obstacle to cooperation with non-Communists who shared some but not all of its goals. If at any given time the party was turning "leftward" and refusing to have any truck with democratic socialists or "bourgeois nationalists," it limited its influence over other movements seeking change; if it was moving to the right and seeking to ally itself with the broader progressive tendencies through "popular front" activities, it aroused suspicions of ulterior motives and conspiratorial aims. Radical black intellectuals who were attracted by Communist opposition to capitalism and racism often found it difficult or impossible to surrender their minds and talents to a party bureaucracy that had little respect for the free play of the intellect and the imagination. If they cooperated with Communists at all, their affiliation as members or "fellow travelers" was likely to be troubled and brief (although there were some conspicuous exceptions, especially in cases where the Communists deliberately and pragmatically refrained from imposing their usual discipline over particularly prominent members or fellow-travelers from the intellectual and artistic world).[1]

The contention that Communists tried to "use" other people for their own ends may be a valid judgment on the motives of some of them, but it cannot be used to predict the actual consequences of their alliances and collaborations. At times when the party lacked a distinctive public agenda of its own—as during the "popular front" era of the late 1930s and again between the German invasion of the Soviet Union in 1941 and the end of the war in 1945—it is an open question as to whether it was furthering its own ultimate goal of proletarian revolution or providing a boost to the liberal or social democratic reformism that it disdained at other times. It is difficult, for example, to know who was using whom when the American Communists put their skills as labor organizers at the service of John L. Lewis and the new industrial unionism of the CIO (Congress of Industrial Organizations) in the mid-1930s. The result was a strengthened and more progressive labor movement but not one that the Communists could readily dominate. Lewis, a committed non-Communist, thought that he was using the Communists, rather than vice versa, and he may have been right.[2]

In their efforts during the 1920s and '30s to appeal to blacks and to progressive whites troubled by the flagrant racial injustices of American and South African society, Communists enjoyed a special advantage. No one else in those decades was fighting so intently and assertively for the abolition of segregation and the complete equality of the races. In the United States, the NAACP shared the same goals, but its gradualist, nonconfrontational approach of working within the existing legal and political system made it seem comparatively conservative and accommodationist. Its concentration on civil rights issues and reluctance to confront the economic sources of black disadvantage provided an opening for the Communists to contend that blacks were being oppressed, not merely as members of a racial or national minority, but also as workers and peasants exposed to an acute form of capitalist exploitation. What was more, the Communist party seemed to provide a unique model for what later generations would call "racial integration" or "nonracialism" and even for the further stage of black incorporation into previously white organizations and institutions that would become known decades later as "affirmative action." The party actively recruited blacks, elevated them to leadership positions in a somewhat preferential fashion, and made strenuous efforts to root out all traces of racial prejudice or "white chauvinism" in comrades of European extraction. Nowhere else in either the United States or South Africa could one find blacks and whites interacting in such an unsegregated and apparently egalitarian way as at Communist party functions. This aggressive anti-racism undoubtedly hindered the party substantially in its efforts to instill proletarian consciousness in prejudiced white workers, but it gave blacks new hope for liberation from white supremacy and created sympathy for the Communists, not merely among the relatively small number who actually joined the party, but also among the much larger group who could not accept Communist discipline and the view of the world that went with it but were intensely grateful for the evidence the party provided that some whites could overcome their racism and treat blacks as equals. The image of the Soviet Union that enjoyed wide currency in the black communities by the 1930s—as the only place in the world where Europeans treated people of color as equals—seemed to give credibility to the view that Communists were quintessential anti-racists.[3]

If the Communists' greatest advantage in their efforts to appeal to blacks was their uncompromising anti-racism, their principal liability was the way that they shifted direction on questions involving the black struggles, as on other issues, in response to changes in the party line prescribed by the Soviet Union and the international Communist movement. In the 21 years between 1927 and 1948, five major shifts in the party's strategy revealed a pattern of opportunistic zig-zagging between go-it-alone revolutionary purity and cooperation with bourgeois or petit bourgeois movements devoted to reforming or democratizing capitalism without overthrowing it. These twists and turns meant that non-Communist movements for black civil rights or self-determination were viewed as implacable foes during left turns and as potential allies when the party moved in the opposite direction. Those who were denounced one day and embraced the next could hardly be expected to develop a high opinion of Communist reliability and dedication to principle.

There was nothing ideologically inconsistent in giving priority to calculations of Soviet national interest. Most Communists accepted the proposition that what was best for the Soviet Union was also best for the international Communist movement and ultimately for all of humankind; they regarded Soviet power and success as a sine qua non for the revolutionizing of the world. Problems and conflicts developed, however, when local Communists or their sympathizers saw special circumstances in their own countries that made a strict application of the international line of the moment impractical or counterproductive. They consequently clashed with Soviet and Comintern bureaucracies that refused to acknowledge such national differences and demanded uniformity and homogeneity, not merely in ideology but even in short-term political strategy and tactics.[4]

Historians disagree on whether Moscow's undeniable efforts to control local struggles in a doctrinaire and sometimes obtuse manner means that the Communist parties of countries like the United States and South Africa were simply alien impositions—agents of Moscow and tools of the Soviet Union—or something more (and better) than that. Revisionists, seeking to break with Cold War anti-Communism and provide a usable radical past, have recently argued that the American Communists, despite their handicaps, succeeded in contributing significantly and creatively to popular struggles or grass-roots movements, including the black freedom struggles that came to fruition in the 1960s. From this perspective, Communist organizers, instead of being condemned for manipulating African-American grievances for their own ends, should be seen as catalysts for mobilizing the black community to battle oppression and injustice on its own terms.[5]

It is possible to accept much of this argument without ignoring the sometimes disastrous effects of Comintern meddling and second-guessing. Occasionally Communists did adapt well to local circumstances and managed to adjust their preconceived ideology and rhetoric to the folk traditions and values of the people they were organizing. Sometimes the shift in the party line opened promising new possibilities and opportunities for agitation. At other times Comintern supervision was relatively lax and permitted considerable variation and deviation within a broad policy orientation. Hence the history of Communist activity in the United States and South Africa cannot be reduced to a footnote in the history of the international Communist movement. But it is also true that some shifts in the party line undermined promising and potentially fruitful endeavors and made the party less influential and effective than it would have been if local perspectives had prevailed. Hence one cannot overlook the international context or Soviet foreign policy and treat the Communists as simply another American or South African movement of social protest. A good history of Communist concern with the race question in the United States and South Africa must be a study of the interactions between an authoritarian international movement and democratic struggles against racial oppression within these nations that at times drew important support and inspiration from Communists.

Treating the encounter of blacks with the Communist party in the United States and South Africa as if it were a single narrative works quite well (at a certain level of abstraction) for the 1920s and '30s. The Comintern directives being implemented in both places at the same time—as well as a perception on the part of Communist strategists that the racial problems of the two societies were similar in character and amenable to the same types of solutions—led to parallel sequences of ideological and organizational initiatives. But the stories have rather different outcomes. After being close to the center of the American black freedom struggle in the 1930s, the Communists found themselves by the late 1940s on the periphery. In South Africa something nearly opposite was the case. From the doldrums of the mid-1930s when the party almost withered away, it moved into a strong and enduring alliance with what was becoming the main liberation movement, the African National Congress. By the '50s South African Communists were an influential and seemingly indispensable part of the struggle against apartheid, while in the United States they watched from the sidelines in diminishing numbers while anti-Communist liberal reformers, black and white, took control of the campaign against Jim Crow. Differences in the extent and effect of government repression will not explain this divergence. South Africa under the Afrikaner nationalists was even more repressive of civil liberties than America during the heyday of McCarthyism. In 1950, at a time when a few Communist leaders were being tried in the United States under the Smith Act, the South African party was banned and driven underground. A closer examination of the asymmetrical ups and downs of the two Communist parties and of their changing relationships with other movements and tendencies may yield a deeper explanation of how the differing political and social contexts in the two countries affected the long-term prospects for Communist involvement in the struggle against white supremacy.

Apprehending the precedents and intellectual resources that Communists brought to their encounters with American and South African racism requires a look at how international socialist movements, and those that emerged in United States and South Africa, dealt with the problems of white supremacy and racial inequality before the period of intense Communist concern that began in the mid- to late 1920s. The classic Marxism of the nineteenth century derived from an effort to make sense of the class conflicts that had developed in Europe between the industrial working class and those who owned the means of production. Socialism was born in societies that were, or appeared to be, racially and culturally homogeneous, and its early adherents had difficulty coming to terms with ethnically divided societies. Racial or ethnic consciousness on the part of oppressed peoples was not taken seriously as an historical force in its own right; it had to be either a vanishing remnant from the precapitalist past or a form of "false consciousness"—an ideological construction designed to obscure the underlying reality of class domination. It was bound to disappear with the maturation of capitalism and the full proletarianization of working people.

Marx, Engels, and their nineteenth-century followers could not ignore the ethnic nationalisms emerging on the periphery of Western Europe, and Marx himself came to favor the independence of Ireland from England—but not, as Lenin later noted, "from considerations of 'justice to Ireland,' but from the standpoint of the interests of the revolutionary struggle of the proletariat of the *oppressing* i.e., *the English nation* against capitalism." Marx's calculation was that English workers could not feel solidarity with their comrades in Ireland and thereby develop true class consciousness so long as their nation oppressed the Irish. Such a logic did not lead, as it might have done, to a call for the liberation of Asian and African colonies from European domination. Although Marx was highly critical of British rule in India, he concluded that it was historically necessary. The progress of humanity required "a fundamental revolution in the social state of Asia"—a transformation from "Asiatic despotism" to capitalism, and this was precisely what colonial domination was providing. "Whatever may have been the crimes of England [in India]," he wrote in 1853, "she was the unconscious tool of history in bringing about that revolution."[6]

A late twentieth-century reader is likely to be struck by Marx's profound Eurocentrism, at how deeply his theories were imbued with nineteenth-century notions of inevitable progress from lower to higher forms of civilization. Europe was the model for the rest of the world, in socialism as in all else, and its historical mission was to bring uncivilized or backward people up to its standards. Once they had been modernized, colonial subjects would lose their ethnic consciousness and fight for liberation, not as members of oppressed nationalities, but rather as part of the international proletariat. Marxists, for the most part, repudiated biological racism, characterizing it as a bourgeois ideological device to divide the working class along ethnic lines. To the extent that movements against Western imperialism or white domination relied on a sense of cultural distinctiveness or ethnic particularism rather than on universalist and internationalist assumptions, Marxists would find them theoretically unsound and could support them only opportunistically in the hope of eventually correcting their consciousness. Given its nineteenth-century Eurocentric assumptions, classic Marxism seemed singularly ill-equipped to deal with what W. E. B. Du Bois in 1903 called "the problem of the twentieth century"—"the problem of the color line."[7]

Early-twentieth-century socialists in both the United States and South Africa faithfully reflected European Marxist thinking and made little effort to adjust their class analysis to take account of the racial diversity and inequality in their own societies. As Eugene Debs, the leader of the American Socialist party, wrote in 1901: "We have nothing special to offer the Negro, and we cannot make special appeals to all the races. The Socialist Party is the Party of the whole working class regardless of color." The race problem, he contended, was created by the capitalists; workers and socialists had no responsibility for solving it other than appealing to blacks as fellow workers to join them in the class struggle. When that struggle was won, "the race problem [would] disappear." In theory, then, socialists advocated racial fraternity and should have welcomed blacks into their ranks. In practice, they often condoned discrimination and failed to make an issue of the racially exclusionary policies of the American labor movement. Occasionally, there were calls for at least recognizing that blacks suffered from special disabilities or disadvantages that would have to be addressed before they could be expected to join the broader working-class struggle, but in general socialists rejected such arguments and continued to focus on the alleged commonalities of class, ignoring the extent to which racism distorted or inhibited class consciousness.[8]

In South Africa, as in the United States, organized socialism began among whites who identified themselves primarily as champions of the white industrial working class. Given the vast economic and political inequalities that existed between black and white workers, South African socialists found it harder than their American counterparts to conceive of blacks and whites having common interests and belonging, even potentially, to the same social class. "Under Socialism," wrote the *Voice of Labor* in 1912, "the native should not be driven from his kraal in order to be exploited. He should remain there with his own kind and develop along his own lines." In other words, industrial labor should be a white monopoly, and Africans should be restricted to peasant pursuits in their own territories or reserves.[9]

Socialists were not responsible for the territorial segregation mandated by the Natives' Land Act of 1913, but some of them saw it as a step in the right direction. Driven by a desire to protect themselves from the lower-wage competition of nonwhite workers, champions of the white working class in South Africa could become extreme segregationists, arguing in effect that they would accomplish the tasks of disengaging the races more effectively than a bourgeoisie whose racial separatism was inevitably qualified by its desire for the profits to be gained by driving down wages. In 1922, a massive and bloody strike on the Witwatersrand registered the intense opposition of white workers to the plans of the gold mining industry to economize by replacing some of its highly paid white miners with lower-priced Africans. The marriage of labor radicalism and racism that characterized "the Rand Rebellion" was summed up by the signs strikers carried—"Workers of the World Unite and Fight for a White South Africa." The government repressed the strike with a massive use of force that included bombing planes, but the election of 1924 saw the coming to power of a coalition of Afrikaner Nationalists and the right-wing socialists of the South African Labor party. Subsequent "color bar" legislation made white workers virtually immune from African competition and gave the white working class a commitment to white supremacy and segregation that they have never lost.[10]

Challenging the American Socialist party and the South African Labor party from the left in the 1920s were the newly founded Communist parties inspired by the Russian Revolution. Communists were able to move beyond the "nothing special to offer" or downright segregationist approaches of the prewar socialists because the character of the Russian Revolution and Lenin's revisions of Marxist doctrine based on the Russian experience encouraged new ways of thinking about ethnicity and nationality. Contrary to the expectations of Marx and his Western European followers, revolutionary socialism triumphed first in a nation that was just beginning the transition to capitalism, rather than in one where it had come to maturity. Marx had assumed that the most "advanced" nations would take the lead, for how could you have a proletarian revolution without a fully developed proletariat? In Russia, the Bolsheviks apparently came to power because of the support they received from peasants and minority nationalities. The proletariat alone could never have provided the necessary base. One of Lenin's great contributions to Marxist thought came from his effort to explain this surprising development and use his explanation as the basis for a new and improved model for world revolution. It was Leninist doctrine, rather than anything that can be derived directly from Marx, that provided American and South African Communists with good reasons to be vitally interested in questions of race and nationality and some ideological guidelines for solving them.

Lenin argued that capitalism had entered a new stage, not clearly foreseen by Marx, in which the class struggle had moved beyond the internal politics of the industrialized nations on which Marx had focused to the international arena created by Western imperialism. Imperialism, as the "highest stage of capitalism," meant that the viability of the capitalist system now depended as much on the exploitation of colonies and dependencies as on the exploitation of domestic workers. In fact some of the fruits of imperialism were now being used to improve the conditions of the domestic proletariat in a effort to cool its revolutionary ardor. Noting the beginnings of national liberation movements in Asia, Lenin hypothesized that good revolutionary opportunities might exist at points where imperial domination was most fragile—the "weakest links" in the world capitalist system.

As the Russian Revolution demonstrated, you did not need to have an advanced proletariat in order to have a Communist revolution. All you needed were the kind of opportunities for seizing power that World War I had provided for Russia and a Communist party to take advantage of them. In Leninist thought the party rather than the proletariat in whose name it acted was the real agent of revolution. If it lacked a proletarian mass to galvanize into revolutionary action, the party should work with other materials such as the one the Bolsheviks had found at hand in Russia—peasants with a hunger for land, minority nationalities thirsting for independence, or at least the right to self-determination, and radical intellectuals with a mission to set the world right or bring freedom to their people.

National liberation movements in European colonies should be supported by Communists throughout the world even if their leadership and ideology was not purely Communist; for victories against Western imperialism could weaken the world capitalist system and bring it closer to collapse. As the Bolsheviks had done in Russia, Communists must affirm that every national group has the right to secede from a nation or empire into which it has been incorporated against its will. As participants in fronts or blocs working for national liberation, Communists should exercise as much control as possible and dominate them if they could. But tactical compromises would be necessary; peasants could not be turned into class-conscious proletarians overnight, and independence-minded "national bourgeoisies" could be allies in the short term, even if ultimately they had to be overthrown to make way for socialism.[11]

The most problematic element in the Leninist strategy for abetting and—if possible—coopting nationalist revolutions was the question of how much cooperation there could be with movements that aimed at independence on the basis of ideologies that were antithetical to Communist doctrine. How could Communists retain their autonomy and remain loyal to the international workers' movement if they gave full support to "bourgeois national movements"? At the Second Comintern Congress of 1920, Lenin advanced the argument that Communists in the non-Western world should generally support movements for national independence even if they were led by bourgeois elements. He was challenged by the Indian Communist M. N. Roy, who contended that the native bourgeoisie was unreliable and would in the end compromise with imperialism. Lenin subsequently clarified his position by acknowledging the need to distinguish between colonial bourgeois movements with a revolutionary anti–imperialist potential and those that lacked it. But it was left to Stalin to lay down specific guidelines.[12]

In his 1925 address to the university that the Soviet Union had established for the training of Asian and African revolutionaries, Lenin's successor noted that "the national bourgeoisie" in colonial areas had a tendency to split between a wealthy strata willing to compromise with imperialism and a petite bourgeoisie prone to support a nationalist revolution. The role of the Communists would depend on how far this process had gone. If it had not yet begun, as in Morocco, Communists should "do everything to create a united national front against imperialism" and should not even organize themselves into a separate party. Where it was occurring, but where "the compromising section of the bourgeoisie cannot yet become welded with imperialism," as in Egypt or China, Communists must move from "the policy of a united national front to the revolutionary party of workers and peasants." Here the Communists would have their own party but would form a "bloc" with "the party of the revolutionary bourgeoisie." As part of such a bloc, they must retain their freedom to agitate and propagandize on their own terms and be in a position to work for "actual leadership of the revolutionary movement." In a third case, such as India, where "the compromising section of the bourgeoisie" had "come to an agreement with imperialism," the task was to create "a revolutionary anti–imperialist bloc" under "the hegemony of the proletariat" that would expose the "treachery" of the native bourgeoisie.[13]

To give movements for national self-determination a place in the dialectic of world revolution, Communists needed to develop a working conception of what constituted a nation. Stalin, the architect of a Soviet nationalities policy that proclaimed the group rights of territorial ethnic minorities and even gave them—on paper if not in reality—the right to secede from the Soviet Union, provided as early as 1913 the definition upon which Bolshevik policy would be based. A nation, according to Stalin, was "a historically evolved, stable community of language, territory, economic life, and psychological make-up manifested in a community of culture." Such entities had the right to self-determination and should not be subjected to other nations against their will. Ultimately, of course, "the national question" was subordinate to "the social question," and the right to self-determination could be overridden by "the right of the working class to consolidate its power." But in the capitalist-dominated parts of the world nationalism was a legitimate and necessary response to imperialist exploitation.[14]

Whatever the merits of such theories for anti–imperialist movements in colonies occupied by Europeans but lacking substantial white settler populations, they would be difficult to apply to the problem of race discrimination in the United States or even to white minority rule in South Africa. Stalin's definition of a nation did not seem to fit African-Americans at all, given their apparent lack of a separate or distinguishing "language, territory, and economic life." A case could be made for distinctiveness of "psychological make-up" and "culture," but was that in itself enough to base a claim to self-determination? Lenin, however, had suggested another way of turning the American race question into a national question. In some comments on the United States written before the Russian Revolution but not actually published until 1935, he stated categorically that American Negroes "should be classed as an oppressed nation, for the equality won in the Civil War of 1861–65 and guaranteed by the Constitution of the republic was in many cases curtailed in the chief Negro areas (the South) in connection with the transition from the pre-monopoly capitalism of 1860–70 to the reactionary, monopoly capitalism (imperialism) of the new era. . . ." Here the definition of a nation depended not on its intrinsic characteristics, which remained unspecified, but on the nature of its relationship to a dominant group. The emphasis was on the character of oppression rather than the cultural prerequisites for nationhood. Marxist-Leninist theory had thus been stretched to include the oppression of nations as well as classes, but a further extension acknowledging that racial or ethnic inequality *within* nations had distinctive characteristics that required a new category for analysis was apparently beyond its capacity. Lenin's logic, one might conclude, was that a form of domination that went beyond the economic exploitation of class rule and designated a racial or ethnic group for special political and social liabilities must be national in character—there was simply no other possibility. Communists in general were left either to follow Lenin's suggestion and pursue the implications of regarding African-Americans as an oppressed nation, or revert to something like the normal socialist assumption that blacks were simply proletarians with dark skins divided from their fellow workers by capitalist propaganda and manipulation.[15]

Applying Leninist or Stalinist conceptions of oppressed nations and struggles for self-determination to South Africa might appear at first glance to be relatively easy. Here a white settler minority, the product of European colonial expansion, was oppressing a native majority. But a literal interpretation of Stalin's definition would yield not one nation but several. Not only were Africans divided into ethnic groups with differing languages and cultures, but one segment of the white community—the Afrikaners—had earlier made their own fierce bid for national liberation from British imperialism, a struggle that had won the sympathy of many European socialists. By the 1920s, Afrikaner nationalists had rallied from defeat in the Anglo-Boer War and were mobilizing politically to challenge the domination of English-speaking mining magnates backed by London capital—or so it appeared to some white leftist observers of the election of 1924. The question of who constituted a nation qualified for self-determination in South Africa was not so easy to decide in the 1920s as it would later become. If there were one or more white nations as well as several black ones in South Africa, how could self-determination for all oppressed nations be achieved without carrying the idea of group segregation to its logical extreme and dissolving the South African union into ethnic states?[16]

Rise of the Black Self-Determination Policy

Beyond the theoretical obstacles to making the United States and South Africa conform to Leninist or Stalinist models of national oppression, there was also the practical problem arising from the fact that the founding members of the American and South African Communist parties were radicals of European extraction whose initial expectation had been that they would revolutionize the white workers first and then go on to bring the gospel of Marx and Lenin to people of color. In China, India, and Egypt, the first Communists had also been members of the oppressed group, a circumstance that obviously made it easier for them to get involved in movements for national liberation. The initiating role of whites in the United States and South Africa meant that the party would have to overcome black suspicions of would-be liberators who were also members of the dominant race. The question of how a white-dominated party could take the lead in the movement to liberate blacks from racial domination was difficult to resolve. In South Africa the obvious answer was to "Africanize" the party, but this took time. In the United States, the Communist party could not become a predominantly black organization without giving up its claim to speak for the American working class as a whole. How to accommodate black needs and interests within a mostly white party was a major challenge for American Communists.

At the beginning of the 1920s, Communists in both the United States and South Africa concentrated mainly on organizing white workers. The predominant view was that most blacks were not yet proletarians and could therefore scarcely have acquired the requisite class consciousness. Special programs for blacks that appealed to a sense of racial or national oppression were considered incompatible with Marxism. But in 1922 the Fourth Congress of the Communist International made a high priority of the recruitment of African-Americans into the party. Since black Americans were more "advanced" than blacks in colonial areas, they could be trained, not only for work among blacks in the United States, but also for revolutionary roles elsewhere in the African diaspora and even in Africa itself. The congress concluded that a Communist-directed "world Negro movement" could be organized, with the United States, as "the center of Negro culture and the crystallization of Negro protest," playing a central role. The new "spirit of revolt" that black Americans had demonstrated since the world war had placed them "in the vanguard of the African struggle against oppression." The "Theses on the Negro Question" of the Fourth World Congress did not actually mention the Garvey movement, but clearly the growth of this populist Pan-African movement in the United States provided inspiration for the new departure. If Communists could not dominate the Garvey movement itself, they could set up a rival Pan-African movement. Accordingly, the Comintern recommended that "a general Negro Conference or Congress" be held in Moscow.[17]

The Comintern's image of African-Americans as the vanguard of the black race—which American Communist writers and orators promptly took up and promulgated—was the new version of an old tradition in Pan-Africanist thought. Indeed it bore a striking resemblance to the notion, long current in missionary circles, that African-Americans would play a key role in the conversion of Africa to Christianity. Now, however, the conversion would be to Communism, and salvation would be achieved in this world rather than in the hereafter.

Taking its cue from the Comintern, the American Communist party by 1924 was making a major bid for black support and seeking to involve itself in African-American politics and protest activities. In that year it participated in a congress of black organizations called the Negro Sanhedrin. But moderate to conservative black organizations blocked Communist efforts to commit the Sanhedrin to a radical protest agenda; in any case the conclave failed to found a new black federation that Communists might have influenced. The year 1924 also saw an article in the journal of the Communist International, apparently written by an African-American Communist, taking the party to task for "ignoring the question of racial antagonism" and thus allowing "the Negro liberation movement in America to take a wrong path and get into the hands of the Negro petty bourgeoisie which has launched the nationalist slogan 'Back to Africa.' " In an effort to show heightened concern for black liberation and to facilitate recruitment of African-Americans, the party founded an organization of its own in 1925—the American Negro Labor Congress, which was designed to attract intelligent and politically informed black workers and turn them into dedicated Communists. The ANLC did not prosper, however, and black membership in the party remained minuscule through the late 1920s. In 1929, the party had no more than 150 to 200 black members.[18]

The South African party also began to make special efforts to recruit blacks in the mid-1920s. After supporting the Rand rebellion, despite its racist overtones, and even endorsing the nationalist-labor coalition in the election of 1924, the party awoke to discover that a government backed by white labor was intensifying discrimination against Africans. On their own initiative, the leading white Communists (who were as yet receiving little guidance from Moscow) decided to concentrate on recruiting blacks and acquiring influence in black protest politics and labor organizations. When the Hertzog government proposed and enacted laws protecting white labor from African competition, Communists had to acknowledge that white workers were becoming a privileged caste and were virtually beyond the reach of left-wing agitation.

Africans, who were more blatantly exploited, seemed a better target. By the late '20s, the great majority of party members were Africans, although whites still held the top leadership positions. At the party's Annual Conference of 1929, twenty black and ten white delegates claimed to represent a mostly black membership of about 3000. Black Communists, encouraged by the party to take leading roles in other organizations, were influential in the ICU until Kadalie expelled them in 1926. They then concentrated on the African National Congress, and in 1927 Eddie Khaile, the first black member of central committee of the CPSA, was elected general secretary of the ANC. To recruit Africans, white Communists made use of their own version of the mission school—night classes taught by white Communists that offered literacy in English along with indoctrination in Marxism. With many times the black membership of the American party, the South African party of 1929 might be said to have recruited a respectable nucleus on which to build.[19]

These Africans, it is worth noting, were not attracted to a revised version of the Communist message tailored to their special concerns as an oppressed nation. The early converts, like early African converts to Christianity, were asked to embrace a whole new way of looking at the world that involved a universalist conception of salvation rather than the particularities of race and ethnicity. Once again, a portion of the African population demonstrated its willingness to break decisively with the past and adopt Western conceptions of progress and liberation.

In 1928, the Communist International, responding mainly to the American failure to attract blacks, made a bold move to combine the class struggle with the kind of nationalistic aspirations to which Garvey had appealed. The new emphasis on "Negro self-determination" as a necessary component of the struggle against capitalist imperialism was an unwelcome surprise to most black Communists in the United States and South Africa. Having made the difficult adjustment to a "class" perspective, they were understandably reluctant to revert to a conception of the struggle based in part on the ethnocentrism that they had forsworn when they became Communists.

There was, however, a precedent within African-American radicalism for a synthesis of Marxism and black nationalism that the defenders of the new policy could have invoked. The African Blood Brotherhood for African Liberation and Redemption, a secret revolutionary organization founded by Cyril Briggs in 1919, had initially put forth such a program. Briggs was from Nevis in the British West Indies, and most of his supporters came from New York's West Indian Community. At first the Brotherhood supported the mass movement headed by the charismatic Jamaican Marcus Garvey and apparently sought to constitute a radical element within Garvey's Universal Negro Improvement Association. But the Brotherhood was also enthusiastic about the Russian Revolution and the prospects that it opened for the liberation of Africa from imperialist domination. "The important thing about Soviet Russia," according to a Brotherhood manifesto, "is not the merits or demerits of the Soviet form of government, but the outstanding fact that Soviet Russia is opposing the imperialist robbers who have partitioned our motherland and subjugated our kindred." Seeing no necessity to choose between Garveyism and Leninism, the Brotherhood by 1920 was moving into the Communist orbit while trying to keep a foothold in the UNIA. Unlike Garvey, who concluded that African-Americans could escape white oppression only in an independent Africa, Briggs experimented with the notion that blacks would achieve some form of political independence within the United States. In his journal *The Crusader* he called for self-determination for blacks in one section of the nation, while also advocating cooperation with radical white workers to overthrow capitalism.[20]

Expelled from the UNIA when Garvey rejected alliances with the left in 1921, the Brotherhood became in 1922 the black unit in the Workers' Party, the federation of pro-Soviet ethnic organizations that was then the visible or above-ground expression of the Communist movement. When the party abandoned its ethnic federalism to become a unitary and centralized organization in 1925, the Brotherhood was dissolved. Its former members became the original cadre of blacks within the party, but the new stress on integrated class action under white, mostly American-born leadership meant that their nationalist or separatist inclinations had to be suppressed or subordinated. Except for Harry Haywood, who became a member of the Brotherhood in 1922, none of the veterans of this Pan-Africanist revolutionary movement was initially sympathetic to the self-determination policy announced in 1928.

The true origins of the policy shift that led in 1928 to a call for "self-determination for the black belt" in the United States and the establishment of a "native republic" in South Africa are obscure, because little direct evidence survives on how the decision was made, except for transcribed debates with a predetermined outcome. Some historians have concluded that the whole project was a personal whim of Stalin's—an effort to view the problems of blacks in the United States and South Africa through the distorting lens of Soviet nationalities policy. Since the policy apparently had little or no local support and was not adapted to local circumstances, it allegedly made it more difficult than ever to attract blacks to the movement. Historians more sympathetic to the policy change not only see it as an accommodation to the legitimate aspirations of black nationalists but also tend to argue that it was not so much an imposition from Moscow as a creative and productive response to input from black South Africans and African-Americans.[21]

The only detailed first-person account of the origins of the policy shift comes from the 1978 autobiography of Harry Haywood, the leading black American advocate of the self-determination policy. Haywood was in a position to know what was going on, and he understood the South African as well as the American side of the issue. His version of events, although presumably self-serving, is plausible and must be taken seriously. When he wrote his autobiography, Haywood was no longer a member of the party and under no pressure to provide a particular view of its history, but he was still on the far left and was not apologizing for anything he had done earlier. Read critically, his account brings us as close as we are likely to get to an understanding of the origins of the Communist conception of black self-determination.[22]

Born Haywood Hall, Harry Haywood followed his older brother Otto Hall into the African Blood Brotherhood in 1922, into the Young Communist League in 1923, and into the Communist party itself in 1925. In 1926, the brothers were sent to Moscow for special training and were among the first African-Americans to be enrolled at the University of Toilers of the East, the scene of Stalin's discourse on national liberation movements the previous year. Haywood later claimed to have learned in the Soviet Union that "national culture could be expressed with a proletarian (socialist) content, and that there was no antagonistic contradiction, under socialism, between national culture and proletarian internationalism." The brothers apparently had an audience with Stalin shortly after they arrived. Haywood does not describe the encounter in his autobiography, but Otto Hall recalled that Stalin told them that "the whole approach of the American party to the Negro question is wrong. You are a national minority with some of the characteristics of a nation." To Hall this sounded like Jim Crow, and he steadfastly opposed the application of Leninist-Stalinist self-determination theory to the United States until it was a fait accompli. Haywood on the other hand was open to persuasion, although in his autobiography he attributes his change of viewpoint not to hints he might have received from Stalin but to his discussions with N. Nasanov, a Russian Communist who, under the name of Bob Katz, had spent time in the United States as a representative of the Comintern. Nasanov had, on the basis of his own observations and also perhaps because he was aware of Lenin's unpublished views, come to the firm conclusion that African-

Americans were members of an "oppressed nation" and not simply a racial minority. Haywood protested initially that he did not see how they could fulfill Stalin's criteria for nationhood; for one thing, they lacked a territory of their own. But, under the influence of Nasanov, he came to see the value of such a perspective. It responded to the "yearnings of millions of blacks for a nation of their own"—feelings which Garvey had tapped but misdirected. Communists, he concluded, needed to endorse black nationalism to make sure that the cause did not fall into the hands of utopian visionaries; in the United States its emotive force needed to be focused on "the struggle against the main enemy, U.S. imperialism." In an effort to give form and shape to black American nationalist aspirations, Haywood himself then proposed—or so he claims—that Communists support "a right to self-determination in the South, with full equality throughout the country." Whether the slogan of "self-determination for the black belt" was really Haywood's invention or not, he quickly became its foremost advocate, glorying in the way that it elevated "the black movement to a position of equality" with the class movement of white workers in "the struggle against the common enemy—U.S. capitalism."[23]

In 1927 Haywood was appointed to the Negro Commission that was to prepare resolutions on the "Negro Question" for the Sixth International Congress in 1928—an undertaking deemed important enough to be chaired by Stalin himself. Arguing fervently for the new departure against the staunch opposition of his brother and other African-American Communists, Haywood appears from his own account to have been neither a puppet of Stalin nor the true originator of the self-determination policy. He admittedly responded to suggestions from above for a general shift in direction but did so with the fervor of a convert, and in the process may have given the policy a practical application to American conditions. His subsequent unwillingness to abandon the stress on black self-determination when it had outlived its usefulness to the Comintern strongly suggests that he was not simply an opportunist seeking to curry favor in high places.[24]

South Africa enters Haywood's account of the origins of self-determination when he is assigned with other black Americans on the Negro Commission to consult with James La Guma, a black Communist from the Cape Province, about how the emerging self-determination doctrine should be applied to South Africa. La Guma played a role in the construction of "the native republic" thesis for South Africa similar to Haywood's in the invention of "self-determination for the black belt," and it is worth tracing his career to the point of collaboration with Haywood in 1928. Under South Africa's scheme of racial classification, La Guma was Colored, rather than African or "native," being of French-Malagasy origins, but such designations meant less in the black politics of the 1920s than during other periods of South African history. At that time Coloreds were active in the African National Congress, especially in the Cape Province, and in the Industrial and Commercial Workers' Union (ICU). La Guma's career as an activist began as an organizer of African diamond miners in South West Africa, where he led a strike in 1918 and an anti-pass protest in 1919. The following year, he helped found a branch of the ICU there, and in 1921, he returned to South Africa to head the ICU branch in Port Elizabeth and serve as one of Clements Kadalie's lieutenants. Having joined the Communist party in 1925, he was expelled from the ICU in 1926 as a result of Kadalie's ban on Communists on the executive committee. Along with the other black Communists purged from the ICU, he then became active in the ANC, becoming executive secretary of the Cape Town branch in 1927. In that same year, he was chosen, along with ANC president J. T. Gumede, to be a delegate to the conference of the League Against Imperialism in Brussels.[25]

The conference, dominated by Communists seeking to establish an international front against Western colonialism, adopted a resolution proposed by La Guma and Gumede calling for "the right of self-determination through the complete overthrow of capitalism and imperial domination." When the conference was over, the two South Africans were invited to visit the Soviet Union. Gumede, who never became a member of the party, returned to South Africa as a staunch supporter of the Soviet Union's international policies. As a nonwhite Communist in good standing, La Guma did more than marvel at Soviet achievements; he became an active participant in discussions of the future of the South African party and on how the new line on blacks in the United States should be applied in his own country. In discussions with Comintern officials, including Nikolai Bukharin, La Guma either advanced the argument that the immediate aim of the South African struggle should be an independent black republic or allowed himself to be persuaded of the virtues of this new approach. As an ANC leader, it is quite conceivable that he had already seen the virtues of collaborating with "progressive petty-bourgeois" elements in the congress to achieve black self-determination as the first step toward socialism. But, as in the case of Haywood, some doubt must remain as to who really initiated the policy. La Guma was also consulted by the Negro Commission and had extended discussions with Haywood. The latter, who found La Guma very impressive, recalled that they uncovered "striking parallels between the struggles of U.S. blacks for equality and those of Native South Africans. In both countries the white leadership of their respective parties underestimated the revolutionary potential of the black movement."[26]

The Negro Commission, guided by Stalin and Bukharin, endorsed the arguments made by Haywood and La Guma and voted to recommend to the Sixth Congress that "Negro self-determination" must be a party goal in both the United States and South Africa. The "Theses and Resolutions on the Negro Question," adopted by the congress treated black-white relations as part of the broader "colonial question," but distinguished between cases such as the United States, in which "the compact Negro masses constitute a minority in relation to the white population," and "the Union of South Africa, where the Negroes are the majority in relation to the white colonials." In the United States most blacks were tenant farmers in the South, living under "semi-feudal and semi-slave conditions," but increasing numbers were migrating to the North and becoming unskilled workers. Communists must work to raise the class consciousness of this new proletariat and simultaneously "fight for the acceptance of Negro workers into all organizations of white workers, and especially in the trade unions." They must also seek to organize "the masses of peasants and agricultural workers in the South." In order to accomplish these tasks, Communists must agitate for full legal and political equality for blacks and root out "white chauvinism" in their own ranks. The call for work among rural southern blacks was new; the American party had previously assumed that its work began only after blacks had left the countryside and joined the urban-industrial proletariat. But agitation among peasants and farm workers required a new slogan, beyond the usual "Black and White, Unite and Fight": "In those regions of the South in which compact Negro masses are living, it is essential to put forward the slogan of 'the Right of Self-Determination for Negroes.' " What this meant exactly was not clearly specified, and further discussions and resolutions would be necessary before it was definitely decided that blacks had the right to establish an independent black-belt republic if they chose to do so.[27]

The similar resolutions concerning South Africa were a bit clearer on the meaning of self-determination. Communists should continue their promising work among the black proletariat and fight for their admission to white unions. Rural oppression should be addressed (as in the American black belt) by calling for "the confiscation of the land of the landlords." In addition, "The Party must determinedly and consistently put forward the slogan for the creation of an independent Native republic with simultaneous guarantees for the rights of the white minority, and struggle in deeds for its realization."[28]

The congress and the executive committee of the Communist International approved the black self-determination policy over the strenuous objections of most of the delegates from the United States and South Africa. The Americans, including all of the black delegates except Haywood, argued that blacks were not an "oppressed nation" with a right to self-determination but rather a "racial minority" that could be incorporated into the struggle only when it became proletarianized. What the party could offer blacks in particular was support in their struggle for civil rights, for success in that campaign would lower barriers to their joint action with white workers. The South African delegation, which was severely criticized for racial chauvinism because of its all-white composition, took a similar stand, arguing in addition that the African nationalism the new slogan sought to accommodate scarcely existed in South Africa. Since black organizations such as the ANC were seeking incorporation into the colonial state rather than independence, party leader S. P. Bunting argued, a strategy aimed at cooperation with a revolutionary nationalist bourgeoisie was misguided, for no such entity existed in South Africa. Although Haywood and La Guma provided a modicum of local backing for the new policy, the conclusion of many historians that the self-determination policy was imposed by Moscow on reluctant local party organizations appears unchallengeable.[29]

Between 1928 and 1930, Communists in the United States and South Africa debated the precise meaning and application of the new slogans. Although the problem that Bunting identified—the lack of a national revolutionary movement for Communists to support—remained unresolved, the South African party at least knew that the black republic was not a mere propaganda device, for they had been ordered to "struggle in deeds for its realization." White Communist leaders such as Bunting and Edward Roux swallowed their misgivings and struggled dutifully to implement the new policy. In the United States, however, the requirement in the resolutions of 1928 to "put forth the slogan of the Right of Self-Determination" was not followed by orders to struggle for its implementation, but rather by a renewed emphasis on explaining "to black workers and peasants that only their close union with the white proletariat and joint struggle with them against the American bourgeoisie can lead to their liberation from barbarous exploitation"—a statement that seemed to deprive the slogan of nationalist implications. It was possible to view self-determination as a mere propaganda slogan designed for local agitation in the South and not a major deviation from the party's main stress on interracial solidarity.[30]

In 1930, the secretariat of the Communist International responded to the confusion of American Communists about how seriously the new line needed to be taken. The first indications were that they did not have to take it seriously at all. Two American Communists cabled from Moscow that "self-determination is an agitation slogan, while the aim of immediate action is the concrete struggle against Jim Crowism and other discriminatory actions of the bourgeois state." "State separation," or the idea that a southern black republic might be allowed to secede from the American Union, was, according to the cable, not an immediate issue and could be played down. On the basis of this information, the American party deleted the words "to the point of separation" from its own statement on "the right of self-determination."

Later in the year, however, the Comintern's Negro Commission stiffened the resolution, making it clear that self-determination was more than "an agitation slogan" and that mass actions should be organized to support the right of black-belt Negroes to form a government of their own and secede from a capitalist America if they so desired. At the same time, however, the Comintern stressed the fact that this black nationalist agenda applied only to the southern states and that interracial proletarian revolution was still the program for the North. At this point, partly because some of those most resistant to the new policy had been on the losing side in a factional struggle and had been purged from the party, opposition to the new departure (as interpreted by Harry Haywood) ceased abruptly.[31]

What, ultimately, was behind the shift and what did it signify? Odd as it may seem in some respects, the main concern of the Communist International was with the black movement in the United States; South Africa's need for a black republic was clearly an afterthought. The United States was emerging as the center of world capitalist power, and all opportunities to weaken it needed to be seized upon. Furthermore, African-Americans were viewed not only as likely catalysts for an American Communist revolution, but also as potential leaders of a Pan-African Communist movement against European colonialism. It was no accident that during the deliberations of 1928 African-American Communists were encouraged to criticize white South African Communists for their alleged racial chauvinism and reluctance to view the South African struggle in the context of the movements for national independence that Communists were encouraging or supporting in Asia and elsewhere in Africa. To the extent that South Africa attracted independent attention, it looked to Moscow like an appendage of the British empire, despite its de facto independence as a white settler commonwealth. Mobilizing African peasants to fight for their own republic would have an effect similar to the support of militant Indian nationalism—it would weaken British capitalist imperialism. Hence Stalin's support of the self-determination policy was meant to support the enemies of his enemies. But it was meant to do more than that: the opportunistic support of moderate, "bourgeois" nationalist movements that Lenin had sometimes recommended would have meant cooperating with groups that were reformist in character and therefore unsympathetic to efforts to mobilize workers and peasants for imminent revolution. The "left turn" in overall Communist policy that also occurred in 1928 precluded such alliances.[32]

The left turn was dictated by Stalin himself and was apparently tied to his suppression of opposing factions within the party, his decision to collectivize agriculture within the Soviet Union, and his fears of a new intervention by the Western imperial powers. The premise of the policy—that the time was ripe for revolutionary action throughout the world—may have seemed dubious in 1928, but it acquired some plausibility from the onset of the world depression two years later. A result of the left turn was the rejection of alliances with democratic socialists, who were now labeled "social fascists," and with reformist nationalist or civil rights movements that were dominated by the "bourgeois" elements of the oppressed group. Consequently, working in tandem with organizations such as the NAACP, the UNIA, or the ANC was ruled out, and in fact Communists were enjoined to denounce such movements and discredit them in any way they could.[33]

The new sectarianism posed no immediate problem for American Communists, because the NAACP and the Garvey movement had kept them at arm's length. But in South Africa it meant failing to pursue to the fullest the opportunity to gain influence in the ANC opened by the elections of Gumede as president-general and Khaile as general secretary in 1927. According to the South African Marxist historian Martin Legassick: "It is unfortunate that the injection of a 'national revolutionary' analysis into the South African Marxist scene coincided with the leftward swing of the Comintern in 1928 and after." He presents evidence to show how the new sectarianism prevented the party from fully capitalizing on African nationalist sentiments and condemned it to a decade of isolation and ineffectiveness. A case in point was the fate of the party's effort in 1929 to establish a mass organization called the League of African Rights. All black organizations protesting segregation and discrimination were invited to participate on a platform of grievances that did not mention revolution or the native republic. The ANC, the ICU, the Cape Native Voters' Convention, and a number of church groups responded to the call. Gumede, the friend of the Soviet Union who then headed the ANC, was chosen president, and the other offices were held by CP members. But this promising effort to put Communists or their supporters at the forefront of an inclusive movement for African rights was torpedoed by the Comintern. It cabled the party in late 1929 ordering that the league be disbanded because its program and membership were bourgeois and reformist rather than proletarian and revolutionary.[34]

In 1930, the American party set up a similarly named front organization called the League of Struggle for Negro Rights. But in this case there was no objection from Moscow, for the platform of the league closely followed the Comintern resolutions on the Negro question, including the right to self-determination for the black belt. Meant to be a radical alternative to the NAACP, the league attracted some black intellectuals, including the poet Langston Hughes, who served for a time as president; but its obvious character as an extension of the Communist party prevented it from developing mass support within the black community. Although American Communists adjusted more easily to the left turn than their South African counterparts, the revolutionary orthodoxy of the early 1930s did not make it easier to take a prominent role in black political movements.[35]

Nevertheless, the American party did make some gains among blacks during the early depression years, a time when the South African party was sliding toward virtual oblivion. This contrast of fortunes under the same mismatched policies of black self-determination and left sectarianism cannot be attributed to differences in how well the Comintern policy worked in the two contexts. Indeed it can be argued that the American Communists succeeded to the extent that they did *despite* Comintern directions that were ill-suited to American conditions, while their South African counterparts foundered despite being encouraged to move toward a valid and appropriate recognition of the revolutionary potential of African nationalism. Hence the twists of the international party line and the myopia of Moscow cannot provide a full explanation of what happened to Communism in these two multi-racial environments during the early '30s. Attention must also be paid to local conditions and how effectively Communists in each country adjusted to the constraints and opportunities that they faced.

In South Africa, the "native republic" thesis failed to bear fruit partly because of a bewildering series of changes in its interpretation. In 1930 the Comintern interpreted it to mean that the South African revolution would be a two-stage affair—the independent native republic would precede the establishment of socialism. Since cooperation with the African bourgeoisie was now ruled out, "the only class capable of uniting the national revolutionary front is the native proletariat." But the visit of the American Communist Eugene Dennis as Comintern representative in South Africa resulted in the rejection of the two-stage model for one that anticipated a simultaneous national and Communist revolution. In 1934, party leader Lazar Bach announced that the native republic was a plural rather than a singular concept; it meant "a voluntary association of national republics—Sotho, Tswana, Swazi, Zulu, Xhosa—in a federation of independent native republics." This formulation was remarkably faithful to Stalin's official conception of how to resolve the national question in the Soviet Union but flew directly in the face of the ANC's commitment to an inclusive (nontribalized) African nationalism and was hence unlikely to win the support of those blacks most actively involved in the struggle against white supremacy. It also, ironically enough, anticipated some features of the 1950s "homelands" policy of the Afrikaner Nationalists.[36]

But the party's decline in the early '30s cannot be attributed exclusively or even primarily to such ideological gyrations. More important were two other developments: a series of purges that deprived the party of its most experienced leaders and the state repression of the Communist movement made possible by new legislation permitting the government to arrest and imprison suspected agitators almost at will or exile them to a part of the Union other than the one in which they had been active. As factionalism, purges, and state repression decimated the party, discussions of the true meaning of the "native republic" thesis became increasingly academic. South Africa of the early '30s in fact provided little scope for radical action of any kind. The depression failed to inspire much new militancy because it could not make the impoverished conditions of blacks much worse than they already were, and, in any case, rebounding gold prices made the economic crisis less severe and long-lasting in South Africa than in most other industrializing countries. The main effect of the depression was to unify the whites politically. In 1933 the two principal white political parties formed a coalition government, and a year later they merged to form a single "United party" committed to economic stabilization and the preservation of white supremacy.[37]

In the United States, the situation was rather different. The depression released liberal and radical impulses that had been held in check by the prosperity and conservative, pro-business political climate of the 1920s. The limited economic gains that blacks had made in finding industrial employment and establishing small businesses during the '20s were wiped out in the early '30s. Since the new party line authorized mobilizing blacks on the basis of their class grievances in urban-industrial areas, the Communists had no qualms about taking the lead in organizing blacks to demand unemployment relief, resist evictions from ghetto apartments, and demonstrate for the hiring of blacks on public works. Because of its conspicuous role in agitating such "bread-and-butter" issues, the Communists built up a substantial reservoir of good will among urban African-Americans. In the course of these actions, black Communists sometimes violated the sectarian imperatives of the left turn by working in cooperation with non-Communist groups, but they did not try to conceal who they were and what they stood for.[38]

Following the orders of the Comintern, the party also expended considerable energy and ingenuity on the difficult and dangerous task of organizing sharecroppers and farm laborers in the southern black belt. According to historian Robin Kelley, they had more success than has previously been acknowledged. In working with rural blacks, he argues, Communists did not try to impose ideological orthodoxy, but rather demonstrated a surprising degree of tolerance for the culture of the people they were dealing with. For one thing, they did not insist that black Christians renounce their faith in favor of Marxist atheism. The "self-determination" thesis was translated to mean gaining rights to the land, a cause that black farmers could readily appreciate. Drawing on folk memories of what had been attempted and promised during Reconstruction, Communists appealed to the deeply rooted belief of many rural African Americans that they had a right to the land they worked and to political control of communities in which they were the overwhelming majority. The Communist-led Sharecropper's Union of Alabama had no chance of reaching the bulk of black farm workers in the state, but it persisted in some counties for several years despite horrendous violence and intimidation and may have had as many as 10,000 members at its peak in 1935. W. E. B. Du Bois of the NAACP condemned such organizing as irresponsible because it endangered black lives in a hopeless cause, but the deadly danger and apparent futility of the endeavor could not dampen the zeal of fearless Communist organizers.[39]

The party's biggest success in winning credibility among blacks throughout the nation resulted from its prominent role in the defense of the Scottsboro boys. After nine black teenagers were condemned to death in 1931 for allegedly raping two white women in a railroad car in Alabama, the Communist-backed International Labor Defense (ILD) rushed to their aid and managed to gain control of the defense, beating back the efforts of the NAACP to take charge of the case. The ILD provided the boys with a competent defense for their appeals and subsequent trials, and the Communist party made the injustice of the conviction into an international cause célèbre, using it as the occasion for rallies and marches throughout the world. Although party literature described the Scottsboro defense as an application of the "self-determination" slogan (what happened to the boys showed what lack of self-government meant), many blacks interpreted the campaign as simply a more militant and effective pursuit of the reformist goals that the NAACP had been advocating since its founding.[40]

Successful Communist agitation among blacks in the early '30s can be attributed to a growing conviction, arising especially from the Scottsboro defense, that the Communists were the boldest and most effective advocates of equal citizenship for African-Americans. There was no mass increase in black membership during this period, but many blacks, perhaps a majority, developed a favorable image of the party. In contrast to South Africa, the party could agitate freely in at least part of the country—the northern states—and, after resolving a major factional quarrel in the late '20s, was relatively unified and competently led. Contrary to what one might have expected, the American Communist party came through the period of the self-determination slogan and the left turn (1928–34) in better shape than its South African counterpart and seemed to be demonstrating a greater capacity to challenge racial injustice in ways that blacks found inspiring.

Blacks and the United Front, 1934–1939

In May 1934, *Pravda* announced an abrupt reversal of policy on the question of whether Communists could cooperate with the non-Communist left. Alliances with socialists and social democrats (heretofore condemned as "social fascists") were now permissible. Hitler's coming to power in Germany and the likelihood that he would attack the Soviet Union forced Stalin to repudiate the sectarianism of the left turn and adopt the slogan "united action by the working people" against fascism. European socialists had earlier responded to the Nazi triumph by calling for an alliance with the Communists, and Stalin was now ready to oblige them. By the time of the Seventh Congress of the Communist International in July 1935, the "united front of the working class" had become a "people's united front" against fascism that might include, in addition to working-class parties and movements, "progressive" segments of the bourgeoisie. The overriding purpose was to mobilize the "democratic capitalist" nations against Hitler and on the side of the Soviet Union; in pursuit of such cooperation against a common enemy, Communists were compelled to play down or conceal the fact that their ultimate objective was the overthrow of capitalism itself and not merely the defeat of its ultra-right or fascist manifestation.[41]

In the United States, the effort to establish a popular front led Communists to revise their views of Roosevelt and the New Deal. Initially denounced as an American variety of fascism, the New Deal was recast by 1936 into the role of useful ally in the struggle against fascism, although such support did not bar Communists from trying to move it further to the left. In South Africa, the popular front policy sanctioned a revival of the effort to work with the pro-segregationist South African Labor party that had been abandoned in the mid-1920s. In both cases, the new line meant deemphasizing black self-determination. This de facto withdrawal of the concessions to black nationalism made in 1928 was part of the broader policy of putting national struggles against imperial domination on the back burner, if not abandoning them entirely. Since the Soviet Union was now seeking the support of colonialist nations like Britain and France, overt support for national liberation movements had to be deemphasized. Inevitably, some blacks and Asians who had joined the Communist movement primarily because of its opposition to imperialism became disillusioned and left the party.[42]

The popular front policy enabled American Communists to work closely with the kind of middle-class black reformers whom it had previously condemned as lackeys of the ruling class. In 1935, the International Labor Defense came to an agreement with the NAACP and the American Scottsboro Committee (a non-Communist black coalition) to cooperate in efforts to appeal the conviction of the Scottsboro boys. In ghettos like Harlem and the South Side of Chicago, campaigns to increase black employment in public agencies, works projects, transit systems, and retail stores serving blacks that were initiated and led by non-Communist black leaders received the full support and cooperation of the party. In Harlem, for example, Communists developed a close working relationship with the Reverend Adam Clayton Powell, Jr., who was emerging as the most prominent black politician in the country. By jettisoning talk about a black republic and endorsing the reform agenda of the northern black middle class, Communists gained respectability among the black liberals and moderates who had previously considered the party too radical and divisive.[43]

But the party failed in its attempts to draw the most prominent black civil rights organization into a working alliance. Recalling past Communist behavior and wary of being manipulated by those who did not fully share its cautious, pragmatic, and legalistic approach to black amelioration, the NAACP rejected the party's calls for a working alliance and forced it to conduct rival campaigns against lynching and Jim Crow in the southern states. Before 1939, however, it appeared that a Communist-backed coalition might win this competition and replace the association as the vanguard of the civil rights movement.

The most ambitious Communist bid to exert leadership in black politics was through its involvement in the National Negro Congress, a federation of black and interracial organizations that sought to put pressure on the New Deal to include blacks in its plans for national recovery and incorporate racial justice into its reform agenda. Instrumental in the founding of the congress were John P. Davis, an influential black lawyer and political organizer, A. Philip Randolph, head of the Brotherhood of Sleeping Car Porters and the nation's most prominent black labor leader, and Ralph

Bunche, a leading black political scientist. Of this distinguished group, it was Davis who was closest to the Communist party, although there is no evidence that he ever became a member.

The congress's first meeting in February 1936 attracted 817 delegates representing 585 organizations from 28 (mostly northern) states. The NAACP, which suspected Communist influence in the new body, refused an invitation to be officially represented, but individual NAACP officers were in attendance. One of them noted that the delegates tended to be a younger, more working-class group than normally turned up for NAACP meetings. Randolph was elected president with the support of the participating Communists, despite the fact that he had fought Communists in his union and was a supporter of the Socialist party, and Davis was chosen for the key administrative post of executive secretary. The congress quickly established local branches throughout the northern states and seemed on the verge of becoming the first successful mass organization of Negroes since the heyday of Marcus Garvey's Universal Negro Improvement Association.

Tensions quickly developed, however, and by 1938 the congress had lost the support of black churches and businessmen, who were put off by its espousal of economic radicalism. If it could not speak for a unified black community, the congress nevertheless had the apparent capacity to put together a working coalition between whites on the left wing of New Deal or the labor movement and virtually everyone in the black community who stood to the left of the NAACP—a substantial group in the late '30s given the NAACP's failure to come to terms with the desperate economic plight of blacks.[44]

The most substantial contribution that Communists made to the cause of African-American liberation was undoubtedly their success in influencing organized labor to open its ranks to blacks. As organizers for the Congress of Industrial Organizations, the new coalition of industrial unions that challenged the conservative hegemony of the American Federation of Labor, Communists insisted on a color-blind recruitment policy that resulted in an upsurge of black membership in interracial unions. Sometimes, in their desire to placate the white membership of unions they influenced or controlled, Communists made tactical concessions to race prejudice, but in general they did much to further the cause of industrial integration.[45]

But the popular front era was not a time of unqualified Communist success in gaining influence within the African-American struggle. Blacks did not rush to join the party itself; African-American membership increased but remained disproportionately small in relation both to total party membership and to the total number of politically active blacks. When the party ran for office under its own banner, even the prominent presence of blacks on the ticket did not lead many African-Americans to desert the major parties. When black leaders accepted Communist support they generally did so on their own terms; their willingness to cooperate with Communists did not mean that they were ready to accept Communist leadership or discipline. Such alliances were inherently unstable and could not survive a major shift of the party line.

In the South, the popular front policy actually weakened the party's effort to gain support among black sharecroppers and farm workers. After 1935, the emphasis of party work in the South was on making common cause with white liberals. Through its involvement in such organizations as the Southern Conference on Human Welfare, the party helped to bring New Deal liberalism and progressive reform to the most conservative part of the country. Opposition to lynching and poll taxes could command the support of an advanced guard of southern liberals, but a frontal attack on Jim Crow, to say nothing of a call for the confiscation and redistribution of land in the black belt, was a cause that was likely to be beyond the pale for even the most progressive southern whites of the 1930s. In their efforts to draw whites into a liberal-labor coalition, the party virtually abandoned its organizational activities among poor blacks and left them at the mercy of landlords and the biased bureaucrats and local officials who normally administered New Deal programs in the rural South.[46]

In South Africa, the popular front policy of the Communists coincided with the revival of black protest activity that was provoked by the prospect of a new government assault on African rights. The 1932 union of Prime Minister J. B. M. Hertzog's National party and Jan Christiaan Smuts's South African party had been premised on the understanding that Smuts and his followers would abandon their opposition to Hertzog's longstanding plan to strip Cape Africans of the limited suffrage that was "entrenched" in the constitution of 1909. (Since the repeal of entrenched clauses required a two-thirds vote of both houses of parliament, Hertzog needed more than the majority that he had won in the election of 1929.) By 1935, the Prime Minister, with the backing of the new United party, was ready to make the constitutional changes necessary to remove an anomalous remnant of nineteenth-century Cape liberalism from the South African political system. Hertzog proposed that Cape Africans be removed from the common voters' roll and that a Natives' Representative Council be established to advise the government on matters relating to the African population. (Africans throughout the Union would indirectly elect twelve members of the Council and four would be chosen by the government.) Instead of the Cape suffrage being extended throughout the Union, as Africans had once hoped, the government had opted instead for a further restriction of black voting rights.[47]

The main vehicle for African protest against Hertzog's Representation of Natives bill was a new organization—the All African Convention. The African National Congress, which might otherwise have been expected to act in defense of African rights, had never had a large following among those who were now threatened with disfranchisement. The elite Cape Africans who had qualified as voters on the common roll had generally preferred to act within their own provincial organizations, such as the Cape African Voters' Convention. The Cape Convention, under the leadership of Professor D. D. T. Jabavu, claimed equality with the ANC when it came to issuing the call for a gathering of Africans to protest Hertzog's suffrage proposals. Among those who answered the joint call of the two organizations to meet in Bloemfontein in December 1935 were leading African members of the Communist party. The first meeting of the convention was far from radical in its pronouncements; as the leadership of Jabavu might have led one to expect, "Cape liberal" moderation appeared to prevail. The convention professed its loyalty to the British Crown and endorsed traditional methods of petition and publicity to avert a further erosion of African political rights. But in 1935 Communists had no better way to relate to African politics than through participating in the convention; as an inclusive coalition of protest groups, it looked like a promising arena for popular front activities. At the founding meeting, black Communists such as J. B. Marks, Edwin Mofutsanyana, and John Gomas featured prominently in the deliberations.[48]

Besides constituting itself as a permanent body, the AAC voted to send a delegation to the government to set forth African objections to the suffrage proposals. Hertzog received the delegates and offered them a compromise that was cleverly calculated to create confusion and division among African leaders. He proposed that the Cape voters removed from the common roll be permitted to vote on a separate roll for three white members of parliament and that Africans throughout the Union be permitted to choose, through a complex indirect process, four white members of the virtually powerless upper house or Senate. This proposal for a new version of political segregation that at first glance might have appeared to offer more than the old one caused some disagreement among the AAC delegation. Some of them, then or later, gave the impression that they could live with the compromise or at least considered it worthy of further consideration. Hertzog seized upon such vacillation and exaggerated it, proclaiming that Africans had been properly consulted and that their objections to his original proposals had been met. Parliament proceeded to pass the amended proposals by an overwhelming majority.[49]

When the AAC reconvened in June 1936, its members were in no mood to endorse the compromise, and much anger was expressed at the new system of representation; but it was now a fait accompli, and the main task was to determine whether the convention would work within the new structure or boycott it. Jabavu's presidential address expressed the prevailing mood of outrage and militancy by making a connection between suffrage restriction in South Africa and the Italian invasion of Ethiopia. Abandoning the Cape liberalism that he had long defended and that had set the tone for the first meeting of the AAC, Jabavu now implied that liberation would be achieved as part of a Pan-African struggle against white rule rather than through the triumph of fair play and "equal rights for every civilized man" within a British imperial context. One of the resolutions passed by the assembly called for a Pan-African conference which would include "overseas people of African descent." More insistently than ever before, a broadly representative body of Africans showed that it was ready to conceive of its liberation in Pan-African terms. But when it came to action, as opposed to rhetoric, pragmatism and moderation carried the day. Despite the objections of a group of independent radicals, mostly from Cape Town, the convention voted to participate in the new system of representation rather than boycott it. This accommodation to the political segregation that it rejected in principle won the support of both Communists and ANC representatives.[50]

For a time Communists viewed the All Africa Convention as the most promising locus for popular front activities among Africans and chose to concentrate their energies there, rather than found a new organization that they could have more readily controlled. Adapting to the realities of South African segregation, the party carried on separate front activities among Europeans, establishing an all-white People's Front in 1936. After endorsing the AAC's decision to take part in elections for the Natives' Representatives Council and the parliamentary seats set aside for whites elected to represent Africans, Communists participated eagerly in public campaigns for these powerless offices. During the late '30s and early '40s, black Communists repeatedly ran for the council and white Communists for the parliamentary seats. None was elected during this period, but the campaigns served to keep the party in the public eye and gave it a forum from which to seek influence among Africans. This emphasis on direct involvement in the only electoral politics available to Africans caused the Communist party to abandon efforts to make the AAC its main organizational ally. By 1938 it was clear that the convention was incapable of establishing the strong local branches that were a necessary base for functioning successfully within the new system of representation and that the African National Congress had stronger local roots and organizational capacities. As a result, black Communists shifted most of their attention and energy away from the AAC and began to work intensively within ANC branches. The party by this time required its black recruits to join the ANC or, if they were deemed not yet ready to become full-fledged Communists, to join the congress as a first step in their political education. Partly as a result of the effective role of Communists as recruiters, organizers, and political coordinators, the congress itself began to revive from the doldrums of the mid-1930s. While the Communists were gaining influence in the ANC, the AAC was becoming a forum for those Africans and Coloreds who favored a total boycott of the new representative system. By the beginning of the 1940s the black left was split into those who favored working within the system in order, ultimately, to undermine it—Communists who were also ANC supporters—and those who refused to compromise with segregation—Trotskyists and independent radicals who viewed Communists as de facto collaborators with white supremacy.[51]

The developing Communist alliance with the ANC represented more than opportunistic collaboration with anyone who could be considered anti-fascist. It could also be justified by the Leninist doctrine that Communists in colonial societies should work for a revolutionary front against imperialism. In 1928 the Comintern's special resolution on South Africa had proclaimed that "our aim should be to transform the African National Congress into a fighting nationalist revolutionary organization against the white bourgeoisie and the British imperialists. . . ." The sectarianism of the left turn hindered fulfillment of this aim, and the ANC's recapture by conservative elements in 1930 (when Gumede was ousted from the president-generalship) had made it seem like a pipe dream. But the interpretation of the "native republic" slogan that was officially promulgated by the Comintern in 1930 pointed unmistakably to collaboration with any nationalist movement that seemed potentially revolutionary. The two-stage model for a South African revolution made the triumph of a bourgeois African republic the precondition for a Communist victory. One African Communist who strongly endorsed the two-stage model when it was first promulgated and who continued to defend it even after it was out of favor was Moses Kotane. More than any other single individual, Kotane transformed the theoretical connection between "bourgeois" African nationalism and Marxist-Leninism into a practical reality. He was primarily responsible for cementing an enduring alliance between the Communist party and the African National Congress that would survive all future changes in international Communist policy.[52]

Kotane came from a Christian African family of Tswana origin that farmed its own land in the Western Transvaal. As a young worker in Johannesburg in the 1920s, he enrolled in a Communist night school and soon mastered the rudiments of Marxist theory. After joining the ANC in 1928 and finding it insufficiently active and militant, he was accepted into the party the following year and was almost immediately elected to its political bureau. Regarded as the most promising of the party's African recruits, he became a full-time party official and union organizer and was sent to Moscow for a year of special training. But in 1934 Kotane quarreled with the white leadership of the party over the meaning of the "native republic" slogan. When party general secretary Lazar Bach concluded that the nationalist and proletarian revolutions would occur simultaneously, Kotane openly disagreed. As he wrote at the time: "In Europe class consciousness has developed immensely, whilst here national oppression and discrimination confuse the class war and the majority of the African working population are more national conscious than class conscious. . . ." Accordingly, the cause of "an independent African republic" could not be advanced by the same kind of propaganda used to justify "the proletarian dictatorship." Nationalism, in other words, had to be given priority over revolutionary proletarianism in Communist appeals to Africans.

Bach's viewpoint prevailed in the South African party, and Kotane was dismissed from the Political Bureau. But he was soon vindicated: Bach was called to Moscow (whence he never returned), and the word came down that Kotane's position was the correct one in light of the new popular front policy. As Kotane's biographer Bryan Bunting has pointed out, Kotane always had a preference for popular fronts, because he sensed that the African "bourgeoisie" was not really a bourgeoisie at all but a part of the oppressed masses. As victims of national and racial oppression, the black educated elite had every reason to fight for a black republic, and Communists should support them to the hilt.[53]

In the late '30s, Kotane became very active in the ANC. As chairman of the organization in the Western Cape, he was successful in turning one of the weakest congress branches into one of the strongest. In 1939, he was chosen to be general secretary of the Communist party, replacing another African, Edwin Mofutsanyana (a loyal functionary who lacked Kotane's charisma and close connections to the nationalist movement). Despite his conspicuous role as the leader of the Communist movement in South Africa, he continued to be active and influential in the ANC and cooperated closely and amicably with its moderate, non-Communist leaders throughout the 1940s. Kotane was an effective link between mainstream "bourgeois nationalism" and Communism, because he himself was as much a nationalist as a Marxist-Leninist, if not more so. Had he merely been using nationalism in a manipulative fashion he could not have played this mediating role so effectively. "It should not be forgotten," he said in a speech of 1939, "that all people have national pride within them and that Communists as we might be, we still feel and suffer the position of the races of our origin or are proud of the achievements of our races. I am first a Native and then a Communist. I came to the Communist Party because I saw in it the way out and the salvation for the Native people."[54]

By 1939, therefore, an African with strong nationalist sentiments and an undeniable dedication to the cause of national liberation as an end in itself led the Communist party of South Africa. Indeed it can almost be said that the Communist party was prepared to subordinate its own ideology and ambitions to the cause of a non-Communist African democratic republic in ways that went beyond what Lenin would probably have found acceptable. Kotane came close to saying to African nationalists that he and his party would support them to the fullest without trying to impose their own ideology on the nationalists' movement—which would have been directly contrary to Stalin's injunction that Communists in anti-imperialist fronts should always seek to dominate. Whether one chooses to interpret Communist motives in such a benign fashion or not, it was clear in 1939 that the party was finding a secure and respected place in mainstream African protest politics.

The party was also having some success in organizing interracial and black trade unions. The one area where it was making little headway, and even regressing, was in its effort to confront the special problems of rural blacks. In 1930, two African Communists—Elliot Tonjeni and Bransby Niobe—had made some progress in organizing farm workers in the Western Cape before being banned from the province. Their work, which has not been studied closely, might very well have important similarities to the contemporaneous Communist work among Alabama sharecroppers in the United States. But there was little or no effort to follow up these initiatives. During the popular front period of the late '30s, the Communists directed almost all of their attention to the problems of urban working-class or petit bourgeois Africans. Yet peasants and farmworkers still made up a majority of the total African population of the Union. As in the United States, work in rural areas was difficult, dangerous, and likely to provoke violence. Consequently it did not fit well with the popular front emphasis on establishing alliances with "progressive" whites or with its generally moderate and cautious approach to black protest. From the perspective of some of the most militant black liberationists, the trouble with the Communist party of both countries in the late '30s was that it was not radical enough.[55]

Communists in the United States and South Africa had by 1939 established roughly similar relationships to the broader black struggles against white supremacy. Their most obvious contribution was to make the ideal of a racially integrated or nonracial society seem plausible and attractive to a large number of blacks. No other group in South Africa both preached and practiced racial egalitarianism. The liberal interracial movement there, as in the American South, remained paternalistic and hierarchical in its basic assumptions. Benevolent whites generally did things *for* blacks rather than make common cause with them. In the United States, the NAACP rivaled the party in its insistence on an integrationist ideal and in fact condemned the Communists' "black self-determination" slogan as a form of racial separatism. But, especially after the party jettisoned this slogan in the mid-'30s, Communists and the organizations they influenced or controlled provided a better model for interracial fraternity than the association. By the late '30s, the NAACP had evolved into a virtually all-black organization and functioned more as a black pressure group and ethnic defense organization than the interracial neo-abolitionist movement that some of its founders had envisioned. Only the Communist party and its fronts gave the appearance of blacks and whites shoulder to shoulder in a common cause. The responses of blacks who visited the Soviet Union or heard reports of race relations there graphically demonstrate how attractive the interracial ideal was to many blacks in both societies during the 1930s and how Communists were able to capitalize on it. Observing or hearing that the Soviets treated blacks as equals and mixed freely with them was often the main consideration that led blacks to join the party or at least develop strong sympathies with the Soviet Union and the Communist movement. The later charge of white supremacists in the two societies that Communists were "behind" the idea of racial integration had a grain of truth in it. In the 1930s, at least, no one had done more than the Communists to convince blacks that a nonracial society was possible.[56]

In neither case, however, did Communist interracialism lead to a massive influx of blacks into the party or into closely affiliated organizations, even during the popular front era when the stringent requirements for membership or collaboration that had existed during the "left turn" were relaxed. One reason why sympathy for Communist racial ideology and gratitude for Communist actions did not translate more frequently into active affiliation was the gap between the world-view of ordinary Africans or African-Americans and that of the party. Communists stopped attacking black Christianity after 1934, but their atheism and secularism remained unmistakable and continued to be a barrier to whole-hearted involvement of large numbers of blacks in their movement. Historian Robin Kelley has argued persuasively that some black sharecroppers in Alabama were able to merge their religious culture and their social radicalism as they struggled under Communist leadership for a better life on the land. But the autobiography of the Alabama black Communist Hosea Hudson suggests that durable commitment to the Communist movement required a kind of conversion experience and the acceptance of a new gospel. Black Communists like Hudson might carry some of their traditional religious culture into the new church, as did slaves converted to Christianity on the plantations of the American South and Africans baptized by European missionaries. But becoming a committed and active Communist, like becoming a Christian earlier, was often a painful and disruptive process; it required the acceptance of new gods and a new way of life, which normally meant a weakening of communal ties and relationships. Only a minority were prepared to substitute Marx and Lenin for Christ or the dictatorship of the proletariat for the millennium.[57]

Beyond the ideological barriers were some practical obstacles making blacks think twice about joining the "reds." For most Africans or African-Americans joining the Communist movement or revealing sympathy for it was dangerous or at least disadvantageous. In South Africa and the American South, black Communists during the 1930s were in danger of losing their lives to white vigilantes or law enforcers—or at best, as the nightmarish experiences of Angelo Herndon revealed, spending much time in hellish jails. Even in the freer atmosphere of the northern states an open affiliation with the extreme left might still mean the loss of a job or a brutal beating at the hands of the police. As was often pointed out at the time, it was hard enough to be black without being red as well.[58]

The vulnerability of African-Americans was one reason W. E. B. Du Bois held himself aloof from the Communist party during this period and advised other blacks to do the same, despite his personal attraction to Marxism and admiration for the Soviet Union. Du Bois believed that a proletarian revolution was unlikely in the United States and that even if one should occur blacks would not necessarily benefit. Modifying significantly the orthodox Marxist interpretation of history as a struggle of economic classes, Du Bois believed that white workers were so strongly prejudiced against blacks that they would fail to treat them as equals even in the context of a conflict against white capitalists. It would therefore be folly for blacks to tie their fortunes to the Communists. His own proposal for a separate black cooperative economy, which was a principal reason for his break with the more consistently integrationist NAACP in 1934, resulted from his belief that anti-capitalist blacks had to go it alone and try to make a life for themselves in a hostile racial and economic environment. In his view the Communist faith in the ability of black and white workers to "unite and fight" was based on a serious underestimation of the strength and autonomy of white racism. Other prominent African-American intellectuals who were influenced by Marxism, such as the political scientist Ralph Bunche and the economist Abram Harris, were more hopeful than Du Bois that black and white workers could find common ground and learn to cooperate in their mutual self-interest, but they shared Du Bois's belief that a social revolution was extremely unlikely in the United States. Consequently they remained independent of the party and advocated a reformist social democracy.[59]

In South Africa, there was no prominent black intellectual like Du Bois to raise troubling questions about the fit between Communist aims and the ethnic or national interests of blacks. Even if there had been, he would probably not have given so much weight to the racism of white workers. In South Africa, unlike the United States, the revolution of a black proletariat without any support from whites was conceivable. Inhibiting the educated black elite from embracing Communism were commitments to the Christianity learned in mission schools and lingering hopes of a peaceful transition to a more democratic society. It was easy, however, for ANC liberals to work quite closely with black Communists such as Kotane, who appear to have made few ideological demands on the more moderate members of the congress.[60]

Parallels between the two narratives of Communist involvement in black politics are striking, but the more deeply we probe the more we become aware of variations and contrasts. In 1939 the American party was a predominantly white organization with a handful of African-Americans among the leadership but more for show than as an independent source of power within the organization. Harry Haywood, whose commitment to a synthesis of Marxism and black nationalism led him to defend the self-determination policy after it had been quietly jettisoned during the popular front era, was ousted from the leadership, although he remained a party member. The Communist party of South Africa was overwhelmingly black in membership and had an African nationalist in its top position. The relatively small number of whites in the party were overrepresented in the leadership, much as blacks were in the United States. But the apparent symmetry of the contrast breaks down in the face of evidence that the whites in the South African party had an influence over policy far beyond what their numbers would justify. What was evolving in fact was a symbiotic relationship between a small, relatively privileged white party and a larger black one. Since many of the whites were wealthy professionals and most of the black Communists were self-educated workers, the whites inevitably served as patrons of the black party, providing legal assistance, meeting places, and most of the funds that the party needed to operate. Whether this patron-client relationship led to a de facto white domination of the party and skewed its ideology and actions to reflect some kind of European ethnic interest would became a matter of intense debate within the black liberation movement in the 1940s and '50s. But there can be no doubt that the blacks in leadership positions, such as Kotane and Edwin Mofutsayana, had a closer and less ambiguous relationship to the larger black movement than their African-American counterparts.[61]

The South African party had another crucial advantage over the American when it came to attracting black support: its claim to be the cutting edge of racial egalitarianism for South Africa was more persuasive than the CPUSA's pretensions to leadership in the American struggle against racism. In South Africa a liberal, as opposed to a Marxist, nonracialism or anti-racism scarcely existed. The Joint Council movement, which continued during the '30s to bring white liberals and moderate Africans together to discuss racial amelioration, took segregation and white domination for granted and merely strove to make the system more humane. Like the Commission on Interracial Cooperation in the American South, it was inescapably paternalistic in its ideology and operations.[62]

In the United States, as a whole, however, the 1930s saw the strengthening of the liberal racial egalitarianism that had inspired the founding of the NAACP a quarter-century earlier. The association itself continued to nibble away at the legal basis of segregation in a series of successful Supreme Court challenges to the blatant forms of inequality, such as unequal treatment by states in higher education and all-white primary elections. In 1937–38, it mounted a campaign in favor of federal anti-lynching legislation that, while ultimately unsuccessful, drew the attention of the country. Within the New Deal interracial liberals, black and white, came to wield significant influence, although they were never able to induce the Roosevelt administration—concerned as it was over the southern white constituency of the Democratic party and the loyalty to the New Deal economic policies of its representatives in Congress—to come out for civil rights laws.

The first lady, Eleanor Roosevelt, emerged as a conspicuous national voice calling for an end to racial discrimination and segregation. Although she was unable to persuade her husband to alter his political priorities, she did give weight and legitimacy to the growing sentiment among moderate to liberal whites in the northern states that racial segregation was wrong and must ultimately be ended. At the same time mainstream liberal social scientists were coming to a consensus that human differences in intelligence, temperament, and behavior that had previously been ascribed to racial genes were actually the product of environment. The scientific racism that helped justify Jim Crow policies was beginning to lose intellectual credibility by the beginning of the Second World War. As the Communists lost the virtual monopoly on the integrationist, anti-racist position that they had enjoyed in the late '20s and early '30s, their position in the forefront of the black struggle for equality became precarious. To the extent that they were willing to cooperate with liberals, and liberals with them, Communists could continue to play a role in the politics of black liberation from Jim Crow. But should a deep rift develop between liberal and Communist goals or priorities, the latter were in danger of being pushed to the margins of the struggle for equality.[63]

The Second World War and the Parting of the Ways

The Stalin-Hitler pact of August 1939 brought an abrupt end to the Popular Front and again put Communists at odds with liberals and social democrats. Now that the Soviet Union was no longer immediately threatened by Nazi Germany, the Comintern ordered national parties not merely to abandon the alliances they had formed to resist the spread of fascism but to denounce the military mobilization of the Western allies against Germany aggression as preparations for "an imperialist war." In the United States this meant opposition to the defense build-up of 1940–41 and aid to the allies—also the position of right-wing isolationists and German sympathizers. In South Africa it entailed the more extreme step of adopting a pacifist stance in a nation that was actually at war. Britain's declaration of war on Germany in September 1939 caused a split in the ruling United party; the dominant faction led by Smuts managed to persuade a parliamentary majority to endorse a war in defense of the British Empire, despite strenuous opposition from Hertzog's Afrikaner nationalist wing of the party and from the emerging Purified National party (which had opposed Hertzog and the United party in the name of an uncompromising Afrikaner republicanism). Thus American and South African Communists were asked not simply to condone Nazi aggression in Western Europe but also to put themselves on the same side as the most reactionary elements in their own societies.[64]

The American party responded to the Comintern's about-face with alacrity and moved resolutely to commit the mass organizations that it influenced to the new peace position. Among these were the National Negro Congress and the Southern Conference for Human Welfare. To the extent that Communists were successful in maneuvering these organizations into line behind the new Soviet policy, pro-preparedness liberals and New Dealers angrily withdrew and accused the party of using domestic reform movements as a front for their cynical machinations on behalf of a foreign power. To dedicated Communists, of course, unconditional support of the Soviet Union and its national interests could be justified as a way of protecting the wellspring of world revolution. The result of the shift was a narrowing of the base of Communist support and a decline of its influence within the African-American community, as in other non-Communist constituencies that had been wooed during the Popular Front era. But party endorsement of the Stalin-Hitler pact was not as important to the party's loss of black support as it is sometimes represented as being. Front activities involving black organizations and movements had begun to unravel or at least show signs of stress before the pact was announced. By the late '30s non-Communist community leaders in northern ghettos were learning how to steal the party's thunder by organizing their own protest demonstrations, and some of them began to wonder if Communist involvement was really essential.[65]

The principal focus of Communist interest, the National Negro Congress, was already foundering by 1937–38, mainly because it had failed to hold the black churches and middle-class organizations within its orbit. The increasingly visible role of the Communists and the CIO in the congress—which also meant more prominence for white radicals—was not so much the result of a left-wing conspiracy to rule or ruin as a simple matter of necessity given the organization's shrinking base of support among blacks. The economic and trade unionist emphasis of the congress was not peculiarly Communist; it had also been pushed by social democrats such as Bunche and Randolph. But even at the height of the depression most blacks probably did not view their problems primarily in economic or class terms. Popular boycotts based on the slogan "Don't Buy Where You Can't Work" reflected national rather than class consciousness and were difficult for Communists and socialists to support. The inability of congress leaders such as John W. Davis to appeal to the black sense of nationality or peoplehood meant that the organization never really had a chance to fulfill its ambition to become the voice of the black community. As one faction after another dropped out because it found that Congress did not adequately address its needs or embody its attitudes, the congress came to rely increasingly on the CIO and white radicals, especially those in the Communist party, for funding and organizational work.

The congress was already a shadow of its original self when the Communists and CIO delegates at the 1940 meeting pushed through their peace resolution. (The role of the CIO was not simply the result of Communist influence; John L. Lewis, the mineworkers' chief who headed the CIO, was an isolationist). But for A. Philip Randolph, the democratic socialist president of the congress, the passage of the peace resolution was the last straw. Randolph had refrained from making a prowar resolution to reflect his personal view that the allied cause against Hitler should be actively supported, for he believed that the organization should concern itself exclusively with the group interests of African-Americans and that divisive foreign policy issues and hidden agendas had no part in such a movement. His resignation was a major turning point in the history of the black protest movement, because it turned one of the most effective and charismatic black protest leaders into an uncompromising anti-Communist.[66]

If the pact delivered the coup de grace to the party's efforts to exert leadership in the black community, it did not have much effect on black party membership. Most of the black members weathered the storm, and there were no defections by prominent leaders. Furthermore, the Communist-influenced left-wing labor unions that welcomed blacks into their ranks actually flourished during the pact period. One Communist-backed organization, the South Negro Youth Congress, also did relatively well between 1939 and 1941. The end of the Popular Front also meant that the Communists could once again talk about black self-determination without fear of losing the white liberal support that was now gone in any case. The SNYC did not revive "self-determination for the black belt," but it did manage to inject a strain of black cultural nationalism into its basic appeal for revolutionary solidarity with white workers.[67]

Nevertheless, the party's stance during the pact period did result in a serious setback for its ambitions to lead the black struggle for equality. In 1940 and 1941, Communists were confronted with the emergence of a militant new movement that deliberately excluded them and ended up stealing much of their thunder. After resigning as president of the National Negro Congress in 1940, A. Philip Randolph took up the cause of fighting discrimination against blacks in the defense industry and armed forces of a rearming nation. Cooperating with the NAACP and the National Urban League, he conceived the idea of having thousands of African-Americans "march on Washington" to demand equal employment in industries with government contracts and integration of the armed forces. The resulting March on Washington Movement was difficult for Communists to support, because it was premised on acceptance of war preparations that they strongly opposed and was virtually impossible for them to join, because Randolph effectively excluded them by limiting membership to blacks. (There were of course black Communists, but they did not generally act independently of whites or participate in racially exclusive organizations.) Randolph was far from being a thoroughgoing black nationalist; he conceived of separate action as necessary to build black self-confidence and insure that African-Americans would lead their own movements, but the goal of his protests was a thoroughly integrated society.[68]

Communists at first greeted the March on Washington Movement with silence; they subsequently vacillated between antagonism and qualified support, attempting to make a distinction between marching for jobs—something they could support—and the leadership of the movement, which in their eyes was guilty of "war mongering" and "black chauvinism." With the Communists carping from the sidelines, Randolph began to organize the march and upped the number of blacks that he hoped to have in the streets of Washington from 10,000 to 100,000. Although it is not all clear in retrospect that Randolph could actually have mustered these numbers, President Roosevelt became alarmed at the possibility of a race riot in the highly segregated nation's capital and agreed to meet with the leaders of the movement. When the president agreed to issue an executive order setting up a commission to encourage "fair employment practices" in defense industries, Randolph declared victory and called off the march, despite the fact that he had failed to achieve his second goal of integrating the armed forces. The March on Washington Movement of 1941 was a landmark in the history of black protest, because it was the first successful effort to use mass mobilization in the streets, or at least the threat of it, to change government race policy. It foreshadowed the growth of a militant, nonviolent civil rights movement from which Communists would be, for the most part, excluded.[69]

The changes in the party line resulting from the Stalin-Hitler pact were much less damaging to the South African Communist party. Although it alienated some white anti-fascists, the shift to an antiwar position actually helped the party's efforts to extend its appeal among the black population. Attempts in the late '30s to establish a popular front involving white liberals and laborites had not been very successful, and it cost the party little to abandon them. When South Africa declared war on Nazi Germany, the party was thrown into momentary confusion because opposing the war seemed to put them on the side of their own homegrown "fascists"—the Afrikaner nationalists. It was not until after the British Communists had come to the painful decision to sacrifice themselves on the altar of Stalinism by opposing a war for the defense of their homeland that the South African party followed suit, trying in the process to make it clear that withdrawing from the international anti-fascist coalition did not absolve them from fighting against the extreme right within South Africa. Although the decision cost the party some white members, it actually helped them in their efforts to recruit blacks. The return to a strong anticolonialist line appealed to some African nationalists who had been disenchanted with the Popular Front's downplaying of anti-imperialism.[70]

In addition, the peace policy enabled the party to extend its influence among South Africa's Indian minority. Following the lead of the nationalist movement in India itself, many South African Indians were opposed to going to war in defense of the British empire. Indian Communists such as Yusuf Dadoo and H. A. Naidoo capitalized on these sentiments and provided leadership for antiwar activity, not merely within their own ethnic group but among the black population as a whole. Dadoo was the leader of the Communist-backed Non-European United Front, an organization formed in 1938 as a popular front for Indians, Coloreds, and Africans that subsequently served as a rallying point for opposition to the "imperialist war." From the aroused Indian community of the pact years, the party recruited a number of capable leaders to add to the Africans and whites already in positions of prominence, thereby adding a new dimension to the Communists' reputation for nonracialism and giving Indians a permanent role in the struggle against white supremacy.[71]

South African Communists also maintained good relations during the pact period with African organizations that they did not actually control, even when these organizations came out in favor of the war. In contrast to the rigidity of their American counterparts, who in effect ruined the

National Negro Congress in order to put it on record against the war, South African Communists continued to work cheerfully within the African National Congress despite differences on the war issue. The ANC, somewhat in the spirit of E. Philip Randolph and the March on Washington, endorsed the struggle of the Western allies against fascism on the condition that blacks be allowed to contribute to the war effort on an equal basis with Europeans. Black Communists within the ANC made little protest and do not appear to have made any strenuous efforts to influence the congress's attitude toward the war. African Communist leaders such as Moses Kotane and J. B. Marks seem to have been equally loyal to the party and the ANC and were unwilling to divide and weaken the latter by seeking to make it agree on all matters with the former.[72]

Hitler's invasion of the Soviet Union in June 1941 forced Communists to change abruptly from strongly opposing the war to fervently supporting it. Their unconditional backing for a war to defend the Soviet Union encouraged them to moderate their efforts on behalf of workers and blacks. Communists in both the United States and South Africa adopted a "no strike" policy for the duration and thus forfeited their position in the forefront of labor activism. The American party also backed off somewhat from its customary demand that blacks immediately be granted equal voting rights and liberated from Jim Crow laws and practices. In the October 1941 issue of *The Communist*, black party leader James Ford did not call for a wartime suspension of agitation for equal rights, but he did warn that "it would be equally wrong to press these demands without regard to the main task of the destruction of Hitler, without which no serious struggle for Negro rights is possible." This reluctance to rock the boat on race issues in ways that might impede the war effort meant that Communists began to be viewed in the black community as less militant on civil rights than the March on Washington Movement, which continued to threaten a mass demonstration in the capital to protest the persistence of discrimination, or even the NAACP, with its "double V" policy of seeking simultaneous victories over fascism abroad and racism at home.[73]

The party's opposition to black militancy was manifested in its savage attacks on A. Philip Randolph and the March on Washington Movement. Before the German invasion of the Soviet Union, Randolph and his movement had been condemned for supporting American militarism; now they were condemned for impeding the war effort by encouraging disruptive protests. The historian Maurice Isserman has recently argued that the conventional view that Communists abandoned the struggle for black liberation during the war is an exaggeration and that by 1944 the party's prestige within the black community had begun to revive; but he does concede that the party did limit "its struggle to those areas that it believed benefited the war effort," such as pushing for the more effective utilization of black personnel in war-related activities. Few blacks left the party during the war and toward the end black membership began to increase, but its modest gains pale alongside the nine-fold increase in NAACP membership between 1941 and 1945. Wartime moderation may not have made party work among blacks totally futile, but it had deepened the gulf between the party and the mainstream of the black protest movement that had opened during the period of the Stalin-Hitler pact. Many African-American activists came out of the war agreeing with Randolph's verdict on the party: "The history and record of this cult shows that it conforms with rigid fidelity to the rapidly changing, unpredictable climate of Soviet Russia, without regard to the national interests of any other group. When the war broke, the Communists who had posed as the savior of the Negro dropped him like a hot potato."[74]

The South African party did somewhat better in gaining and holding the respect of blacks during the war, because it did not allow the moderation of the national party to prevent vigorous involvement in local protests. Like the CPUSA, the CPSA eschewed strikes and mass protests for the duration and focused on issuing calls for the equal inclusion of blacks in the military and war industries, arguing that discrimination hindered the effort to defeat the Axis powers. African Communists such as Kotane and Marks worked closely with Alfred B. Xuma, the moderate president of the ANC, to build up the African Nationalist movement and prepare it for more forceful actions in the postwar years. In 1944, when the war was virtually won, the ANC and the Communist party embarked on a joint campaign against the pass laws, which mainly took the form of circulating petitions condemning the pass system and presenting them to the government. Sometimes described as a reawakening of radical protest, the campaign was actually a relatively tame affair. Unlike the ANC anti-pass campaign of 1919 and the Communist-led pass burnings of 1930, the 1944 protest did not involve civil disobedience and seemed to be based on the naïve assumption that the government might be persuaded to repeal the pass laws merely because it received evidence that Africans found them oppressive.[75]

Despite its lack of a militant stance in national affairs, the party increased its black membership during the war and gained added legitimacy in the eyes of Africans outside the party. In part, these gains resulted from the simple fact that the government eased up on the persecution or repression of Communists as a consequence of the alliance with the Soviet Union. During this brief and special moment in South African history Communism became almost respectable. The Friends of the Soviet Union, an organization founded by white Communists at the beginning of the war, had by its end enlisted the mayor of Johannesburg and the minister of justice as official patrons. In 1944, one white Communist in Johannesburg and two in Cape Town were elected to city councils. But the Communist gains in black membership and support, while clearly facilitated by the party's new aura of legitimacy and respectability, were due principally to the role that party members came to play in the internal politics of African townships. This local political activity revolved around the elected advisory boards that were analogous to the Natives' Representative Council on the national level. The Communist decision before the war not to boycott the shadow representative institutions that the government had set up as an alternative to black participation in the actual governing of the country now began to bear fruit. During the war African urban townships grew very rapidly as hundreds of thousands moved from rural areas to the cities in response to the wartime demand for labor. Since the authorities generally failed to provide the infrastructure to meet the most basic needs of this growing population, desperate overcrowding, serious food shortages, lack of water, and other severe economic hardships resulted. Communists elected to township advisory boards or serving as officers of residents' associations often took the lead in protesting these conditions to the authorities and demanding their alleviation. By putting their skills as organizers and agitators at the service of communities striving to satisfy their elemental material needs, Communists earned the respect and gratitude of a large number of urban Africans. According to historian Tom Lodge, "the early forties were the years in which the Communists successfully established themselves as a popular force in local urban politics."[76]

The American Communist party failed to find an equivalent sphere of activity during the war years; in the United States there was no set of grass-roots institutions and organizations to which Communists had access that could serve as a springboard to influence in local black communities. The Fifteenth Amendment to the Constitution made it impossible to institute the kind of separate-and-unequal political segregation that was one of South Africa's special contributions to the history of white supremacy. Although most blacks were effectively disfranchised by theoretically color-blind electoral restrictions, the legal fiction that they possessed the right to vote on the same basis as whites had to be maintained. Hence southern African-Americans had no officially recognized elected councils that were authorized to petition white authorities for the redress of grievances. In politics, if not in public facilities, exclusion rather than legalized separation was the rule. Virtually the only institutions that provided a forum for the discussion of community problems were the churches, and normally Communists were not welcome there.

Electoral politics in the urban North did provide an opportunity for Communists to put themselves and their positions before the public; in New York the party and its allies in the CIO took control of the locally competitive American Labor Party in 1944, thereby gaining a firm foothold in the city's politics. But the New York situation was unique; in cities such as Chicago the party had little visibility or impact among blacks because African-American political leadership was under the thumb of Democratic political machines. Where the American two-party system provided the structure for black electoral action, it was more difficult for Communists to exert influence than in the completely segregated and detached politics of the African townships. Ironically, political segregation provided opportunities that an apparently integrated politics could not offer.[77]

Because of its greater effectiveness during the war, the postwar CPSA had a closer connection to the mainstream of black protest than the CPUSA. In 1946, the Communists revealed the strength within the African labor movement that wartime relaxation of restrictions on black unions had allowed them to develop by providing the leadership for a massive national miners' strike. (African Communist and ANC leader J. B. Marks was chairman of the Mineworkers Union.) In the end, the Smuts government crushed the strike, and the movement for the rights of African workers received a blow from which it would have difficulty recovering, but a Communist role in whatever kind of struggle could be mounted seemed assured. The principal threat to its playing a major part came from the group of younger Africans in the ANC, who had founded the Congress Youth League in 1944. The militant Youth Leaguers were at this time promoting a form of African nationalism that seemed to rule out cooperation with the Communists. As these advocates of black self-determination gained influence in the congress, Communists faced some danger of being outflanked and relegated to the sidelines. But this challenge did not have to be faced directly until the Youth Leaguers came to power in the ANC in 1949.[78]

The American party staggered into the postwar era with a number of handicaps, some self-inflicted and others arising from the inherent difficulties of its situation in the United States. Unconditionally supporting a war for the survival of the Soviet Union had dictated a position on the struggle for black rights that was less militant and confrontational than that of the March on Washington Movement or the NAACP. After the war the party turned sharply to the left, once again in response to dictates from Moscow, and in the process condemned the deposed leadership of Earl Browder for lapses of revolutionary fervor that included the failure to press the cause of black equality. "Self-determination for the black belt" was revived as a party slogan, and Harry Haywood was restored to prominence as the party's principal spokesman on "Negro liberation." Communists and the organizations they influenced, especially the Civil Rights Congress, were conspicuous in their condemnations of Jim Crow, disfranchisement, and lynching, and the party fortunes in the black community revived to some extent between 1945 and 1948. But the anti-Communist NAACP grew even more rapidly in membership and influence, and the party was unable to regain the position of leadership in the black struggle that it had achieved in the late '30s. The failure of all but a tiny minority of African-Americans to vote for the Communist-supported Progressive party in the election of 1948 showed that the party carried little weight in the black community.[79]

Both Communist parties suffered after 1948 from massive repression as a result of the hysterical anti-Communism of the early Cold War era. But the differing positions from which they entered this dark night of persecution meant that the South African Communists were able to weather it without losing their major role in the politics of black liberation, while the American party was being virtually annihilated as a source of radical thought and action in the United States. The contrast of subsequent Communist roles in black freedom struggles—increasingly important in South Africa and decreasingly so in the United States—cannot be attributed to any greater state tolerance of party activities in South Africa; for the new Nationalist regime forthrightly banned and criminalized the party in 1950, at about the same time that a small number of Communist leaders in the United States were being prosecuted under the Smith Act for conspiring to overthrow the government. The credibility of the parties before proscription helped to determine how they weathered it.

Differences in the quality and astuteness of Communist leadership may help to account for the greater ability of the South Africans to turn the war to their advantage. The CPUSA had no leaders with the kind of resonance among blacks and the ability to cultivate linkages between the party and the black community that Moses Kotane and J. B. Marks demonstrated. The closest equivalent was Paul Robeson, the brilliant actor, singer, philosopher of Third World liberation, and champion of the Soviet Union. But Robeson was not a party member. As a prestigious exemplar of black achievement and assertiveness, he provided influential support for some of the party's positions; but, unlike Kotane, he lacked a firm base in organized black protest politics. The most prominent blacks within the party tended to be uncharismatic functionaries with a limited ability to galvanize their compatriots and forge alliances with community leaders.[80]

Even if American Communists had been more adept at showing blacks that they had a genuine and steadfast commitment to their cause, they would probably still have failed to match the success of their South African counterparts. The most significant factor accounting for the contrast of fortunes was that the American Communists faced strong competition from an interracial liberalism that appeared to be the wave of the future, whereas the South African party did not. The partial success of the March on Washington Movement in 1941 and subsequent Supreme Court decisions outlawing white primaries, segregation in higher education, and restrictive real estate covenants made the period between 1941 and 1948 a time when hopes that equal rights for blacks could be achieved within the context of democratic capitalism steadily increased. During the war, the Communist party itself had briefly endorsed the reformist doctrine that racial discrimination could be eliminated without the overthrow of capitalism. But this heresy was repudiated in the left turn of 1945, and Communists reverted to their normal view that no true progress toward equality was possible without a proletarian revolution. When President Harry Truman and the Democratic party endorsed a number of civil rights reforms in 1947 and 1948, the gradualism and liberalism of the NAACP seemed vindicated, and the Communist claim that capitalism and racial equality could not coexist lost credibility.[81]

Black South Africans had much less reason to put their trust in liberal reformism. As earlier chapters have shown, what passed for liberalism in South Africa in the period before the 1940s was the product of British influences and projected an idealistic view of the British empire as a place where all "civilized" subjects of the crown had equal rights regardless of color. A tattered remnant of this tradition was the Cape African suffrage which Hertzog abolished in 1936. By World War II, the notion that Africans could be relieved from racial oppression by some action of the King, the parliament, or the British Empire was a dead letter. South Africa was de facto an independent republic, and its ruling white minority had no inclination to extend rights to Africans, even those who met a "civilization" test. South African white "liberalism" in the 1920s and '30s was analogous to the southern white liberalism of the same period; it objected to flagrant mistreatment of black as a violation of Christian ethics and Western liberal values, but its remedy was a more equitable form of segregation rather than equal rights in a common society. Even this form of liberalism was in retreat in the postwar period. Jan Hofmeyr, Smuts's deputy prime minister and his putative successor, was its last major exponent among South Africa's white political leadership, and his behavior did not always conform to his professed principles. In the election of 1948, the Afrikaner National party made Hofmeyr's alleged liberal tendencies a main campaign issue, claiming that when he succeeded the aging Smuts as prime minister blacks would be made equal to whites. The charge was absurd, but it was effective. The nationalists won control of the government and imposed the more stringent, elaborate, and blatantly unequal form of segregation that became known as apartheid. It was blindingly obvious to most politically conscious South African blacks that they had no basis whatever to expect that their status of noncitizenship and abject legal inferiority would be overcome or even ameliorated by some form of liberal reformism.[82]

Harry Truman and the Democratic party were the unexpected victors in the 1948 American election, partly and perhaps mainly because of the solid support they received from black voters. Overwhelmingly, blacks resisted the attractions of the Progressives, a left-wing party with Communist backing but with a prominent mainstream political figure as its standard-bearer that opposed the Cold War policies and anti-Communist "loyalty" programs of the Truman administration. African-Americans supported Truman in large part because they believed his party's platform when it proclaimed its commitment to equal rights for black Americans and thought that the Democrats might actually be able to deliver on some of their promises. It apparently did not bother them that Truman and the liberals in the Democratic party were strongly anti-Communist or that the Progressives had taken an even stronger stand in favor of equal rights.

In the same year, extreme white supremacists triumphed in a South African parliamentary election in which participation was limited to whites and a relatively small number of Colored voters in the Cape Province (whom the Nationalists would proceed to disfranchise once they were in power). The simple juxtaposition of these basic facts about the elections of 1948 reveals a great deal about why Communists had a future in South Africa's black liberation movement and very tenuous prospects for a significant role in the African-American freedom struggle. In one case, an ascendant anti-Communist liberalism was promising equal rights to blacks in a democratic capitalist society. In the other nonracial liberalism was not a viable option; the majority of whites had opted for a quasi-fascist form of racial oppression, and only the Communists seemed to offer the possibility of blacks and whites fighting together for a democratic South Africa.

In the late 1940s and thereafter, the main challenge to Communist partnership with the African nationalist movement in South Africa came not from interracial liberals but from Africanists who viewed the Communists as too white and too Western to speak for the African masses. Not until the civil rights movement had run its course in the 1960s would a comparable black separatist impulse emerge in the United States to challenge the integrationist liberalism that had dominated black political thought since the '40s. When civil rights laws failed to bring full equality, a disenchantment with liberals and liberalism rapidly set in. By that time, however, the leadership and inspiration that Communists had provided to the cause of black liberation in the late 1930s was a dim memory. During the late '60s and early '70s, Marxist-Leninist perspectives would again inform the thinking of some black radicals—it was a major element, for example, in the ideology of the Black Panther party—but (with the conspicuous exception of Angela Davis) no prominent black activists and intellectuals formally affiliated themselves with the remnant of diehard loyalists that now comprised the American Communist party.[83]

· 6 ·

"We Shall Not Be Moved": Nonviolent Resistance to White Supremacy, 1940–1965

The Gandhian Tradition

People committed to a militant struggle against colonial or racist domination during the middle decades of the twentieth century had one clear alternative to the Marxist-Leninist path of violent revolution. It was the method of resistance pioneered by Mohandas K. Gandhi in South Africa and India that has been variously called "passive resistance," "nonviolent direct action," and "militant nonviolence." During the 1940s and '50s in South Africa—and for a few years longer in the United States—advocates of nonviolence tried to demonstrate the superiority and efficacy of their method of protest. They argued for it, tried it out on a small scale, threatened mass action, and in a few notable instances led thousands of demonstrators in direct action against oppressive laws or policies.[1]

The term nonviolence is difficult to define with precision, but there seems little question that it refers to a range of protest or resistance activities that fall between the straightforward use of physical force and the mere expression of dissatisfaction in conventional, legally authorized ways within officially constituted bodies or channels. Boycotts, strikes, mass marches or demonstrations, and planned civil disobedience are all forms of nonviolent resistance. Nonviolent actions do not necessarily signify a nonviolent ideology. Movements that are not committed to nonviolence as a philosophy or moral imperative often make use of such methods. Furthermore, there is a natural tendency for people whose grievances are not being addressed through conventional politics and legalistic appeals to engage in nonviolent action out of frustration, or the pragmatic need to try something else, rather than out of the conviction that it represents a morally superior form of struggle. The practical function of nonviolent doctrine or ideology—as set forth by a Gandhi or a Martin Luther King, Jr.—might be to increase the morale and motivation of nonviolent protesters and prevent or inhibit their normal tendency to escalate the conflict to the level of violent resistance when nonviolent campaigns do not bring the results that are hoped for. It would be wrong, however, to assume that only committed pacifists can lead effective nonviolent movements. It is quite possible to regard violence as self-defeating within a particular historical context—and to act consistently on that conviction—without rejecting the use of force or violence in all conceivable circumstances.

It is also extremely difficult, if not impossible, for believers in nonviolence to act consistently on the Gandhian injunction that they should do no harm to their enemies. Effective nonviolence can rarely be simple moral suasion, in which the protesters take all of the suffering on themselves and thus shame their opponents into behaving decently. It is difficult to dispute the judgment of Reinhold Niebuhr that "so long as [nonviolence] enters the field of social and physical relations and places restraints on the desires and activities of others it is a form of coercion." Boycotts come to mind as examples of such coercion. The actual record of nonviolent campaigns supports a conclusion that success usually results from the protesters' ability to coerce or disadvantage the target group. If, to reverse Clausewitz's dictum, diplomacy is war carried on by other means, then perhaps nonviolence is simply a use of force that avoids doing bodily harm to a political opponent.[2]

But there may be third parties with a stake in the outcome of conflicts between a dominant group and aggrieved subordinates who will be susceptible to an appeal to conscience because their liberal or humanitarian ideals are not at war with their interests and may even be compatible with them. If nonviolence is more likely to coerce than convert the oppressor, it may nevertheless convert an interested and influential third party with a capacity to intervene on behalf of the protesting group. The absence of such a third force reduces the likelihood that nonviolent resistance against entrenched privilege will remain nonviolent.[3]

For mid-twentieth-century black protesters in the United States and South Africa nonviolence was not merely a set of abstract propositions and possibilities; it was also a form of resistance against white or European domination that had been previously employed in the struggle for Indian independence from British rule. The ideas and actions of Gandhi thus became an inspiration and touchstone for nonviolence elsewhere. Gandhi first developed and employed the form of nonviolent resistance that he called *satyagraha* ("truth force") among the Indian minority in South Africa between 1906 and 1914. His use of civil disobedience in an effort to relieve Indians of some special disabilities under which they labored in Natal and the Transvaal was not a frontal assault on racial inequality in South African society. Gandhi's concern was exclusively with the rights of Indians as an expatriate community; not only did he pay no attention to the plight of Africans, but in fact he aimed explicitly at making Indians part of the "civilized" and privileged settler minority.[4]

But Gandhi's ethnocentrism may offer a clue to the kind of historical circumstances that tend to give force and effectiveness to nonviolent resistance. Gandhi's campaigns, in India as well as in South Africa, were always nationalistic in the sense that they were defined as the actions of a specific people or nation in the making—an "imagined community," to use Benedict Anderson's phrase. The universalistic character of Gandhi's conception of *satyagraha* could not conceal the fact that its function was the liberation of a particular community through its own efforts. At times Gandhi's conception of who could and should undertake militant nonviolence was culturally specific to the point of chauvinism: in 1940 he suggested that the only nation with the capacity to create a nonviolent state was India.[5]

The case against a universalistic application of *satyagraha* might seem to be further strengthened by a recognition that Gandhi's nonviolent philosophy was rooted in the Hindu religious ideals and practices that were associated with the strictest kind of virtue, specifically in the doctrine of *ahimsa*—the refusal to do harm to any living thing. For Gandhi a true *satyagrahi* must refrain from eating meat and animal products as well as refuse to resist physically when attacked. Gandhi's identification of the pursuit of truth with vegetarianism, celibacy, and a host of other ascetic practices obviously limited the degree to which Westerners or Africans could embrace the total way of life that he considered essential to the consistent practice of nonviolence. Had Gandhi been a true sectarian who felt no sympathy or affinity for those who advocated similar methods of resistance from a different religious or philosophical standpoint, his doctrines would have gained little international currency. But Gandhi believed that all of the great world religions were compatible and that they all prescribed the same basic morality. (Religious toleration was of course essential to his project of uniting Hindus and Moslems behind Indian nationhood.) Most significant, he found the nonviolent philosophy perfectly summed up in the Sermon on the Mount. Having himself been influenced by Tolstoy's Christian anarchism, he concluded that Christians need only obey their savior's injunction to "turn the other cheek" and love their enemies in order to be authentically nonviolent.[6]

Gandhism could therefore find an ideological port of entry into American and South African freedom struggles through the nonviolent tradition in Christian thought; his doctrines were obviously compatible with the radical pacifism implicit in perfectionist Christianity. But Gandhi brought something new to the pacifist tradition—an element of political realism or pragmatism that had previously been lacking. One of Gandhi's most remarkable attributes was his rare combination of high idealism and political astuteness. Unlike those pacifists who had thought that their duty was done when they had taken a personal stand abjuring war or violence,

Gandhi organized and led a political movement that mobilized masses of people to press the British government for relief from very specific injustices and exhibited great tactical skill in doing so. Although historians may doubt that Gandhian nonviolence was actually responsible for Indian independence, there is no doubt it proved an effective way to achieve more limited objectives and that part of its justification stemmed from its effectiveness. Unlike earlier Christian pacifists, therefore, Gandhi's morality could be validated by its results as well as on the basis of its conformity to a set of moral absolutes.[7]

Indeed a central feature of Gandhian nonviolence was that it offered resistance leaders a wide choice of tactics short of violent insurrection or terrorism and made achievement depend to a considerable extent on choosing the method most appropriate for the circumstances at hand. As Gandhi, and later Martin Luther King, Jr., would demonstrate, successful nonviolent leaders had to be skillful politicians and masters of tactical maneuver as well as charismatic exemplars of principles and virtues prized by their followers. Among the forms of resistance used by Gandhi and later employed by passive resisters elsewhere were the one-day community strike (*hartal*), the mass protest march, the consumer boycott of goods produced in an unjust or undesirable way, noncooperation with official bodies that were deemed illegitimate because of their unrepresentative character, and finally civil disobedience—the deliberate and open violation of allegedly unjust laws and the courting of arrest for the offense. Sometimes nonviolent resistance is made synonymous with this latter and extreme form of *satyagraha*, but Gandhi made it clear that "all Satyagraha is not Civil Disobedience" and that the latter is "like the use of a knife to be used most sparingly if at all."[8]

Gandhi considered civil disobedience as a kind of last resort only to be employed after everything else had failed, because he was acutely aware of its tendency to provoke or inspire violence both by the enforcers of the law and by undisciplined members of the protesting group. "At the same time that the right of civil disobedience is insisted upon," he wrote in 1922, "its use must be guarded by all conceivable restrictions. Every possible provision should be made against an outbreak of violence or general lawlessness. Its area as well as its scope should be limited to the barest necessity of the case." But he was also aware, as he wrote to some English sympathizers in 1930, that even the most circumscribed civil disobedience "may resolve itself into violent disobedience. . . . But I know that it will not be the cause of it. Violence is already there corroding the whole body politic. Civil Disobedience will be but a purifying process and may bring to the surface what is burrowing under and into the whole body."[9]

A more serious problem for the theory of nonviolence than the danger that it might lead to its opposite was the tendency of passive resistance to turn into a form of compulsion or coercion that was difficult to distinguish on moral grounds from the application of physical force. When some of his followers proposed in 1933 to block access to temples so that the orthodox Hindus who were opposing his campaign against untouchability would be shown the error of their ways, Gandhi vetoed the suggestion on the grounds that it represented compulsion rather than the peaceful persuasion and appeals to conscience that were at the heart of *satyagraha*. But some of the boycotts and strikes that he did support relied more on economic coercion than on the "truth force" that they could exert on the minds of the oppressors. The conflict between the religious ideal of nonviolence and use of it as a form of coercive pressure politics would be a problem for theorists of nonviolence in the United States and South Africa, unless they were willing to accept the distinctly un-Gandhian notion that this form of resistance was purely and simply a pragmatic response to oppressive situations in which violent resistance was not feasible or likely to be productive.

Gandhi himself, after years of translating *satyagraha* into English as "passive resistance," decided by 1920 that a better phrase was "direct action" and that the older term should be retained to indicate a movement of people who were nonviolent not out of religious or ethical conviction, but because they were simply too weak or too cowardly to engage in violence—the form but not the spirit of *satyagraha*. He also came to recognize that there might be a gap between leaders and followers, with the former acting out of religious conviction but the latter pursuing self-interest. Drawing a distinction between individual civil disobedience based on conscience or altruism and "mass" nonviolent action in which mundane and impure motives were likely to be involved, he appeared to be making the spiritually elitist argument that only an exceptional few can rise to the full heights of *satyagraha* and that the followers were like an army pledged and conditioned to follow the generals even though they did not fully understand the cause for which they were fighting. "It is sufficient in mass civil disobedience," he wrote in 1921, "if the resisters understand the working of the doctrine"— which in context means that they know how to behave as good soldiers of what Gandhi elsewhere called "the nonviolent army."[10]

When Gandhi returned to India in 1914 after more than twenty years in South Africa, he left behind the Natal Indian Congress, a cautious and accommodationist organization dominated by merchants that was quite willing to consider the limited victories Gandhi had won as sufficient reason to rest on its laurels. The South African Indian community did not in fact engage in militant nonviolence again for more than thirty years. The South African Native National Congress (later the African National Congress), which was founded in 1912 when Gandhi's nonviolent campaign against Indian disabilities was in full swing, was sufficiently taken by his example to include in its constitution, along with more conventional methods of seeking redress of grievances within the channels provided by the South African government, an endorsement of "passive action" as a means of protest. In 1919, the congress branch in the Transvaal actually undertook "passive action" against that most gnawing and persistent of African grievances—the pass system. Thousands were arrested on Witwatersrand for publicly turning in or discarding their passes. But the campaign could not be sustained for long in the face of government repression, and it failed to abolish or significantly modify the pass system. The congress would not use nonviolent direct action again until the 1950s.[11]

The failure of the ANC to employ nonviolent resistance as a method of protest for more than three decades can be attributed in part to its leadership's fear that it could not control masses of Africans in volatile situations. In 1914, Congress founder Richard Msimang wrote in a pamphlet describing the new organization that its "leaders have had and are having a difficult and responsible task in leading a large mass of people whose method of resisting or displaying dissatisfaction is and was resort to arms." There was clearly nothing in the cultural traditions of Zulus, Xhosa, Sotho, or Tswana that could make a virtue out of refraining from violent resistance to conquerers or oppressors. Furthermore, it was a common belief that nonviolence required a Gandhi, a charismatic leader who would show the way through his personal suffering, and that no such figure was on the horizon. The Non-European Conference of 1930, in which the ANC participated along with Colored and Indian organizations, recommended "passive resistance" if proposed parliamentary legislation restricting nonwhite protest meetings were to be passed. But the president of the conference, the Cape Colored leader Abdul Abdurhaman, contended that such action was ill-advised and premature, because "there is not a single man in South Africa who could make a success of passive resistance. You must have a leader who is prepared to make sacrifices, such as Gandhi in India. We have not such a man."[12]

In 1939 a prominent ANC leader, the Reverend S. S. Tema, visited India and asked Gandhi directly what the ANC could do to become as successful as the Indian National Congress. The Mahatma's forthright answer suggested that a whole new cultural orientation would be needed before mass nonviolent action could be undertaken. In his view, the Western-educated elite that dominated the ANC had lost touch with the masses of Africans who were still attached to their traditional ways of life and were thus incapable of acting effectively and unselfishly on behalf of the people whom they claimed to represent. "You must not be afraid of being 'Bantuized,' " Gandhi admonished Tema, "or feel ashamed to carry an assegai or of going about with only a tiny clout around your loins. . . . You must become Africans again." This advice highlights an element of Gandhism that is often overlooked—mass nonviolence needs to express the culture of a community, and its leaders must be in tune with the basic beliefs and values of the masses. One does not have to endorse Gandhi's anti-Western populism to recognize the validity of his criticism of the ANC leadership of 1939 as incapable of directing militant mass action because "it has not sprung from the common people." But Gandhi's conception of nonviolent resistance had its own undemocratic aspect.

What the South African Indian community required, he told a group of visiting students that same year, was a single leader "who will be so pure, so cultured, so truthful and so dignified in his being that he will disarm all opposition." Gandhi's own sense of personal mission—some might call it a messiah complex—and his belief that nonviolence could not succeed elsewhere unless it had an equally virtuous and charismatic leader was a troubling legacy for the nonviolent tradition. Tensions between the concept of nonviolent resistance as a democratic, grass-roots movement and the alternative view that it is called into being by a messianic leader would persist.[13]

The eve of World War II saw a revival of interest in passive resistance in South Africa, as a more activist leadership took charge of the ANC and radical elements in the Indian community began to call for action against new efforts to discriminate against Asians. At the same time, A. Phillip Randolph in the United States was proposing that African Americans undertake mass nonviolent action to influence national race policy—a major new departure for the civil rights movement.

As historians August Meier and Elliott Rudwick have shown, Randolph did not propose what he called "nonviolent direct action" to a community that had never practiced such a form of resistance. In fact there was a long history of nonviolent protest on the part of African Americans and their allies. Boycotts of segregated facilities and disobedience of Jim Crow laws and customs had been a feature of the antebellum abolitionist campaign against racial discrimination in the northern states and had also been part of the spontaneous response of southern African Americans to white attempts to impose streetcar segregation during Reconstruction and again around the turn of the century. Twentieth-century efforts to extend school segregation from the South to the North had frequently been met by boycotts of public education by black students that often achieved their objective of keeping the schools formally integrated. In the 1930s, more militant forms of nonviolent protest came to the fore. In urban ghettos, economic nationalists picketed stores that sold to blacks but did not hire them. The Communist party instigated a range of nonviolent actions, including mass marches of the unemployed, demonstrations and sit-ins at relief offices, obstruction of evictions, rent strikes, and even a march on Washington on behalf of the Scottsboro Boys. But none of these actions gave rise to a nonviolent *movement* or even envisioned one. Either the protest was a spontaneous and localized expression of black anger at the growth of segregation or it was simply one tactic of a group or organization that had no special commitment to nonviolence. None of the actions of the '30s appeared to have drawn inspiration from Gandhi and the Indian freedom struggle. Communists in fact tended to be highly critical of Gandhi, describing him as a tool of capital and a traitor to the Indian people because his "reformist" idealism clashed with their revolutionary materialism.[14]

Nevertheless, some African Americans before the 1940s evinced a strong interest in Gandhi and his method. As early as 1922, James Weldon Johnson, the executive secretary of the NAACP, wrote: "It will be of absorbing interest to know whether the means and methods advocated by Gandhi can be as effective as the methods of violence used by the Irish. . . . If noncooperation brings the British to their knees in India, there is no reason why it should not bring the white man to his knees in the South." Other black activists and intellectuals were skeptical or unsure that Gandhism would work for blacks in an American context, but the black press frequently heralded Gandhi's virtues and achievements, and, as Sudarshan Kapur has recently demonstrated, calls for a black American Gandhi sometimes appeared in editorials. But one needs to be wary of overestimating the extent to which black Americans were ready for militant nonviolence before the 1940s. Some of Gandhi's strongest defenders were moderates or conservatives who viewed his rejection of the use of force as supporting a quietistic alternative to the Marxist-inspired radicalism that was spreading during the 1930s among the younger generation of educated African-Americans. They clearly did not view themselves as preparing the way for militant mass action. One writer in the *Chicago Defender* in 1933 found that the teachings of Christ, the accommodationist interracial philosophy of Booker T. Washington, and the nonviolence espoused by Gandhi were perfectly compatible.[15]

It was A. Philip Randolph, the African-American labor leader, who made a link between the depression-era readiness of blacks to engage in more militant and confrontational forms of protest and the diffuse but persistent African-American fascination with Gandhi and Gandhism. Randolph's March on Washington Movement, discussed above in relation to Communist involvement with black causes in the 1930s and '40s, was the first serious effort to create a nonviolent mass movement of African-Americans. The Committee on Racial Equality (CORE), founded in 1942 as an action-oriented offshoot of the pacifist Fellowship of Reconciliation, was more consistently Gandhian in theory and practice and pioneered such important nonviolent tactics as the sit-in and the freedom ride. But during the 1940s and '50s CORE was neither a black-dominated organization nor a prospective mass movement. Interracial but predominantly white in its early years, it had no solid base in the black community and represented little more than the individual consciences of the small number of radical pacifists who were its members. Two of them—James Farmer and Bayard Rustin—were African-Americans who would later play important roles in the civil rights movement, but during the '40s they were known more for a principled but unpopular pacifism than for effective race leadership. Randolph, on the other hand, was a mainstream black leader whose prominence and popularity during the 1940s made him the most influential and prestigious black spokesman of that decade.[16]

Randolph was less than an ideal candidate for the role of American Gandhi; for he totally lacked the religious inspiration that Gandhi considered indispensible for *satyagraha*. Although the son of an African Methodist Episcopal minister, he had early lost his faith and become a life-long atheist. His convictions came mainly from the strain of Marxist thought that had produced twentieth-century democratic socialism. Something of an economic determinist, he believed that human beings acted out of mundane, material concerns more than out of conscience or idealism. As a labor organizer and founder of the most important black trade union, the Brotherhood of Sleeping Car Porters, his inspiration for militant action derived most directly from the confrontational tactics recently used by labor in its struggles with management. Randolph was impressed with how, during the late 1930s, the United Auto Workers had carried the strike weapon to the verge of nonviolent revolution in the sit-down occupations of Detroit auto plants. More generally he hailed the effectiveness of "great masses of workers in strikes on the picket line." Not only was Randolph no pacifist or unconditional believer in nonviolence, but the major campaigns that he launched were directed primarily toward the inclusion of blacks on an equal basis in the emerging American "warfare state" of the 1940s. A final problem complicating any characterization of Randolph as a pioneer of African-American nonviolent resistance was the fact that during the 1940s he did not actually lead, or even participate in, a militant nonviolent action. A master of self-promotion, publicity, and bluff, he achieved his victories simply by threatening mass demonstrations; no one knows whether he could actually have produced one.[17]

There is no evidence that Randolph's call for a march on Washington in 1941—his initial campaign to pressure the government into giving blacks equal access to the defense jobs that were opening up—was directly inspired by Gandhi's campaigns in India. But in one respect Randolph was closer to the spirit of Gandhi than the more orthodox Gandhians in pacifist groups like the Fellowship of Reconciliation. His decision to exclude whites from the March on Washington Movement was motivated in part by his fear that white Communists would try to take over the movement as they had apparently taken over the National Negro Congress during Randolph's presidency. But a deeper motive was his sense that African Americans needed more experience of acting communally in defense of their rights and that group pride and solidarity could be achieved only if they acted independently of whites under their own leaders. "The essential value of an all-Negro movement such as the March on Washington Movement," he explained in 1942, "is that it helps to create faith by Negroes in Negroes. It develops a sense of self-reliance with Negroes depending on Negroes in vital matters. It helps break down the slave psychology and inferiority complex in Negroes which comes with Negroes relying on white people for direction and support."[18]

If Gandhi's campaigns were designed to restore pride and honor to Indians as a national or ethnic group, whether occupying minority status as in South Africa or constituting the bulk of the population as in India, Randolph's March on Washington Movement sought to achieve something similar for African Americans. Randolph was no black nationalist in any ordinary sense. He had little feeling for African roots or the cultural distinctiveness of African-Americans, and his ultimate goal was assimilation of the black population into a democratic and color-blind American society. He favored an all-black protest movement because "Negroes are the only people who are the victims of Jim Crow, and it is they must take the initiative and assume the responsibility to abolish it. Jews must and do lead the fight against anti-Semitism, Catholics must lead the fight against anti-Catholicism, labor must battle against anti-labor laws and practices." In disarmingly simple terms Randolph made a point that Gandhi also expressed: any group seeking liberation from subjugation must make self-assertion and independent action a central part of its struggle, otherwise it will remain dependent on its putative emancipators. Randolph was always willing to accept the help of white liberals and to participate in interracial organizations—as a realist and a pragmatist he acknowledged that the African-American minority could make little headway without outside help. But by limiting exposure to the dangers and potential hardships of nonviolent direct action to blacks themselves, he sought to insure their priority and pride of place in the broader movement for racial justice and equality.[19]

The March on Washington Movement accomplished part of its initial purpose in 1941 when President Roosevelt signed an executive order discouraging racial discrimination in hiring under government defense contracts—the most significant federal action in support of African-American rights since Reconstruction. But Randolph did not disband the movement; he foresaw enforcement problems and wished to continue agitating for the movement's second original goal of desegregating the armed forces. By September 1942, Randolph's vision had expanded, and he contemplated an extensive use of what he called "non-violent good will action" against all forms of racial discrimination and segregation. In his address to the policy conference of the March on Washington Movement he drew explicitly on the Gandhian model: "Witness the strategy and maneuver of the people of India with mass civil disobedience and non-cooperation and marches to the sea. . . . The central principle of the struggle of oppressed minorities like the Negro, labor, Jews and others is not only to develop mass demonstration maneuvers, but to repeat and continue them." He then went on to advocate an ambitious program of action including mass marches, organized pressure to register voters in the southern states, and public protests against segregated public facilities.[20]

Randolph discussed more fully what a comprehensive nonviolent freedom struggle would look like in the essay he contributed to the notable wartime collection of black viewpoints, *What the Negro Wants*, published in 1944. In the light of subsequent nonviolent actions his proposals may not seem very bold or militant. It is noteworthy that none of them actually involved civil disobedience, which may be the reason he no longer explicitly invoked Gandhi as a precedent. He endorsed the tactic that CORE was developing of using small interracial groups to desegregate northern restaurants and recreation facilities, a form of nonviolent pressure that in some cases led to the first sit-ins. To southern blacks, he recommended protests that were largely symbolic, such as occasional one day boycotts of segregated schools, streetcars, buses, and trains. But even such a limited program of direct action found few supporters in 1944—by that time the March on Washington Movement had lost most of its momentum. In the midst of a war that most African-Americans supported, it was scarcely an ideal time to engage in civil disobedience or even to mount mass demonstrations. In addition, the race riots of 1943 in Detroit and Harlem had inspired fears that mass action might provoke violence. Early in the war, the March on Washington Movement had succeeded in sponsoring some giant public rallies, and its existence served to keep pressure on the government to keep its commitment to fair employment practices in war production, but at no time did it go beyond exhortation and actually embark on mass nonviolent action.[21]

The American nonviolent movement of the war years was only a pale imitation of what Gandhi had done in India and only a weak anticipation of what would occur in the United States in the 1950s and '60s. Randolph had publicized the idea of mass protest and showed how the fear of blacks marching en masse and possibly provoking race riots could wring concessions from white authorities. CORE had actually carried out small-scale direct actions against segregated facilities in the North, and as pacifists and close students of Gandhi's philosophy and methods they came closer than Randolph to the spirit of *satyagraha*. But their actions involved only a handful of blacks and made relatively little impression on the black community. Even less widely noticed were the strikes of draft resisters in federal penitentiaries against the racial segregation of prisoners. In 1947, CORE conducted its most ambitious and precedent-setting action—the Journey of Reconciliation—which featured a small number of demonstrators testing recent court orders for the desegregation of southern interstate transportation facilities. Among them was Bayard Rustin, the black pacifist who along with two other CORE members served twenty-two days in a North Carolina jail for violating state segregation laws.[22]

In 1948, when Congress was on the verge of enacting a peacetime draft in response to growing tensions between the United State and the Soviet Union, Randolph and Rustin organized the Committee Against Jim Crow in Military Service and Training, which threatened to mobilize young blacks to resist being conscripted into a segregated army. Testifying before a congressional committee, Randolph invoked the example of the recently martyred Gandhi: "In resorting to the principle of direct-action techniques of Gandhi, whose death was publicly mourned by many members of Congress and President Truman, Negroes will be serving a higher law than any passed by a national legislature in an era when racism spells our doom." Once again Randolph was able to win a victory for nonviolent protest without actually engaging in it. Reacting to Randolph's threat, as well as to his concern about how blacks would vote in the forthcoming presidential election, President Truman ordered the gradual desegregation of the armed forces. Randolph immediately called off his campaign despite the strenuous objections of Rustin, at this time still a dedicated pacifist, who favored resistance against conscription whether or not the armed forces were being desegregated. The Randolph-Rustin split of 1948 revealed the difference between the former's conception of nonviolent direct action as a political tool in the struggle for racial equality and the latter's unconditional opposition to violence in any form. The future effectiveness of a nonviolent movement might depend on its ability to combine Randolph's skill at political maneuver and acute sense of the need to involve masses of blacks in nonviolent resistance with Rustin's more principled and disciplined approach to direct action. Neither large numbers of imperfectly controlled protesters nor small groups of well-disciplined and right minded *satyagrahi* were likely to challenge southern segregation successfully.[23]

By the early 1950s the stirrings of thought and action that pointed toward an application of Gandhi's methods—if not his philosophy—to the black American freedom struggle appeared to have dissipated, leaving little impression. Randolph devoted himself during these years mainly to his perennial struggle from inside the American labor movement to win full acceptance of blacks in unions and labor federations. CORE, after having used direct action to desegregate restaurants and recreation facilities in some northern and border cities, foundered on the question of what to do next. Moving south seemed unthinkable, because demonstrators coming from the North were likely to risk (or even lose) their lives without having much impact in a region where whites seemed firmly committed to segregation and blacks were apparently cowed and quiescent. Consequently CORE declined in the early '50s and was almost moribund in 1954 when the Supreme Court declared school segregation to be unconstitutional.[24]

The civil rights movement was not, as is sometimes supposed, on a steady upward trajectory from the March on Washington Movement to the Civil Rights Acts of 1964–65. The onset of the Cold War and the anti-radical hysteria personified by Senator Joseph McCarthy put a damper on militant protest movements of all kind. Although CORE was staunchly non-Communist, its members were often accused of being "reds" by people who shared the common belief that only Communists believed in racial equality. Since direct action often interfered with customary property rights, it was unacceptably radical to many Americans whatever its ideological coloration. Randolph was protected from red-baiting to some extent because of his long record of opposition to Commmunist influences in black organizations and in the labor movement; but in 1951, at the height of McCarthyism and the Korean War, he acknowledged the final congressional defeat of efforts to establish a federal Fair Employment Practices Commission. Randolph had hoped that the kind of agency established during the war in response to his threatened march could become a permanent office concerned with equalizing black economic opportunities, and for a decade he had devoted much of his attention to this campaign. But the reactionary climate of opinion in the early '50s made such legislation impossible. One can only speculate as to exactly why Randolph did not propose a new march on Washington during the postwar years to push for a permanent FEPC. Perhaps his keen tactical sense told him that the time was not ripe.[25]

But progress of a sort was being made in the struggle for civil rights in the late '40s and early '50s. The NAACP was winning the court victories against segregation and discrimination that culminated in *Brown v. the Board of Education* in 1954. But one effect of the apparent success of the association's legal strategy was to convince white liberals and moderate black leaders that mass protest and direct action were unnecessary. When the extreme repressiveness of McCarthyism declined somewhat with the fall of McCarthy himself, the argument that gradual steps toward desegregation would aid the United States in the Cold War became increasingly persuasive. Desegregation without violence or disorder, if it could be achieved, would show the Africans and Asians whom the United States hoped to secure as allies in the conflict with the Soviet Union that steps were being taken to curb the American habit of humiliating people because of the color of their skins. In the period just before blacks in Montgomery, Alabama, began boycotting buses late in 1955, white Americans who considered themselves liberal or tolerant in racial matters, as well as the large number of middle class blacks who put their faith in the legal strategy of the NAACP, were likely to be unreceptive to the notion that mass nonviolent protest against segregation was either necessary or promising. If anything, it might endanger the gains that were being made.[26]

Nonviolence in South Africa

In South Africa, as in the United States, the war years brought discussions of nonviolent protest as a possible tactic in the struggle against white supremacy. But rather than declining in the early cold war period, the interest in nonviolent action grew, and "passive resistance" campaigns were undertaken, first by the Indian community and later by the African National Congress. In the late '40s and early '50s, mass civil disobedience became the first line of defense against apartheid—the intensification of segregation and discrimination that resulted from the electoral victory of the Afrikaner Nationalists in 1948. A fundamental difference was emerging between the likely future of black people in the two societies if events continued on their present course. The situation of African-Americans undoubtedly improved in some important respects during the period between 1941 and 1955. In fact this was a period of great economic gain relative to whites. Despite the success of racially conservative elements in blocking civil rights legislation in Congress, the Supreme Court was continuing to make decisions that chipped away at the edifice of legalized discrimination. In South Africa, however, the legal and political status of Africans, Indians, and Coloreds came under increasing attack from white supremacists in the late 1940s and early '50s; as a result the kind of gradualist legal methods and reformist tactics that were bearing some fruit in the United States were shown to be totally ineffectual in the South African context. Consequently, nonviolent direct action was embraced out of desperation.[27]

The South African Indian community, a relatively small but locally significant minority, was the seedbed of nonviolent thought and action in the 1940s. (Because of restrictions against their movement Indians remained concentrated mainly in Natal, especially in the city of Durban, with a smaller contingent in the Transvaal). Here the influence of Gandhi was direct and personal. After his return to India in 1914, the Mahatma kept an eye on South African affairs and from time to time provided advice and guidance for South African Indian leaders and organizations. Gandhi's approach to the rights of Indians in South Africa was to seek to have them included within the privileged or "civilized" segment of an ethnically divided society—also the aim of the relatively conservative leaders who dominated South African Indian politics in the 1920s and '30s. An agreement negotiated between the governments of India and South Africa in Cape Town in 1927 and endorsed by Gandhi provided for assisted repatriation to India for those who wished to return to their homeland and promises of improved conditions for those who chose to remain in South Africa; it also provided limited assurances of Indian privileges relative to Africans, although it fell far short of establishing their equality with whites.

By 1939, it was evident that the Europeans were unwilling to open the way for the eventual social and political equality of Indians and were in fact preparing to make new assaults on their economic rights as a property-owning and trading community. At the same time, a younger generation that had been born in South Africa and was committed to remaining there—unlike some of their elders who had been born in India and had a sojourner mentality—was coming to maturity and beginning to contest the leadership of the old guard. Among this younger group was a radical element that not only favored reviving the Gandhian method of militant nonviolent protest against the new assaults on Indian rights but also proposed transcending the ethnocentric Gandhian legacy by making common cause with Africans against the white racist regime.[28]

Dr. Yusuf M. Dadoo, leader of the young radicals in the Transvaal, attempted in 1939 to organize Indians for militant action against a legislative proposal for their residential and commercial segregation. Dadoo was a Marxist who had helped form the Non-European United Front in 1938. In 1939 he joined the Communist party and would remain a faithful member of the CPSA for the remaining forty four years of his life, eventually becoming its chairman. But Dadoo saw no conflict between his Marxism and the employment of Gandhian rhetoric and tactics. He professed great respect for Gandhi and acknowledged his leadership of the Pan-Indian cause. When Gandhi advised Dadoo to suspend the proposed action of 1939 while the former attempted to negotiate a settlement, the latter complied. After the war broke out, the plans for passive resistance were shelved for the duration, although Dadoo did manage to get himself jailed for his antiwar agitation during the period of the Stalin-Hitler pact. (Conveniently for his dual allegiance, Communists and Gandhian Indian nationalists could agree in 1940 to oppose a war that involved the defense of the British empire.) In a statement issued on the eve of his trial in early 1941, Dadoo addressed prospective passive resisters in language that was clearly inspired by Gandhi rather than Marx: "The path before passive resisters is one of suffering. They must be armed with the weapon of truth and so steeled in the school of self-discipline that they will be able to endure the trials of the struggle with calm dignity, unflagging determination, uncomplaining stoicism, ungrudging sacrifice and unswerving loyalty to the cause. Such an attitude of mind and behavior will disarm all opposition and open the road to the vindication of justice."[29]

During the war, young radicals also gained influence in the organizational life of the larger Indian community of Natal. Here Indians confronted a persistent campaign by white politicians to stem what they regarded as Indian intrusion into white residential and commercial areas. The agitation led the Union government to pass a law in 1943 temporarily banning real estate transfers between whites and Indians. The Indian elite split on the question of how to respond to the threat of segregation. The older and more conservative element that still dominated the Natal Indian Congress favored efforts to reach a compromise with white authorities, allowing perhaps a restriction of residential rights in return for an acknowledgment of the trading rights that were of crucial importance to the merchant elite that constituted the established leadership.

The emerging radical opposition, which organized itself as the Anti-Segregation Council during the war, was a remarkable coalition between a small group of young professionals, the first generation of South African Indians to achieve that status, and the vigorous trade union movement that had sprung up among the Indian urban working class. In 1945 the radicals took control of the Natal Indian Congress and elected Dr. G. M. (Monty) Naicker as president. Like Dr. Yusuf Dadoo in the Transvaal, Naicker was a British-educated physician; unlike Dadoo, however, his political inspiration was exclusively Gandhian—he never joined the Communist party or proclaimed himself a Marxist. In 1951, when he was asked to recall what Gandhi had meant to him, Naicker professed that as a young man he had "made a deep study of the writings of Gandiji" and then when the time came for nonviolent resistance in South Africa, he had heard "the voice of Mahatma Gandhi calling for action."[30]

Under the leadership of Dadoo and Naicker, the Indian Congresses of

Natal and the Transvaal undertook a passive resistance campaign beginning in 1946; before it was over they had managed to attract international attention, culminating in the first United Nations resolution condemning racial discrimination in South Africa. The provocation was the Asiatic Land Tenure and Indian Representation Act of the Smuts government. This law anticipated the Group Areas Act of the apartheid era by denying to Indians the right to reside or trade in urban areas specified for exclusive white occupancy. Dubbed "the Ghetto Act" by the protesters, its enactment was a response to the success of Indians in competing with white merchants as well as an effort to stop a growing Indian middle class from seeking better housing outside of the run-down areas in which Indians had traditionally concentrated. The campaign began when 15,000 people took a pledge to disobey the act in Durban on June 13, 1946; subsequently 1700 volunteers came forward in that year to participate in specific acts of civil disobedience in Natal. Ultimately more than 2000 demonstrators were arrested. The principal action was the occupation of vacant lots in areas set aside by the law for Europeans. Assaults by white hooligans severely tested the nonviolent convictions of the demonstrators, as did the conditions in some of the jails in which they were incarcerated for trespass. Late in the campaign, the Transvaal and Natal Indian congresses combined to repeat a famous march of Gandhi; they sent a mass of demonstrators across the border between Natal and the Transvaal in violation of the laws restricting the movement of Indians from one province to another. Of those arrested during the two years of actions, virtually all (including Dadoo and Naiker) refused bail. Gandhi, who died in 1947 in the midst of the South African Indian struggle, gave full support to the campaign, although he continued to disapprove of proposals for a joint struggle of Indians and Africans that the South African passive resisters were beginning to take quite seriously.[31]

Although it aroused international indignation and led to a United Nations General Assembly vote condemning South Africa, the Passive Resistance Campaign failed utterly to induce the South African government to give up its plans to impose residential and commercial segregation on Indians. The effort was called off in 1948 just as the Nationalists were coming to power on the platform of comprehensive segregation for all non-Europeans. At that point it had become clear that Indians could achieve little on their own and that their movement would have to be merged into the larger black struggle for liberation. The previous year, in the "Doctors' Pact" signed by Dadoo, Naicker, and Dr. A. B. Xuma, the presidents of the two Indian congresses had joined with the president of the ANC to endorse a common program of universal suffrage, equal economic opportunity, equal access to education, and the end of all legalized segregation and discrimination. But the reality of a joint effort would be difficult to achieve; many Indians retained hopes for a special status closer to whites than to Africans, and many Africans were resentful of Indian privileges, suspicious of Indian culture, and distrustful of the Indian traders with whom they often had to deal. In 1949, the precarious state of Indian-African relations in Natal was revealed when Zulus in Durban rioted against Indians, killing a substantial number while the police looked on.[32]

The Indian Passive Resistance Campaign provided part of the inspiration and leadership for the most significant example of nonviolent resistance in South African history—the Defiance Campaign of 1952, formally a joint effort of the African National Congress and the South African Indian Congress. But Indians were junior partners in that campaign and sometimes found that their claims to special knowledge of the philosophy and practice of nonviolence were resented by Africans who thought of themselves as acting on a self-generated impulse rather than following the lead of others. The specifically African roots of the massive nonviolent resistance that was undertaken in the 1950s can be found in the debates that took place in the congress during the 1940s on the question of whether or not to cooperate with the segregated political institutions that the white government claimed gave some representation to Africans. The leadership of the ANC during the Second World War favored working within the system for whatever advantages Africans could get out of it. Communists in the ANC leadership tended to view elected township advisory boards, the Natives' Representative Council, and the provisions for white representation of Africans in parliament as opportunities for agitation and infiltration. Some of the old-guard moderates in the leadership continued to believe that the channels provided by the government for the expression of African opinion could still provide a means of nudging the government and white political opinion in the direction of justice for Africans.[33]

The African National Congress Youth League, founded in 1944 out of the impatience of younger members with the ANC's lack of militancy, has drawn the attention of historians mainly because some of its leaders espoused a mystical Africanist philosophy that anticipated the later Pan-Africanist and Black Consciousness movements. But for our present purposes it is more significant that it challenged the leadership's accommodation to the separate-and-unequal political structure and came out for a boycott of all the existing representative bodies. It wished in effect to base resistance policy on what Gandhi had called "noncooperation." Furthermore, the Youth League from the beginning advocated mass action and civil disobedience. Shortly after its first meeting, the organizers sought and obtained an interview with Dr. A. B. Xuma, the ANC president, in which, according to a summary in the ANC files, the young rebels accused the leadership of following an "erratic policy" as "was shown by the fact that there was no programme of action—no passive resistance or some such action. Dr. Xuma replied that the Africans as a group were unorganized and undisciplined, and that a programme of action such as envisaged by the Youth League would be rash at that stage."[34]

Xuma's postponement of mass action until some distant day when the ANC had organized and disciplined the masses enjoyed the support of Communists in the ANC leadership. Congress's most ambitious wartime campaign—the anti-pass agitation of 1944 which was jointly sponsored by the Communist party—did not involve the civil disobedience that had been a part of some earlier anti-pass campaigns, namely the turning in, discarding, or burning of the hated documents. Instead, the 1944 protesters merely circulated a monster petition protesting the pass system that was then presented to the government (which of course proceeded to ignore it). The Youth League's early anti-Communism was in part a response to the wartime timidity of a Communist-influenced movement that did not wish to impede the war effort by engaging in militant and disruptive action. Its view of Communists as unreliable associates with an agenda of their own paralleled that of the March on Washington Movement in the United States and had a similar justification. After the war, the congress went partway toward the Youth League's goal of a boycott of segregated institutions when the ANC-dominated Natives' Representative Council adjourned indefinitely in 1946 after it became clear that the government would pay no attention to its deliberations. But Xuma and the older generation of leaders continued to believe that mass action was ill-advised because it would lead only to violence and repression. In 1949 the votes of Youth Leaguers prevented Xuma from being reelected to the presidency, and their Programme of Action became the official position of the ANC.[35]

The Programme of Action redefined the aims of the South African liberation struggle and gave it a new tactical direction. For the first time in an ANC policy statement, it called for "political independence" and "the right of self-determination" for "African people." Other Youth League statements made it clear that this demand for "National freedom" did not mean "Africa for the Africans" in an exclusionary sense, but it did imply that the majority would seize power from the racist minority. It thus repudiated the ANC's traditional belief that Africans would be gradually elevated to participatory status by whites who had been persuaded that it was the decent and democratic thing to do. The heart of the action program itself was the establishment of a "council of action" to work for "the abolition of all differential political institutions the boycotting of which we accept and to undertake a campaign to educate our people on this issue, and, in addition to employ the following weapons: immediate and active boycott, strike, civil disobedience, non-co-operation and such other means as may bring about the accomplishement of our aspirations." The first thing to be done was to prepare "for a national stoppage of work as a mark of protest against the reactionary policy of the government."[36]

The methods of resistance proposed were precisely those that Gandhi had used in the Indian independence struggle, including his distinctive way of raising consciousness and testing discipline through the *hartel*, or one-day strike. But nowhere in the Programme of Action or in the surviving Youth League documents that anticipated it was there any mention of Gandhi and his campaigns. It is also noteworthy that the triumphant Youth Leaguers made no reference to the recent Indian Passive Resistance campaign in South Africa itself. This failure to cite Gandhian or Indian precedents for nonviolent resistance makes sense if one recalls that the Youth League was seeking to make the ANC an authentically and distinctively African movement. Its literature emphasized that Africans should not follow the leadership of Europeans or Asians and should not base their struggle on "foreign" ideologies such as Marxism. Gandhism was never explicitly included among the alien ideologies being rejected, and the method of action being adopted was strikingly similar to Gandhi's. But nowhere did the Youth Leaguers reveal any sympathy for the religious or philosphical nonviolence that Gandhi had espoused. Furthermore, the idea that the oppressor could be converted through his exposure to the undeserved suffering of the oppressed was alien to Youth League thinking. For them, even more than for A. Philip Randolph in the United States, nonviolent direct action was simply an available means to exert pressure and disrupt the plans of the white supremacists; it did not imply a repudiation of violence under all circumstances. In Gandhi's terms, the Youth League was proposing passive resistance but not *satyagraha*. But the emerging new leadership of the ANC conformed to the Gandhian precedent in one sense even while it departed from it in others. As we have seen, Gandhi had always conceived of his movement as an authentic expression of Indian culture or the Indian soul. The Youth League was striving to make the ANC a manifestation of the African soul which meant that it could not be based on values identical to those of Indian nonviolent resisters.[37]

A question not really faced then or later by ANC ideologists was where to find, or how to convey, the traditional African cultural values that would inspire people to make the sacrifices and endure the suffering essential to the effective practice of nonviolent resistance. The composite and unifying African nationalism to which the intellectuals in the Youth League referred was in fact a construction of their own which would have to be explained to Zulus, Xhosa, Tswana, and members of the other African tribal or ethnic groups. The role that religious beliefs might play as a stimulus to African nationalism and a source of morale and courage in the struggles to come was not fully explored. Most of the Youth Leaguers were themselves Christians, but they came from a variety of denominational backgrounds. Anton Lambede, the founder of the movement, and A. P. Mda, its president from 1947 to 1949, were practicing Roman Catholics. Oliver Tambo was a devout Anglican, who prepared at one time for the priesthood. Walter Sisulu also had a strong Anglican upbringing, and Nelson Mandela was a Methodist, although not a noticeably devout one. What is striking is that all of the leading Youth Leaguers were affiliated with missionary churches; none was a member of an independent African church or had any experience of separatist "Ethiopian" Christianity. In a remarkable letter written in 1949, Nelson Mandela addressed the question of what role Christianity should play in the African liberation movement and concluded on a note of great uncertainty:

> The question of our attitude toward the Christian religion is a very delicate and complicated affair. . . . It might well be, however, that as the forces of nationalism gather momentum in the political field there might be a corresponding upsurge of the same factors in other fields as well, including the religious field. I might mention the emergence of such denominations as the Bantu Methodist Church, the Bantu Presbyterian Church, and the AME Church, [which] despite the fact that they still embrace the Christian faith is not without significance. The existence of such Religions as Shintoism and Buddhism among the Japanese, Hinduism and Mohammedanism among the Indians, which are religions fashioned and firmly embedded in the National traditions and philosophies of these nations and which extol and deify their national heroes, cannot fail to exercize its influence on the future attitude of the African nation toward the Christian faith.[38]

Mandela wondered whether Christianity was compatible with African nationalism and whether black separatist churches that had not hitherto been active in the liberation struggle might become so in the future. In raising this question about the potential of independent black Christianity to be a vehicle for nationalist sentiment and mythology, he revealed his lack of awareness of the extent to which black churches had themselves provided much of the content for the early development of black nationalism. He was obviously viewing Africanized Christianity from a great distance, mentioning only the relatively orthodox churches of the "Ethiopian" type and registering no apparent interest in the more thoroughly Africanized "Zionist" churches. The gulf between the mission-educated intellectuals of the Youth League and the most dynamic tendencies in black popular religion would make it difficult to mount a nonviolent struggle that was a genuine folk movement and not simply a new and more radical expression of elite protest. In 1951, Dr. J. S. Moroka—the veteran African leader who had been elected president with the Youth Leaguers' support in 1949 because he had endorsed their program—called attention to the movement's lack of a religious impetus. As "a great step toward the realization" of "*African Nationalism,*" he recommended "consideration of the steps to be taken to establish a national church. . . . It is in the fold of your nation—the African nation—that you come nearer to God than in the foreign atmosphere of European churches."[39]

The 1950s was the decade in which nonviolent resistance received its most thorough trial in South Africa. The ANC's embrace of direct action stemmed from its frustration at the fact that its traditional method of passing resolutions and petitioning the government had failed to stop or even slow the flood of segregationist legislation. The triumph of the Youth League with its program for militant action took place at the time the Afrikaner Nationalist regime, elected in 1948, was preparing to elaborate its grand design for separating the races from the cradle to the grave and assuring the dominance of Europeans for all time to come. If the ANC had not responded in some dramatic fashion to the apartheid legislation passed between 1949 and 1952, its claim to be in the forefront of the struggle for African rights would have lost most of its credibility. The ANC did in fact gain significantly in prestige and membership as the result of its new militancy, although it failed to stem the tide of repressive and discriminatory legislation.

The most serious organizational and ideological problem that the National Congress confronted as it began to implement the Programme of Action was whether it should invite the participation and cooperation of groups that did not share the Youth League's original Africanist impulse. The Nationalist government's legislative barrage was not aimed exclusively at Africans. The principal victims of the Group Areas Act were Indians and Coloreds. The Coloreds in the Cape were also the targets of an effort to remove from the South African Constitution the entrenched clause protecting their limited voting rights. White radicals, along with those of other races, found their civil liberties denied by the Suppression of Communism Act of 1950—a measure that created such a broad and flexible definition of statutory Communism that it made the search for shades of red during the McCarthy era in the United States look almost benign by comparison. Sometime between 1949 and 1952, key members of the group of Youth League insurgents who were taking control of the ANC—especially Nelson Mandela, Oliver Tambo, and Walter Sisulu—made a crucial decision for reasons that went unrecorded but can readily be imagined. They decided to downplay the Youth League's original Africanist nationalism and depart decisively from its nativist tendency to have no truck with movements and ideologies of "foreign" inspiration. Under their influence, the congress welcomed support from whites, Indians, Coloreds, Communists, Gandhians, Christians—anyone, in other words, who repudiated apartheid and was willing to work against it.

Rarely has a liberation movement made fewer ideological demands on its supporters. Some in the ANC continued to fear that cooperation with whites, Indians, and Communists would lead to non-Africans taking charge of the movement. There was tension in 1950 over the implementation of the first step in the Programme of Action—a one-day work stayaway. The Communists took the initiative by declaring May Day as the occasion for a one-day general strike to protest the Suppression of Communism Act. Rather than simply falling in behind the May Day *hartal*, the ANC hastily organized its own day of protest for June 26. But in 1951 the ANC national executive, which included the Communists Moses Kotane and J. B. Marks, invited the leaders of the South African Indian Congress and the Franchise Action Council of the Cape (a predominantly Colored group) to discuss taking joint action against "the rising tide of national oppression." The resulting conference of executives backed a plan for all three groups to engage in nonviolent direct action in a coordinated fashion. In the end, the African National Congress and the South African

Indian Congress agreed to joint sponsorship of the Campaign of Defiance Against Unjust Laws. A five-man planning council, chaired by Dr. Moroka and composed of Sisulu and Marks representing the ANC and Dadoo and Yusuf Cachalia representing the Indian Congress, was put in charge of the campaign. Cachalia, who had spent five years in India, was greatly influenced by Gandhi, and both he and Dadoo could draw on their experience in the Indian Passive Resistance Campaign. It was, however, the Youth Leaguer Sisulu who took the lead in formulating plans for civil disobedience.[40]

The Defiance Campaign began in classic Gandhian fashion by making very specific demands on the government and threatening civil disobedience if they were not met by a certain date (February 29, 1952). Demanded were the repeal of the pass laws, the Group Areas Act, the Suppression of Communism Act, the Coloured Voters Act, and the recently passed Bantu Authorities Act. In an effort to appeal to rural Africans, the government was also enjoined to end the unpopular practice of forced cattle culling. Unlike Gandhi, who did not make demands that had no chance of being met and always allowed room for an honorable compromise, the Defiance Campaigners must have known that there was not the slightest chance that the Nationalist government would respond favorably to their ultimatum, even though they had refrained from asking for the equal political rights to which they were committed as an ultimate objective. In South Africa in 1952 the oppressor was inflexible and immovable. There was no viable alternative to creating nonviolent confrontations, even though the protesters were likely to be treated with brutality by the authorities and were unlikely to win the kind of concessions that would increase their belief in the effectiveness of nonviolence.[41]

The plan of action contemplated a three-stage process. In the first, a few "selected and trained" demonstrators would publicly violate certain laws in a few major cities. In the second, larger numbers of passive resisters would go into action in more "centres of operation." Finally, there was "the stage of mass action during which as far as possible the struggle should broaden out on a country-wide scale and assume a general mass character." The small groups of selected volunteers went into action on June 26, the second anniversary of the stay-away of 1950, under the leadership of Nelson Mandela as volunteer-in-chief. Most of the groups were composed of members of a single racial group acting against the law or laws that they found most onerous—for example, Africans violated the pass laws, while Colored and Indian protesters usually defied recently enacted legislation that applied Jim Crow-type segregation to them for the first time. But a few mixed groups went into action, and late in the campaign a small number of whites joined in defying segregation laws. By December more than 8000 people had been arrested. Actions took place all over South Africa but the greatest concentration of them and a majority of all those arrested were found in the Eastern Cape, an area that had a long history of politicization and an especially vigorous black labor movement.[42]

Although the official literature of the campaign made no mention of religious justifications for nonviolence, historian Tom Lodge notes that "a mood of religious fervor infused the resistance, especially in the Eastern Cape." Prayer, hymn singing, and nightly church services helped to give courage and determination to the resisters, and "the verbal imagery of the campaign involved ideas of sacrifice, martyrdom, and the triumph of justice and truth." Although the actual resisters maintained discipline worthy of *satyagrahis* and went to jail without protest, destructive rioting broke out in the wake of nonviolent actions in Port Elizabeth on October 18 and East London on November 9. In one of the riots a white nun sympathetic to the protesters was killed and mutilated by an African mob. This violence and disorder gave the government a pretext for repressive actions that made civil disobedience increasingly difficult, and the campaign lost its momentum. The projected third stage of country-wide mass action did not come close to realization. The original aim had been to escalate mass action to the point of a black general strike, but a combination of insufficient grass roots organization on the part of the congresses and the government's capacity to crack down effectively on public protest before it reached threatening proportions put this objective well out of reach. In order to prevent mass civil disobedience in the future, the government gave itself new powers to declare states of emergency and made breaking a law for the purpose of protesting against it a serious offense in its own right punishable by a combination of flogging and long terms of imprisonment.[43]

Assessing the place of the Defiance Campaign in the history of black resistance in South Africa is difficult. Was it an abortive Gandhian revolution or merely an extension of the ANC's traditional reformist strategy—a more dramatic and insistent way to bring black grievances to white attention? Albert Lutuli, the chief of a Christian Zulu community in Natal who was dismissed from his office in November 1952 because of his support of the ANC and its Defiance Campaign, gave eloquent expression to the reformist interpretation. He described "Non-Violent Passive Resistance as a non-revolutionary and, therefore, a most legitimate and humane political pressure technique for a people denied all effective means of constitutional striving." The campaign then in progress "may be of nuisance value to Government, but it is not subversive since it does not seek to overthrow the form and machinery of the state but only urges for the inclusion of all sections of the community in a partnership in the government of the country on the basis of equality." Since Lutuli would soon be elected president-general of the ANC, his interpretation—even if it does not accurately convey the original intentions of the campaign—would be influential as a retrospective view of what it was all about.[44]

A more radical interpretation of what was intended can be inferred from the plan to escalate actions to the point of mass noncooperation and something like a black general strike. Had this occurred, the campaign would have had more than "nuisance value" for the government. Naboth Mokgatle, a rank-and-file Communist, was critical of the campaign because it operated at the outset with small numbers of trained volunteers (and thus had to turn away some potential recruits), rather than encouraging the masses to engage in spontaneous action. He recalled in his autobiography how he told a meeting attended by Nelson Mandela that the right way to "break the apartheid machine" was "to throw in its spokes, its wheels, and all its parts everything they could—sand, rags, stones—to jam it. By that, I told them, I meant that hundreds and thousands of volunteers should flood police stations, courts, and prisons." There may have been validity to Mokgatle's claim that a slow-developing campaign gave the government a chance to respond effectively, but it is also clear that the original plan seemed calculated to produce as its end result just the breakdown Mokgatle was calling for. (Except in the highly unlikely event that the government met the campaign's demands at an earlier stage.) No one knows what would have happened if the economy had been brought to its knees and the authority of the government effectively undermined. The white settler regime could not have withdrawn from South Africa as the British did from India. Perhaps the safest conclusion as to what was anticipated from the Defiance Campaign is that its instigators did not really know what would happen but agreed that something new had to be tried because the old protest methods had proved unavailing against the onslaught of apartheid.[45]

Although the campaign failed to jam apartheid or even to slow it down, it did serve to give the ANC new legitimacy and credibility as a vehicle for popular black protest. Although Indian Congress leaders played a major role in directing nonviolent activity and in putting their own bodies on the line, the response of the masses of Indians was disappointing—much less than had been seen in the passive resistance of 1946–48—and virtually all the rank-and-file resisters were Africans mobilized by the ANC. As a consequence, the contribution of Indians as forerunners and co-sponsors of the campaign tended to be forgotten, and the ANC's position at the forefront of the anti-apartheid struggle was solidified. A left-wing rival, the Non-European Unity Movement centered in the Western Cape had criticized the ANC for more than a decade because of its failure to implement a policy of total noncooperation with the "herrenvolk," but it never recovered from the do-nothing image that it acquired from its opposition to the Defiance Campaign.

The ANC's membership on the eve of the campaign has been estimated at between 7000 and 20,000, a large proportion of whom must have been members of the educated elite or "African bourgeoisie." The actions of 1952 brought a dramatic upsurge of membership until the organization could claim 100,000 members. If this figure was not inflated, it had to have included many working-class adherents. Membership soon dropped back to about 30,000 during the inactive period following the campaign, but supporters greatly exceeded the hard core willing and able to pay dues. Considered as a recruiting device, therefore, the campaign was a great success. More difficult to measure was the change of consciousness that resulted. But it seems safe to conclude that it helped many Africans to lose a sense that white domination was an unfortunate and painful fact of life that could not be changed and to acquire hope that liberation was possible. As one young activist reported at the time, "there has been a transformation in the thinking of Africans—a revolutionary transformation." Never again could the regime argue with any plausibility that Africans were content with their situation.[46]

Nevertheless, the results of the campaign did little to encourage a belief that South Africa could achieve racial justice through nonviolent action. The apartheid regime was not the British government in India. (Even in India most nonviolent actions were successfully repressed, and Gandhian resistance was not the most important factor behind the British decision to grant independence in 1947.) The nationalist regime of the 1950s was hard-bitten, determined, and fanatical in its devotion to white supremacy. Its draconian response to the civil disobedience of 1952 foreshadowed the total ruthlessness of its subsequent efforts to repress African resistance. One of its most effective devices, first put into effect at the time of the Defiance Campaign, was to disable the resistance by lopping off its leadership. This was accomplished initially by banning leaders from public activity, later by accusing them of serious crimes and tying them up in long trials, and ultimately by imprisoning them after conviction for subversive activity or without trial under emergency laws or regulations. Under such repressive circumstances, it was virtually impossible to organize public protest demonstrations, to say nothing of mounting mass campaigns of civil disobedience.[47]

Although outright civil disobedience was difficult and dangerous after 1952, some types of nonviolent action were still possible. In the mid-'50s, the ANC participated in a number of boycotts or noncooperation campaigns, such as resistance to forced removals of Africans from the Western Areas of Johannesburg to what is now Soweto, a boycott of the schools to protest the new curriculum mandated by the Bantu Education Act of 1953, and boycotts of buses in Evaton and Alexandria in an effort to roll back fare increases. The bus boycotts had the greatest success, for here the issues were economic rather than political, and the adversaries were private bus companies rather than the state. But these were essentially spontaneous local initiatives; the ANC supported them but was unable to harness them to a broader political movement. Most historians agree that the school boycott and attempts to resist the forced removals near Johannesburg were conspicuous failures. They revealed that the collapse of the Defiance Campaign and the subsequent repression had left the ANC incapable of carrying out sustained and coordinated nonviolent campaigns.[48]

The most impressive nonviolent actions of the mid to late '50s were the mass demonstrations of African women against legislation extending the pass laws to them for the first time. Earlier efforts to force women to carry passes, going back to 1913, had been met with refusals to comply that had forced the authorities to back down. In the 1950s, the women resisted again. In 1955 1000 to 2000 women, predominantly African but including whites, Indians, and Coloreds, demonstrated at the seat of government in Pretoria under the leadership of the interracial Federation of South African Women (FSAW). Closely allied to the ANC and its Women's League, the FSAW had been formed in 1954 to draw attention to women's issues and coordinate women's protests. Spontaneous pass burnings in 1956 showed the depth of feeling on the pass issue among African women. In the first seven months of that year an estimated 50,000 women participated in 38 demonstrations. The campaign culminated with another mass rally in Pretoria, this time with 10,000 to 20,000 women participating. Demonstrations continued through 1957 and 1958. Especially dramatic was the mass refusal of female servants in Johannesburg to register under the new law in 1958, an action that included efforts to blockade government offices. But after many women were arrested for obstructing the issuance of passes, the ANC took direct charge of the campaign and called off civil disobedience. By 1959 the anti-pass campaign had dissipated, succumbing to government repression and the movement's loss of faith in the efficacy of nonviolent resistance.[49]

Under the circumstances that prevailed, it is not surprising that Congress's commitment to nonviolence began to wane. An editorial in the December 1958 issue of the ANC's journal *Liberation* reviewed a decade of struggle and repression, concluding that "one of the most important concepts behind the 1949 plan of work [the Programme of Action], which is related to the Gandhian 'satyagraha' idea, is clearly inadequate to meet the changed conditions of 1959." The editorial lamented that the Defiance Campaign's aim of filling the jails with protesters had become the policy of the government, which was trying 156 leaders of the resistance for treason. But this rejection of nonviolent direct action did not lead to a straightforward endorsement of violent resistance. "There is," the editorial concluded, "only one power that can end this system: the systematic, massive, enlightened and determined organization of the people in their national liberation organizations and trade unions. And for such organization, Congress needs a new sort of plan; a plan not based upon emotional platform appeals and heroic gestures, but upon relentless work, day and night, throughout the land, in town and country . . . . ." The plan referred to was probably the "M Plan" conceived by Nelson Mandela in 1953. It was essentially a method to counter repression by establishing a closely knit cellular organization that would allow the Congress to function and grow outside of the glare of public scrutiny. Implicit in the M Plan, however, was the possibility that ANC cells in a Nazi-like South Africa would function like the underground movements in occupied European nations in World War II and engage in sabotage and other acts of clandestine violence.[50]

The ANC's propensity to engage in public nonviolent resistance and thereby expose its leadership to prosecution and incarceration had thus pretty much exhausted itself even before the crisis of 1960 and the banning of public dissent against apartheid. The killing by police of 70 demonstrators from the breakaway Pan Africanist Congress as they tried to turn in their passes at the police station in Sharpeville revealed something that the ANC already knew —that efforts at mass civil disobedience were likely to lead to deadly violence and that the blood shed would be that of the protesters.

Even ANC president Albert Lutuli, a devout Christian and strong believer in the possibility of reaching the white conscience and achieving a reconciliation of the races, had to admit reluctantly by 1964 that he could not criticize those in the congress who had chosen the path of violent resistance. In 1952 Lutuli had tried to make Christian nonviolence the keynote of the ANC's struggle when he had concluded the speech resigning his chieftainship with the ringing affirmation that "the Road to Freedom is Via the CROSS." Although not as charismatic or eloquent as Gandhi or Martin Luther King, Jr., Lutuli might have been the prophetic leader of a nonviolent crusade if he had been allowed to play that role. As it was, he was either banned or on trial for virtually the entire period of his presidency. He could not attend meetings, lead protests, or make public statements. Lutuli's Nobel Peace Prize, the first to be awarded to an advocate of nonviolent protest against racism, was not so much a recognition of anything he had been able to accomplish as a tribute to what he stood for—his seemingly impossible dream of a peaceful transition to democracy and racial equality in South Africa. Chief Lutuli had sought in vain to make the advocacy and practice of nonviolence in South Africa live up to Gandhian ideals (in 1955 he criticized "our campaigns" for failing to embody adequately "the gospel of 'Service and sacrifice for the general good' "); but even if he had succeeded and if the South African struggle had been more deeply penetrated by the spirit of *satyagraha*, it is doubtful that the outcome would have been much different.[51]

The ANC remained nonviolent (if not purely Gandhian) for as long as it did partly because of its deep underlying commitment to a democratic multi-racialism. In 1955, the Congress Alliance—a federation of African, Indian, Colored, and white groups under the general leadership of the ANC—adopted and promulgated "the Freedom Charter," which signified the organization's decisive break with the Africanist tendency to see the future of South Africa in orthodox black nationalist terms. When the Charter proclaimed that "South Africa belongs to all who live in it, black and white," it meant that the enemy could not be defined as the white race and that liberation would have to be something other than victory in a race war. A commitment to the ultimate reconciliation and democratic coexistence of whites and blacks in South Africa prevented Congress from endorsing terrorism and the indiscriminate killing of whites even after it went underground and began to engage in sabotage. The view of Frantz Fanon and his followers that violence against white colonizers was a necessary catharsis and the very essence of black liberation would never influence the ANC in a significant way. What is remarkable is not that the congress turned to a highly selective use of violence in 1961 when it joined with the Communist party to form the clandestine military organization *Umkonto We Sizwe*, but that the violence subsequently perpetrated was so limited and resulted in so few white fatalities.

If the ANC turned to armed struggle after Sharpeville, it did not thereby license itself to engage in that savage and unconditional form of violence that has become so common in the modern post-colonial, post-imperial world—the no-holds-barred, zero-sum, eye-for-an-eye struggle of racial or ethnic groups for domination or survival. The multi-racialism or nonracialism of the ANC not only limited and contained its use of violence during thirty years of underground resistance, it also made it easier, when the time was ripe, to give up violence and return to other methods of struggle. Consequently the ANC's change in tactics between the mid-'50s and the early '60s was not a decisive shift from one absolute to another. It can be better described as a shift from a conditional and expedient use of nonviolence to an equally conditional and expedient use of violence.[52]

Martin Luther King, Jr., and Nonviolence in the American South

On the twenty-fifth anniversary of the Defiance Campaign, E.S. Reddy, an Indian former assistant secretary-general of the United Nations and a longtime observer of the South African freedom struggle, assessed its international significance and especially its effect on the black liberation movement in the United States: "The Defiance Campaign and the subsequent bus boycotts and other acts of non-violent resistance," he maintained, "were an inspiration to the black people in the United States in launching the Civil Rights Movement under the leadership of the late Reverend Martin Luther King, Jr."[53]

Unfortunately, the historical record gives little support to Reddy's claim. The instigators and leaders of the civil rights movement may have had some awareness of events a few years earlier in South Africa, and it would have been surprising if no one during the 1950s had noticed that bus boycotts were occurring almost simultaneously in the United States and South Africa—but a review of the available evidence suggests that nonviolent resistance in South Africa did not significantly influence or inspire comparable activity in the United States. The Defiance Campaign did cause some ripples while it was in progress. The Council on African Affairs, a group of black radicals who sought to influence American opinion on behalf of decolonization, circulated a petition supporting the campaign that garnered 3800 signatures and $835 in donations; and an interracial group of pacifists and democratic socialists formed the Americans for South African Resistance, which raised $5000 to aid the ANC. (A year later it became the American Committee on Africa). But interest did not extend much beyond such small coteries of Africa-oriented activists. A petition to the United Nations on African issues sponsored by the NAACP and 25 other black or interracial organizations mentioned the campaign in passing; but the association's organ *The Crisis*, which commented frequently in 1952 on the rise of apartheid, did not provide its readers coverage of the campaign against it.[54]

Rather than drawing on each other, the two movements both drew directly on Gandhi and the Indian precedent, although not in the same way or to the same extent. The American movement, more than the South African, sought legitimacy through a close identification with the Gandhian legacy of militant nonviolence. In November 1955, Harris Wofford (a young white civil rights advocate who would later advise the Kennedy administration on race policy) delivered a speech on "Gandhi, the Civil Rights Lawyer," that was brought to the attention of Martin Luther King, Jr., shortly after the onset of the Montgomery bus boycott and may have helped to inspire the latter's embrace of Gandhism. In it, Wofford referred briefly to the Gandhian and Indian contribution to the South African struggle against apartheid. If South Africa was noticed at all, it would seem, it was as a parallel branch from the Gandhian trunk rather than as a direct inspiration.[55]

If they have demonstrated anything, recent historians of the American civil rights movement of 1955–65 have shown that it sprang from no single ideological or organizational source and was certainly not the creation of one great man. The movement had its origin in the local struggles of black communities to protest against some of the most aggravating manifestations of the official racial discrimination that characterized all parts of the American South in the 1950s. Formal ideology and linkage with struggles going on in other communities, to say nothing of elsewhere in the world, came only after the fact of nonviolent activity at the grass roots. But King's prophetic eloquence and personal charisma may have been indispensible to the task of providing a broad audience with a compelling explanation and justification for the movement and putting a sense of urgency behind the call for national action on civil rights. Without his kind of leadership, the local struggles might have remained local—there would have been civil rights movements but no Civil Rights Movement.[56]

The issue of who sat where on the municipal buses was the catalyst for organized protest in several communities. The first of these was not Montgomery, but Baton Rouge, where in 1953 there was a successful boycott of the buses to reform (but not abolish) segregated seating by allowing blacks and whites to be seated on a first-come-first served basis with whites starting at the front and blacks at the back, rather than using the traditional system of having a reserved white section at the front that could then be expanded as necessary to make certain that all whites were given seats. The issue that arose two years later in Montgomery was exactly the same, but in this case the boycott lasted for a year rather than for only a few days as in Baton Rouge and thus could attract the attention of the country and serve as the catalyst for a broader movement.[57]

The Montgomery bus boycott did not begin, as is commonly believed, because a previously nonpolitical black woman felt tired and decided on the spur of the moment to refuse the bus driver's command to give up her seat to a white passenger. It is true that Rosa Parks had probably not intended to engage in civil disobedience on that particular day. But as a longtime activist and secretary of the local NAACP, she understood very well what she was doing when she defied the segregation law; and the man she contacted after her arrest, E. D. Nixon, was an important civil rights leader (he had earlier been president of the state NAACP and currently headed the Montgomery chapter) who had been actively seeking an incident that would provide a legal test of the bus segregation policy and an occasion for mobilizing the black community against it. The prime instigator of the boycott, Nixon was a Pullman car porter and union organizer; as a longtime admirer of A. Philip Randolph and an officer of the Brotherhood of Sleeping Car Porters, he had been drawn to the prospect of nonviolent resistance by the March on Washington Movement of the previous decade. His principal associate in the initial mobilization for a protest against the treatment of Rosa Parks was Joanne Robinson, leader of the Women's Political Council, a group that had also been concerned about the bus situation.

It was only after Nixon and Robinson had decided to launch a boycott and had called a meeting of the black ministers of Montgomery to generate support that the Reverend Martin Luther King, Jr., emerged as leader of the protest. King was drafted to be the head of the Montgomery Improvement Association partly because Nixon's work schedule kept him away from the city for several days each week. As pastor of the most fashionable black Baptist church in Montgomery and a recent arrival in town who had not acquired many enemies, King was viewed as a safe consensus choice to head the boycott effort. It was only after he had assumed this position that his talents as an orator and inspirational leader became fully apparent.[58]

That King did not start the boycott does not mean that he was unimportant or that he was merely the creation of a movement made by others. The Montgomery bus boycott turned into much more than anyone could have anticipated at the outset. What started as a call for a fairer system of bus segregation ended up as the first major nonviolent challenge to segregation itself. King and the movement matured simultaneously, each becoming by the boycott's end much more than either had been at the beginning. What made Montgomery different from Baton Rouge two years earlier was the adamant refusal of the white political authorities to accede to the modest original demands of the protesters. What would otherwise have been a boycott of a few days to achieve an adjustment of segregation became an epoch-making confrontation that initiated the ultimately successful assault against Jim Crow in all its forms.

Along with purely local factors that may have made the political leaders of Montgomery more rigidly segregationist than those of Baton Rouge, it seems likely that the differing responses of white power to black boycotts in the two cities was to some extent the result of national developments in the intervening two years. The 1954 Supreme Court decision in *Brown v. the Board of Education* that school segregation was unconstitutional had panicked southern white supremacists and undercut the moderate reformist position that gradual progress toward "separate but equal" would satisfy the nation's recognition of the need for racial fairness. Hence the refusal of the Montgomery city fathers to agree to a plan for separate bus seating that had previously been implemented in several other southern cities was a manifestation of the new mood of "massive resistance" against desegregation that had followed the Supreme Court decision. During the course of the boycott, King became fully aware that white intransigence made it vain to hope that segregation could be humanized or gradually chipped away. But it also showed him that it could be challenged dramatically and perhaps effectively through nonviolent action and that the masses of ordinary black people were now ready to withhold their cooperation from segregation and to accept the hardships and danger of participating in a militant movement for its total abolition.[59]

It was the creative use that King made of these insights that explains much of his success as a leader. Understanding that segregation had to be attacked root and branch, that nonviolence might be an effective weapon against it, and that masses of southern blacks could be persuaded to boycott, march, and even disobey unjust laws, his task was to refine, channel, and coordinate the intense desire for liberation that was welling up in Montgomery and other black communities throughout the South. The new hope for a revolution in race relations stemmed in part from the same Supreme Court decision that had provoked the white South to massive resistance. The combination of the new black hopes and the new white intransigence made nonviolent direct action an emotional necessity for many southern blacks, who saw white supremacists trying to shut doors that had recently opened a little for the first time. The potential friendliness of federal courts to those challenging segregation gave grounds for optimistic assessments of how the conflict might be resolved. But success would be less likely if black protesters engaged in violence. The fact that a few random shots were fired at the buses during the early stages of the Montgomery boycott showed what could happen. If black protesters against segregation engaged in violence or even disorderly and undignifed behavior, any moral and legal advantages that their movement enjoyed before the court of American public opinion might be lost. Elevating nonviolence from a tactic to a principle and making himself an exemplar of the Gandhian ethic, as King attempted to do during the course of the Montgomery boycott, would have been politically astute and highly expedient even if it had not reflected his true beliefs. King was converted to Gandhi's conception of nonviolent struggle at a time when the circumstances that he faced made this kind of commitment highly advantageous for the black liberation struggle.

In his book *Stride Toward Freedom*, King gave the impression that he had been a serious student of Gandhi even before he assumed the leadership of the boycott. But there is in fact little evidence in his papers before 1955 that he had more than a casual interest in the great exponent and practitioner of nonviolence. As a student of Christian theology, he of course wrestled with the practical implications of the Sermon on the Mount and the claims of pacifists that they alone were faithful to its ethical prescriptions. But he absorbed enough of the moral realism of Reinhold Niebuhr to have doubts about unconditional pacifism. By arming for self-defense at the beginning of the boycott, King clearly demonstrated that he had not yet become a thoroughgoing advocate of nonviolence. He became one under the influence of two advisers who had been dispatched to Montgomery by national pacifist groups to encourage the boycotters to learn from Gandhi. One was Bayard Rustin, a founder of CORE and longtime associate of A. Philip Randolph. The other was the Reverend Glenn E. Smiley, a white representative of the Fellowship of Reconciliation, the prominent pacifist group from which CORE had hived off. Although both were dedicated Gandhians, they derived their principles from different sources. Smiley's inspiration was clearly religious and came out of the Social Gospel and antiwar strains in American Protestantism. Rustin, on the other hand, was apparently not a believer; his devotion to nonviolence appears to have been rooted in a humanistic philosophy that condemned violence because it was destructive of the highest human values and possibilities. He also showed an acute awareness that the African-American minority had nothing to gain and everything to lose from the use of violence in its struggle for equality.[60]

It was Smiley who left the fullest report of how King came to embrace Gandhism. According to his recollection of their first meeting, King responded somewhat as follows to a query about whether he had been influenced by Gandhi: "As a matter of fact, no. I know who the man is. I have read some statements by him, and so on, but I will have to truthfully say . . . that I know very little about the man." But Smiley found King receptive to being tutored in Gandhism, and he came away from the interview with a conviction that "God has called Martin Luther King to lead a great movement. . . . King can be a Negro Gandhi."[61]

At times one almost gets the impression that a role had been scripted for King that he had to play whether he really wanted to or not. The expectation in both black and white religious circles that profound racial reform would only come with the emergence of a "Negro Gandhi" obviously put great pressure on him, once he had been chosen for the part, to give a good performance. One historian of the civil rights era, Taylor

Branch, believes that King remained a Niebuhrian rather than a Gandhian even after being being indoctrinated in nonviolence by Smiley and Rustin. The basis of King's nonviolent politics, he argues, was not Gandhi's writings, which he shows no signs of having actually read closely, but rather the discussion of Gandhism in Niebuhr's *Moral Man and Immoral Society* (1932).[62]

Assessing Branch's thesis requires a brief exegesis of Niebuhr's remarkable chapter on "The Preservation of Moral Values in Politics." First of all, Niebuhr attacked the claims of Gandhism that there was an absolute moral distinction between violence and nonviolence, pointing out that the latter, as practiced by Gandhi, sometimes involved coercing and injuring (economically at least) the target of the protest. Since it involved coercion and power, effective nonviolence could not maintain the kind of ethical purity that Gandhi ascribed to *satyagraha*. But Niebuhr went on to argue that there was a difference between actions, whether violent or nonviolent, which were inspired by "moral goodwill" and those that were not, and that "nonviolence is usually the better method of expressing good will" and was "a type of coercion which offers the largest opportunities for a harmonious relationship with the moral and rational faculties in social life."

After making this pragmatic and relativistic defense of nonviolence, Niebuhr went on to discuss the application of this "type of coercion" to American social problems and made his remarkable prophecy that the "emancipation of the Negro race in America probably waits on the adequate development of [nonviolent] social and political strategy." What would happen would not be a conversion of whites to racial justice but rather a demonstration of the fact that "the white race will not admit the Negro to equal rights if it is not forced to do so." If one desired a sophisticated defense of nonviolence that faced up to implications for the power politics of the real world but made it a morally sensitive and responsible doctrine that had special application to the condition of African-Americans, one could find it in Niebuhr. Taylor Branch believes that this is precisely what King did, but that he allowed the media and others to portray him as a pure Gandhian for the public relations advantage that this provided. As direct evidence for this somewhat devious strategy, Branch offers a secondhand report that King once told a colleague that nonviolence was "merely a Niebuhrian stratagem of power."[63]

But it is difficult to hold that Niebuhrian realism was the ideology that King drew from the Montgomery experience without indicting him for hypocrisy. Continually and insistently in the period during and after the boycott, he spoke and wrote of the power of love to disarm and convert the oppressor and of the redemptive quality of undeserved suffering. The goal of nonviolence, as he normally expressed it, was not a Niebuhrian balance of power that could check but not eliminate the human capacity for evil; rather it was "the beloved community" in which the prejudice and greed that lay behind racial discrimination had been eliminated. In an article published in the magazine *Fellowship* in May 1956 while the boycott was still going on, King wrote that his protest depended entirely on "moral and spiritual forces" and that its "great instrument is the instrument of love." The organization King founded after the boycott to encourage and coordinate nonviolent resistance throughout the South—the Southern Christian Leadership Conference (SCLC)—issued a pamphlet explaining its ideas that is, to all appearances, an unqualified celebration of Gandhism: "The basic tenets of Hebraic-Christian tradition coupled with the Gandhian conception of *satyagraha*—truth force—is at the heart of SCLC's philosophy," and its "ultimate aim is to foster and create the 'beloved community' in America where brotherhood is a reality."[64]

It remains true, however, that the bus boycott was more obviously an economic weapon to coerce the oppressor than a demonstration of concern for his soul. As Niebuhr had argued in the case of Gandhi's boycotts in India there was dissonance between the movement's idealistic philosophy and the understanding of power politics that its actions demonstrated. An early statement of King's that came close to recognizing that nonviolence was an exertion of power was his 1957 assertion that "a mass movement exercising nonviolence is an object lesson in power under discipline." By the time of the Birmingham campaign of 1963, King's rhetoric had lost some of its exalted idealism, and he was more willing than previously to recognize that coercion as well as a purely moral force was at work in nonviolent resistance. "Lamentably," he wrote in his famous Letter from Birmingham Jail, "it is an historical fact that privileged groups seldom give up their privileges voluntarily. Individuals may see the moral light and voluntarily give up their unjust posture; but, as Reinhold Niebuhr has reminded us, groups tend to be more immoral than individuals." In this reversion to Niebuhr, King did not quite concede that nonviolence, rather than being a way of awakening a sense of justice and humanity in the oppressor, was simply a form of coercion available to people who could not exert force in any other fashion, but he came very close to doing so.[65]

Does this mean, as David Garrow has argued, that King's experience with white resistance to his nonviolent campaigns cured him of the utopian belief that returning love for hate would change the oppressor's attitudes and make "the beloved community" possible? If we view Gandhian idealism and Niebuhrian Christian realism as two poles to which King was attracted, but to differing degrees at different times, then it would make sense to see him moving from an early phase in which he was closer to Gandhi in his thinking to a later one in which experience had increased his respect for Niebuhr's Christian realism.

Interpreters of King's thought have often pointed to the "Hegelian" or mediative tendency in his thinking, his attraction to contrary points of view and his efforts to reconcile them. In fact synthesizing Gandhi and Niebuhr did not require dialectical gymnastics, because the German-American Protestant theologian and the Hindu lawyer turned resistance leader and holy man shared some basic assumptions that would set them off from those who merely employed nonviolent means because violence was not a safe or practical option. The key point on which they agreed was on the need to act in the present so that in the future there was a possibility of reconciliation with one's current oppressors or enemies. Defining one's enemies as nonhuman or innately malicious and thereby justifying their annihilation could never be morally defensible. Niebuhr believed that "goodwill" might sometimes dictate violence, and Gandhi conceded that violence could sometimes result from peaceful protests. To both men—and to King—glorification of violence as an instrument of group liberation or advantage was rejected not simply because it was impractical under circumstances such as those that the African-American minority faced in the United States, but also because it did not admit the possibility that current oppressors and oppressed could ever be members of the same harmonious community. The hope of humanizing and rehabilitating the oppressors because one would have to live with them in the future did not inevitably require a total abstinence from violence and certainly not from coercive methods of nonviolence. Circumstances and patterns of behavior might have to be somewhat changed before any sense of common humanity could be awakened. But clearly violence and coercion had to be restrained and kept at a minimum; assassinations and other terrorist acts directed at enemies signified a desire to make them vanish forever. Resistance that showed by its restraint that it wished members of the oppressive class, race, or ethnic group to change their ways rather than disappear from the face of the earth offered the hope of a peace that was not the peace of the grave.[66]

There was one sense in which King's cause was closer to Gandhi's than to that of the white Christian moralists—whether realists like Niebuhr or unconditional pacifists like Glenn Smiley and A. J. Muste—who came from the Social Gospel tradition. The apparent contrast between King, the integrationist, and Malcolm X, the black nationalist, has obscured the fact that the southern civil rights movement was preeminently the expression of a group struggling collectively for its rights *as a people* and not merely as individual Christian believers or American citizens. To the considerable extent that the movement, and King as its most representative voice, found nonviolence to be the appropriate protest vehicle of a distinctive people struggling for its liberation, it recapitulated the "national" or "ethnic" impulse of Gandhian struggles in India and South Africa. Historian William Chafe has summed up very effectively the "national" roots of the movement, noting that "its strength was rooted in the collective solidarity and vitality of black institutions" and that "it spoke for a united community." Not often noted is the fact that SCLC, despite its commitment to an integrated society, deviated from the practice of older civil rights organizations by having an all-black rather than an interracial leadership. One need not deny the significance of the extrinsic sources of nonviolent ideology that influenced King and the leadership of

SCLC to recognize that the readiness of large numbers of black southerners to accept the discipline of nonviolence derived in large part from the fact that it was, in the words of Nathan Huggins, "consistent with a traditional black Christian belief and a kind of stoic Christianity." Much as Gandhism in India drew strength and cultural resonance from aspects of traditional Hinduism, southern nonviolence drew heavily on African-American folk Christianity and thus possessed a comparable ethno-cultural authenticity.[67]

Whatever its precise rationale and however it may have changed over time in the thinking of King and others, the insistence on nonviolent methods that King first articulated at Montgomery contributed enormously to the success of the civil rights movement. Nonviolence did not shame many southern white supremacists into accepting blacks as equals. Court decisions banning segregated facilities, such as the one that ended the Montgomery boycott, did not usually result in actual integration because whites often responded with their own boycott of desegregated institutions. But the ultimate banning of Jim Crow by federal legislation in 1964 would not have been possible if northern white racism had not been neutralized. The belief that black protest was strictly nonviolent was probably essential to the process of winning northern public sympathy and ultimately government backing. Sociologist Inge Powell Bell explained this necessity as resulting from the lack of a public consensus behind the ideal of racial equality: ". . . the Negro's claim to equality and his right to use strong methods to attain it were so widely questioned by the prevailing culture that even the members of the movement had to legitimate their activity in their own eyes by denying the extent of the coercion they used and by renouncing the right of self-defense." Whether or not Bell is right about self-doubts within the movement itself, it seems likely that the northern white majority that came to favor federal action against segregation found black assertiveness worthy of sympathy only because it was framed by an aura of heroic nonviolence. Blacks, it seemed, could be regarded as equals only if they showed themselves to be morally superior to whites. Many northern whites also had a fear of black violence stemming from stereotypical reactions to urban crime and disorder that the King's "Negro Gandhi" image must have helped to mollify. From a purely public-relations perspective, therefore, the nonviolent imagery was extremely effective.[68]

Besides providing the movement with a nonviolent theory and practice, the Montgomery bus boycott set a pattern for the broader civil rights movement in several other respects. Despite the founding of SCLC in 1957, the movement did not become a single coordinated campaign with a centralized organization at its head. Several organizations competed or cooperated under the broad banner of a nonviolent movement for civil rights. The NAACP continued to focus primarily on legal actions, but some of its local chapters in the South participated in, or even initiated, nonviolent direct action. The Student Non-violent Coordinating Committee (SNCC), founded in 1960 to coordinate the student sit-ins that began in Greensboro, North Carolina, was inspired by King and SCLC but chose to be independent. In the first year or two of its existence, SNCC proclaimed a somewhat more radical version of Gandhian nonviolent resistance than King himself espoused. Under the spiritual guidance of a charismatic divinity student, James Lawson, who had spent three years in India and knew Gandhism first-hand, SNCC *satyagrahis* broadened the agenda for nonviolent action beyond civil rights and viewed themselves as involved in the early stages of a nonviolent revolution that would not only bring legal and political equality to African-Americans but also radically transform the American social and economic system as a whole. Finally, a revitalized CORE brought its nonviolent experience and expertise to bear in the Freedom Ride of 1961.[69]

But the movement's diversity and decentralization went beyond a pluralism of national organizations. The deepest sources of energy and activism were, as in Montgomery, local black communities marching under a variety of organizational banners, the most important of which in a particular confrontation with white power might represent a purely local entity. Sociologist Aldon Morris has highlighted the importance of these "local movement centers." The supra-local organizations had a complicated and protean relationship to these centers. SCLC and SNCC began as efforts to federate or coordinate these local movements. At a later stage, they might be called in by local movements to add support to a campaign already in progress. Finally, by 1963 and 1964, they were planning and attempting to carry out broader campaigns that might involve entire states or focus on communities selected for strategic reasons in the battle for national opinion. As in Montgomery, the black churches served as the institutional base for the local movement centers.[70]

The central role of the black churches and the black clergy in the nonviolent civil rights struggle was almost inevitable given the facts of black political and social life in the South. No other high status figures or community leaders had as much independence from the white power structure as ministers. Unlike professors or teachers, for example, they were not employed by white-dominated or -funded institutions. Unlike doctors, lawyers, and businessmen they did not engage in activities subject to government control or regulation. Similarly, the churches were the only permanent institutions, except for fraternal organizations, within which southern African-Americans could expect to make collective decisions and govern themselves. Before they were movement centers, the churches were community centers. As the movement gained force and recruited supporters, it quickly became apparent that the only places mass meetings could be held were the churches; no other halls were likely to be at the disposal of black activists.

To say that the clergy and the churches were indispensible is not the same as saying that they originated the movement. The southern black churches did not have a history of social activism or militancy, at least not since the Reconstruction era. Many ministers had accommodated to white domination and concerned themselves more with the souls of their congregants than with their worldly condition. Even during the civil rights era of the late '50s and early '60s, a large number, possibly a majority, of southern black churches and ministers failed to give active support to the movement. The pattern of Montgomery, in which lay activists such as E. D. Nixon and Joanne Robinson instigated the movement and then maneuvered the clergy into the forefront, would be repeated elsewhere. Understanding that the ministers had more autonomy and immunity to reprisal, the nonclerical middle class activists, who had often dominated the southern chapters of the NAACP, stood aside when they could prevail upon ministers to take a leadership role. The new and more strictly church-based movement arose in the wake of the repression and virtual banning of the NAACP in some states of the Deep South. The local NAACP activists were not less militant or inclined to direct action than the ministers; on the contrary, their thinking was often ahead of that of the clergy. What is crucial to explaining the enhanced clerical role is not so much the beliefs and instincts of the clergy as the pressure from within aroused communities for a kind of leadership that only the ministers were in a position to provide.[71]

Nevertheless, ministers such as Martin Luther King, Jr., and some of his lieutenants in SCLC gave the movement much of its staying power and capacity for growth by reviving or reconstructing those aspects of the black religious tradition that could serve the intellectual and psychological needs of a militant nonviolent protest movement. A defining feature of African-American religious belief has been its hopeful eschatology—its expectation of God's intervention into history to free His children from bondage and sinfulness. The nationalist or Ethiopianist variant of black millennialism focused on the special destiny of the Negro race as a people chosen by God, but the orthodox mainstream of southern black Christianity tended to see the salvation of mankind in color-blind, universalist terms. But there was no better way to know if the millennium was approaching than to see if black people were being freed from oppression. Black Baptists and African Methodists could never succumb entirely to the individualistic pietism of many of their white counterparts; for to be black was to feel the need of some form of collective or corporate deliverance. But black ministers and church leaders could be quietistic and racially accommodationist when—as was often the case between Reconstruction and the 1950s—they threw up their hands in the face of overwhelming oppression and decided that black redemption would come only when God willed it. Religiously sanctioned nonviolence restored human agency to sacred history by making confrontation and resistance authentic expressions of black Christianity. When King recognized and articulated the fit between Gandhian nonviolence and what all Christians recognized as the strictest and most sublime conception of Christian ethics—the Sermon on the Mount's injunction to return good for evil—

he politicized and empowered black Christianity. The Christian Millennium, the "beloved community," and the racially integrated American democracy fused into a single vision among those who heard King's sermons and marched with him.[72]

If the religious sanction for nonviolence helped to account for the mass appeal, high morale, and steadfastness of the movement, how can we explain the comments of many of King's followers to the effect that his absolute nonviolence went beyond what they could really accept? There is no doubt that most southern African-Americans believed that they had a right to self-defense and were uncomfortable with an absolute prohibition of violence. Not fighting back when attacked by white racists was difficult and could be justified only if it could be demonstrated that such behavior was essential to the cause. But such an attitude did not preclude admiration for those who conformed to what Christians had to recognize was a higher ethical standard than the one to which ordinary people could be expected to adhere. It was part of King's charisma among black and white Christians that he projected the image of exceptional and almost superhuman virtue, the kind of holiness or saintliness that is often an attribute of religious and moral authority. In noting the problem caused by King's personal repudiation of self-defense, critics have sometimes failed to note two important factors that eased the tension in the movement. First, King did not insist that everyone in the movement endorse his position on self-defense or behave in a thoroughly nonviolent fashion; unless, of course, they were taking part in a nonviolent action and had taken a special pledge forswearing retaliation. More even than Gandhi, King refrained from imposing his entire philosophy of life on his followers. Second, disagreement on the abstract right to self-defense did not prevent a consensus behind the more fundamental proposition that for reasons of expediency *and* morality, *aggressive* violence was ruled out. Blacks could not hope to win an out-and-out race war and mobilizing for one would undermine the hope of someday living in harmony and equality with the white majority.[73]

The nonviolent civil rights movement of the 1950s and '60s is rightly regarded as one of the most successful reform movements in American history. It achieved virtually all of its formal objectives; the SCLC's Birmingham campaign of 1963 and the massive March on Washington under multiple sponsorship later the same year provided impetus for the Civil Rights Act of 1964, which successfully outlawed the segregation of facilities open to the public. In 1965, the protests in Selma culminating in the March to Montgomery led even more quickly and directly to the voting rights legislation that effectively employed federal power to guarantee that southern blacks had free acess to the ballot box. The barest recitation of these well-known facts is sufficient to refute any unqualified assertion that the movement failed. At a bare minimum, it sped up a course of development that was making the United States into a modern state based on equal citizenship under the law. The legal strategy of the NAACP would probably have succeeded eventually in having all forms of legalized segregation and publicly enforced disfranchisement declared unconstitutional, but without the nonviolent movement implementation and enforcement of these decisions would likely have taken several decades rather than one.

A fair and accurate balancing of the movement's achievements against its failures or limitations requires two distinct frames of reference. Viewed strictly in a southern context, the movement lived up to its slogan of "nonviolent revolution." It used a variety of coercive and disruptive methods—sit-ins, mass marches, and demonstrations—that often violated state and local law and went beyond normal American limits on peaceful dissent and reformist agitation. If these methods were nonviolent, they were likely to provoke violence and were to some extent intended to do so; although, as historian Adam Fairclough has shown, careful planning kept the violence to a minimum. The activists of the movement also acted like true revolutionaries when they rejected gradualism and demanded "Freedom Now." Within the context of a southern social order based on legalized racism, the movement was indeed revolutionary in its intentions and achievements. For Southern African-Americans, freedom from the day-to-day humiliation of Jim Crow, government protection of voting rights, and a measure of security against white supremacist terror and intimidation added up to a profound change in social and political conditions.[74]

But if the movement is viewed in a national context its essentially reformist character and the limits of what it was able to achieve become apparent. By bringing southern racial practices into harmony with those of the rest of the country, the national elites who acquiesced in the demands of the southern civil rights movement were in fact promoting the health and safety of the American social and economic system as a whole. An internationally embarrassing blot on the escutcheon of American democracy was removed by withdrawing explicit legal support for unequal racial status, but inequalities resulting from extralegal white racism and the normal operations of American capitalism remained. Consequently the roughly one-half of the black population that resided in the northern states and was not directly affected by the civil rights acts found that its circumstances were not simply unchanged but were deteriorating, as the low-wage, relatively unskilled industrial jobs that had provided earlier immigrants a point of entry in the American dream began to disappear from the American economy. The "de facto segregation" of housing, schools, and economic opportunites that characterized northern cities was, for the most part, beyond the reach of the civil rights laws, and the persistent, unrelieved poverty of a disproportionate segment of the black population in both regions (a legacy of past discrimination as well as current policies) made it clear to anyone who cared to think about it that the civil rights movement had not achieved full racial equality, to say nothing of "the beloved community." Martin Luther King, Jr., became increasingly aware of what had not been achieved, especially after the failure of his 1966 campaign for open housing in Chicago, and he turned during the last two years of his life to the advocacy of a form of democratic socialism and the organization of the poor for nonviolent action on the basis of class or economic status rather than race.[75]

Comparing Nonviolent Struggles

Unlike the transit of people and ideas from the United States to South Africa that characterized the spread of earlier ideologies such as Ethiopianism, Washingtonianism, and Garveyism, links between the American and South African nonviolent resistance movements of the 1950s and '60s were relatively tenuous. The South African Defiance Campaign of 1952 was not very much, if at all, in the thoughts of the Montgomery bus boycotters and subsequent southern nonviolent protesters. Similarly the American movement, once it emerged, does not seem to have had a significant, discernable influence on the thinking of the anti-apartheid movement in South Africa. Compared with the attention given earlier to Booker T. Washington, Marcus Garvey, or even W. E. B. Du Bois, King and his actions did not cause a great stir in South Africa during his lifetime. The popular black press, to the extent that censorship and repression allowed, did report occasionally on developments in the southern segregation struggle, but a review of important ANC literature for the period when King was active failed to yield references to him and the American movement. Apparently the ANC had little inclination to identify itself publicly with the American civil rights movement.[76]

This relative lack of mutual awareness and cross-fertilization was partly an accident of chronology. The Defiance Campaign occurred at a time when African-American interest in militant nonviolence protest was at a low ebb; in 1952 the March on Washington Movement was dead, and CORE was barely surviving. By the time that the Montgomery bus boycott of 1955–56 showed the possibilities of nonviolent protest against Jim Crow in the American South, South African interest in Gandhian methods was beginning to decline, as a result of the failure of the Defiance Campaign and subsequent boycotts to make any dent in the armor of apartheid. The most obvious result of nonviolent activism in South Africa, it seemed, was to provoke the government to greater repression and discrimination. At the time when the first student sit-ins of 1960 signified that the American civil rights movement was about to engage in nonviolent direct action on a broad front in the South, the African National Congress, having been banned and driven underground, was in the process of repudiating its nonviolent past and joining with the Communist party to carry out sabotage and prepare for armed conflict. Militant nonviolence was apparently demonstrating its utility in one case and its futility in the other. The American and South African struggles seem to have diverged decisively. The coming of the rigorous and oppressive apartheid regime at the same time that civil rights for blacks became a mainstream political issue in the United States would have provided ample cause for South African black intellectuals and protest leaders to break their habit of making close analogies between their own struggle for rights and the apparent progress of black Americans toward full equality.

The social ideologies of the two movements diverged during the 1950s even more decisively than did their choice of means to be employed in the struggle for liberation. In 1948, the ANC and the NAACP, then clearly the organizational expressions of the black struggle for equality, were headed by men of similar social and economic beliefs. Walter White, executive secretary of the association, and the American-educated Dr. Alfred B. Xuma, president-general of the congress, were both liberals who believed that racial justice could be achieved through the gradual reform of an essentially capitalistic society. A difference between them, however, was that White would have no truck with Communists, whereas Xuma was willing to cooperate with them in the pursuit of immediate objectives. During the 1950s and early '60s, the NAACP remained an anti-Communist liberal reform group. The direct-action groups that were at the forefront of the southern struggle did not fully share the association's fear of Communist subversion—one of King's closest advisers, Stanley Levison, had a party background, and SNCC at no point excluded Communists from membership. But the dominant assumption within the leadership of the nonviolent movement, at least up to 1963, was that its goals were compatible with democratic capitalism and were indeed a fulfillment of its promises of common citizenship and equal opportunity. It was only when some activists concluded in 1964 and 1965 that substantive black equality would not be achieved through pending or enacted civil rights legislation and required a fundamental transformation of the American economic and social system that broadly radical perspectives began to emerge, especially in SNCC.[77]

In South Africa, on the other hand, the triumph of the militant Youth Leaguers in 1949 led to the decline of bourgeois-liberal ideas and influences in the ANC during the '50s. Although initially anti-Communist, the Youth Leaguers, unlike the old-line conservatives in the ANC, were not advocates of liberal capitalism; for the most part they favored "African socialism," which meant trying to adapt the communal ethos of the precolonial African village to the economic life of a modern nation. The working alliance between the militant African nationalists and Communists that was cemented during the 1950s was premised on Communist support for African nationalism within South Africa and nationalist support for the Communist or Soviet position in world affairs. Hence the ANC journal *Liberation* condemned American intervention in Korea and in 1958 maintained that the United States had inherited from the European colonial powers the role of principal antagonist to the independence movements of Africa and Asia. Soviet suppression of the Hungarian revolution, on the other hand, was applauded. Communist manipulation or an uninformed anti-Americanism will not explain the depth of hostility to the United States found in the ANC literature of this period. It also has to be acknowledged that the Truman and Eisenhower administrations were giving strong support to a racist South African government that they regarded as a bulwark against the spread of Communism in Africa. In the United Nations during the 1950s, the United States generally voted against resolutions condemning South African apartheid out of fear of alienating a cold war ally. After one such vote, a young ANC activist wrote to his father, a leading non-Communist intellectual with longstanding ties to the United States, to express his total disillusionment with American policy. "I must say I am pretty fed up with the U.S.A.," Joe Matthews wrote to Professor E. K. Matthews in November 1952, "their stand is rotten and the Eastern nations have beaten the West on the colour issue. . . . I think America has lost African friendship. As far as I am concerned, I will henceforth look East where race discrimination is so taboo that it is made a crime by the state."[78]

African-American nonviolent resisters of the late '50s did not of course defend American policies toward South Africa and the Third World. In fact they identified from the beginning of their struggle with the independence movements of people of color elsewhere in the world and would have greeted with dismay evidence that the United States government was impeding these movements. When the Montgomery boycott was only two months old, King described what was happening as the local incarnation of a worldwide phenomenon: "The oppressed people of the world are rising up. They are revolting against colonialism, imperialism and other systems of oppression" (of which Jim Crow was obviously one). What King and other nonviolent civil rights leaders did not do until the Vietnam War raised the issue in an acute form in the mid-'60s was to criticize the United States government publicly and insistently for its role in suppressing popular struggles for independence from Western domination. There was a pragmatic justification for this restraint. It was to the United States government and to mainstream political groups that the movement was appealing in its effort to get segregation outlawed and the ballot made accessible. Criticism of the foreign policy of that government, especially when such criticism would have been immediately condemned as echoing the Communist line, would probably have damaged, perhaps fatally, the prospects for civil rights reform. The resulting failure of the movement to question publicly the world mission of American capitalism meant that its discourse could not have the relevance for the South African struggle that the rhetoric and ideology of some earlier African-American movements had possessed.[79]

The lack of sustained interaction on a common ideological wavelength does not obviate comparison between the two movements. The feature that makes comparison fruitful was of course the use of nonviolence as a method of protest and political pressure—in South Africa during the '50s and in the United States between 1955 and 1965. In one crucial respect the situation of southern blacks and black South Africans was the same—neither had the right to vote. Some forms of nonviolence are difficult to reconcile with democratic theory because they seek to nullify decisions made by a properly constituted majority. But in both of these instances the protesters were denied the suffrage and were able to argue that their employment of extraordinary means of exerting pressure was justified by their lack of access to other forms of political expression. Since 1943 the official aim of the African National Congress had been universal suffrage, and in the United States gaining the right to vote was a central goal of the civil rights struggle. John Lewis of SNCC recognized this commonality in his speech at the March on Washington in 1963: "One man, one vote is the African cry. It is ours, too." Attainment of equal suffrage would presumably reduce, if not eliminate entirely, the need for nonviolent direct action, especially in South Africa where blacks would then constitute a majority of the electorate. As Chief Lutuli put it in 1952, "Non-Violent Passive Resistance" is "a most legitimate and humane political pressure technique for a people denied all effective means of constitutional striving." Speaking at the Prayer Pilgrimage to Washington in 1957, King made a litany of the phrase "Give us the ballot," and promised that if it were done "we will no longer have to worry the federal government about our basic rights. . . . We will no longer plead—we will write the proper laws on the books."[80]

The social and economic content of the democracy to be achieved through a nonracial suffrage was not firmly established in either case. The Congress Alliance's Freedom Charter of 1955 had indicated that government ownership of some basic industries would be an essential part of the new order. In the United States, civil rights leaders such as A. Philip Randolph and Bayard Rustin were longstanding advocates of democratic socialism, and King had a strong inclination in that direction that surfaced toward the end of his career. But, as in the South African Communist conception of a two-stage revolution, the cause that required immediate attention was political and civil emancipation, a sine qua non for further progress.

The two movements were also comparable in that they recruited most of their leadership from a similarly situated section of the black community—what a sociologist might describe as the educated elite of a subordinate color caste. Studies of the social composition of the ANC through the 1950s have shown conclusively that the organization was dominated by members of "an African bourgeoisie" or "petite bourgeoisie." Since membership in this group depended more on education and professional qualifications than on wealth or relationship to the means of production and since this class did not function as a middle or buffer group within the larger society, the Marxian terminology is not entirely adequate. Distinguished both by their prestigious and privileged position within their own communities and by their lack of opportunity in the white dominated core society, members of the African elite were pushed inexorably into assuming leadership in the black freedom struggle rather than accepting the meager rewards of staffing some of the organs of "separate development." Paradoxically, apartheid provided new opportunities for independent professional and service activity in African townships and therefore helped to strengthen a group that was well situated to mobilize Africans against that very system. Studies of the social and professional profiles of ANC leaders and activists reveal that most of them came from relatively favored or "bourgeois" backgrounds and had acquired a far better than average education and set of occupational qualifications. But it also shows that their ambitions were limited by the horizons of their own segregated communities, and that their loyalties were with the people whose status disabilities they shared and whom they served on a daily basis. In the 1950s, lawyers were the professionals best represented in the top leadership, replacing to some extent the physicians who had held sway in the 1940s.[81]

The leadership of the southern civil rights struggle possessed similar social characteristics. In his study of the Montgomery bus boycott, Steven M. Millner described the emergence of "a relatively independent black professional class" as crucial to the development of the movement. The educated class that had been produced by the black colleges of the South and then returned to their own communities as teachers, ministers, lawyers, and physicians were in a position almost identical to that of the "African petite bourgeoisie." The main opportunities they possessed were in the service of their own people—with whom they shared the humiliations of Jim Crow and, more often than not, exclusion from voting booths to which uneducated whites often had access. The combination of personal frustration at the lack of wider opportunities and enforced solidarity with less advantaged blacks made members of this class, like their South African counterparts, ready for radical actions on behalf of equal rights. Teachers were the largest element in this educated group, but ministers played the most conspicuous role, because, as we have seen, they had both greater independence and higher status in their communities. In the 1960s, the younger aspirants to middle class status, the students in the black colleges, moved to the forefront of protest. Although many of them had working class backgrounds they shared much of the world view of the African-American educated class and, as students, were relatively free from day-to-day white surveillance and intimidation.[82]

It was a special product of legalized segregation that such elites were not—as is often the case under less stringent and blatantly racist forms of ethnic or colonial domination—subject to detachment and alienation from their communities by a system of rewards and opportunities that allowed a favored few to move into the lower ranks of the governing apparatus established by the dominant group. Where race per se is the main line of division in society, as it clearly was in South Africa and the American South, resistance will normally take the form of a cross-class movement led by members of the educated middle class. This does not mean, however, that less-educated and working-class blacks made little contribution to whatever success these movements achieved. It was the plain folk who sustained the boycotts, often at great personal sacrifice. In the campaigns to desegregate lunch counters and department stores in southern cities, the abuse or arrest of student protesters exposed the problem and raised people's consciousness, but it was the well-sustained consumer boycotts that brought local merchants to the bargaining table. Consumer boycotts played a lesser role in the South African struggle, although there was a relatively effective boycott of potatoes, a crop raised mostly with black convict labor, in 1958–59. South African blacks were so poor, and consumed so little that they could get along without, that the boycott weapon was less available to them than to the relatively more affluent southern African-Americans. Successful bus boycotts to protest higher fares showed what poor Africans could do when their vital interests were threatened and there was a way to fight back. The point needing to be made is not some elitist notion that educated people are more intelligent and virtuous than others. It is rather that freedom struggles fought on the terrain of color differentiation established by centuries of white racist domination were, and had to be, movements of peoples or communities rather than social classes.[83]

The most significant structural difference between the Defiance Campaign and the nonviolent civil rights movement was that the latter grew out of a number of local struggles and was sustained by strong organizations and institutions at the community level, whereas the former was for the most part a centrally planned, from-the-top-down operation. The one area where the Defiance Campaign achieved something like mass involvement was the Eastern Cape, where it was able to build on the firm base provided by a recent history of local mobilization and protest activity. But nothing like the network of "movement centers" that was the source of the American movement existed to buttress nonviolent campaigns in South Africa. Where such centers existed in South Africa they were usually tied to labor organizations and trade unions; in the United States it was the black churches and colleges that did the most to sustain local activism. Since every southern city had relatively prosperous black churches and many had some kind of higher education facility for blacks, such an institutional matrix for community protest was widely available, whereas black unions were well established in only a few places in the South Africa of the 1950s. Furthermore, South African black townships of the 1950s were quite different from southern black communities. Their populations, which included a large number of transients and illegal residents, were much less stable and significantly poorer; there was a proportionately much smaller middle class; there were fewer well established cultural or religious institutions; and there was little black entrepreneurship or business activity. Attempts were constantly being made to establish community associations, but they had much less success than comparable efforts in Montgomery or Birmingham.[84]

Even if the forces opposing each movement had been identical in strength and determination—which of course they were not—there seems little doubt that a centralized movement like the South African would have been easier to repress than the more decentralized and diffuse American movement. Even before the ANC was banned, the government was able to hobble it severely merely by banning or arresting its top leaders. In some states of the American South in the 1950s, the NAACP was rendered ineffectual by state legal harassment that came close to an outright ban. It was partly to fill the vacuum created by persecution of the NAACP that independent local movements developed. These local movement centers were more difficult to suppress by state action, and they flourished in places where the NAACP could no longer show itself. If such strong local communities and institutions had existed in South Africa, the government might have faced a variety of local actions that would have been much more difficult to counter than the nationwide campaign of the ANC in 1952. (In the 1980s, when a stronger matrix of community organizations existed, a more effective nonviolent resistance to apartheid became feasible.) The ANC's efforts, during the mid-'50s, to assume the leadership of local struggles over housing or transportation suffered from its failure to adjust its methods and organizational style to grass-roots initiatives. In the later stages of the civil rights movement, SCLC was sometimes accused of coopting local campaigns and undercutting local initiatives. But its great successes in Birmingham and Selma were the product of a skillful coordination of local, regional, and national perspectives. SCLC's genius was that it could channel and harness community energies and initiatives to make them serve the cause of national civil rights reform.[85]

Besides differing structurally, the two campaigns also diverged in the less tangible realm of movement culture and ethos. As the special prominence of ministers and churches in the American struggle reveals, religious belief and emotion were integral to the movement and helped to define its essential character. As prophet/saint of the movement, King was instrumental in making it a moral and religious crusade, and not merely the self-interested action of a social group. The opposition of large numbers of black churches and churchpeople to nonviolent direct action may belie any notion that African-American Christianity inevitably or automatically sanctions militant protest, but King's creative application and interpretation of the gospel showed that it had the capacity to do so. The South African struggle, unlike the American, did not produce a Gandhi-like figure who could inspire the masses by persuading them that nonviolent protest was their duty to God. There was a reservoir of religious belief and practice that might have been tapped—it surfaced at times in local actions. But the ANC leadership was mainly composed of highly educated men who had gone to mission schools and whose religious beliefs had little connection with those of the working-class Africans who had been attracted to independent Zionist churches by the millions. The rival Pan-Africanist Congress formed in 1959 made a greater effort to draw independent churches into the struggle, but it did not have time to accomplish much before it was banned. What King did that no South African leader had a chance of doing was to fuse a black folk Christianity that was his own heritage with the Gandhian conception of nonviolent resistance to define a cause that stirred the souls of its followers and disarmed the opposition of many whites. The more obviously conditional and pragmatic civil disobedience that characterized the Defiance Campaign did not have the same resonance.[86]

Of course the resonance of the American movement was in part the result of the extensive and usually sympathetic press coverage that it received, and, by the '60s, of its exposure on national television. The Defiance Campaign by contrast received relatively little attention from the white South African press and the international media (which is one reason why it did not figure more prominently in the thinking of African-American passive resisters.) The possibly decisive effects of contrasting press or media treatment suggests that differences in the nature of the movements may tell us less about why they ultimately succeeded or failed than we are likely to learn from examining their external circumstances — what they were up against.

The American protesters faced a divided, fragmented, and uncertain governmental opposition. The most important division among whites that the movement was able to exploit was between northerners who lacked a regional commitment to legalized segregation and southerners who believed that Jim Crow was central to their culture. A key element in the kind of success the movement was able to achieve stemmed directly from its ability to get the federal government on its side and to utilize the U.S. Constitution against the outmoded states'-rights philosophy of the southern segregationists. When King proclaimed that "civil disobedience to local laws is civil obedience to national laws," he exploited a tactical advantage the South African resisters did not possess; for they had no alternative to a direct confrontation with centralized state power. South African black protest leaders had long tried to drive a wedge between British imperial and South African settler regimes, but the withdrawal of British influence beginning as early as 1906 had rendered such hopes illusory. For all practical purposes, South African whites in the 1950s were monolithic in their defense of perpetual white domination. In the United States it was of course federal intervention to overrule state practices of segregation and disfranchisement in the southern states that brought an end to Jim Crow. In South Africa there was no such power to which protesters could appeal against apartheid.[87]

The geopolitical context of the cold war and decolonization of Africa and Asia also cut in opposite ways, ultimately helping the American movement and hindering the South African. In the United States, the competition with the Soviet Union for the "hearts and minds" of Africans and Asians, especially by the early '60s when several African nations had achieved independence, made legalized segregation a serious international liability for the Kennedy and Johnson administrations. As reasons of state were added to other factors working against Jim Crow, the federal government became more susceptible to pressures from the civil rights movement. In South Africa, on the other hand, fears of Communist subversion within the country and of Soviet influence in the newly independent African states of southern and central Africa panicked the white political leadership into pressing ahead with more radical schemes for racial separation and the political repression of the black majority. Underlying these contrary assessments of the dangers of black insurgency was the basic difference between a white majority facing a demand for the inclusion of a minority and a white minority conscious that extension of democratic rights would empower a black majority.[88]

It would be cynical, however, to see nothing in the positive responses of many white Americans to the civil rights movement except self-interested calculations. White America has not been of one mind historically on the place of blacks in the republic. In the North at least, there has been an alternative or oppositional tradition in white racial thought, originating in the antislavery movement, that advocated the public equality of the races and offered a standing challenge, although one that was only intermittently influential, to the deeply rooted white supremacist tradition that was a legacy of African-American slavery. At times, as during Reconstruction and in the 1960s, racial liberals became ideologically dominant and responded to black demands for civil and political equality with major reforms. (But being liberals they had great difficulty in addressing the problem of economic inequality.) In South Africa, by contrast, there was virtually no tradition of white racial liberalism that went beyond a benevolent paternalism and no strong set of democratic assumptions that could be applied to race relations. Nelson Mandela caught this difference when asked by an American journalist in one of his rare prison interviews during the 1980s why he had not followed the example of Martin Luther King and remained nonviolent: "Mr. Mandela said that conditions in South Africa are 'totally different' from conditions in the United States in the 1960s. In the United States, he said, democracy was deeply entrenched, and people struggling then had access to institutions that protected human rights. The white community in the United States was more liberal than whites in South Africa, and public authorities were restrained by law."[89]

Was it therefore inevitable that a nonviolent movement for basic civil rights would succeed in the United States and fail in South Africa? As probable as these outcomes might seem to be, one can readily imagine things turning out differently. It is arguable that without the astute and inspirational leadership provided by King and others the struggle for black civil and political equality would have taken much longer. Any claim that the Civil Rights Acts of 1964 and '65 were inevitable obscures the creative achievements of the movement. For South Africa, the 1961 decision of the ANC to sanction some forms of violence remains questionable; the full potential of nonviolent resistance may not been have been exhausted, and the sabotage campaign that resulted from the decision was, in the short run at least, a disastrous failure that devastated the organization. Historian Tom Lodge has pointed to the relative success of the last mass nonviolent action of the 1960s—the three-day stay-at-home of 1961—and has also noted that one ANC-affiliated organization that was not banned immediately after the Sharpeville massacre of 1960—the South African Congress of Trade Unions (SACTU)—had a capability for politically motivated strikes that was never fully exploited. Clearly the sabotage campaign that became the center of resistance activity in the 1960s exposed the ANC's top leadership to arrest and imprisonment. If nonviolence had its inherent limitations as a resistance strategy under the kind of conditions that prevailed in South Africa, it would be hard to establish from its record of achievement in the 1960s and '70s that the resort to violence, however justifiable in the abstract, represented a more effective method of struggle. Of course the key historical actors, such as Nelson Mandela and Oliver Tambo did not have the benefit of historical hindsight and can scarcely be condemned for trying something different when nonviolent resistance had obviously failed to move the regime and had become more and more difficult to undertake. Furthermore, sabotage at least kept the ANC visible at a time when the government was trying to destroy it. A revival of sabotage actions in the late '70s would be instrumental in restoring the ANC to the forefront of the anti-apartheid struggle.[90]

Martin Luther King, Jr., first showed a strong interest in South African affairs in 1957 when he acted as chairman of an international "Declaration of Conscience" against human rights violations under apartheid. In 1959, he wrote to Chief Lutuli to express his admiration for the ANC president's courage and dignity and to forward a copy of *Stride Toward Freedom.* The Sharpeville massacre in 1960 and the awarding of the Nobel Peace Prize to Lutuli for his espousal of nonviolent resistance heightened King's interest and prompted him to speak out more frequently against apartheid. In a 1962 address to the NAACP national convention, King invoked Lutuli to exemplify his doctrine of nonviolence: "If I lived in South Africa today, I would join Chief Lithuli [*sic*] as he says to his people, 'Break this law. Don't take the unjust pass system where you must have passes. Take them and tear them up and throw them away.' "[91]

King made his fullest statement about South Africa in a speech given in London, England, on December 7, 1964, as he was on route to receiving his own Nobel Peace Prize in Oslo.

> In our struggle for freedom and justice in the U.S., which has also been so long and arduous, we feel a powerful sense of identification with those in the far more deadly struggle for freedom in South Africa. We know how Africans there, and their friends of other races, strove for half a century to win their freedom by non-violent methods, and we know how this non-violence

> was met by increasing violence from the state, increasing repression, culminating in the shootings of Sharpeville and all that has happened since. . . . even in Mississippi we can organize to register Negro voters, we can speak to the press, we can in short organize people in non-violent action. But in South Africa even the mildest form of non-violent resistance meets with years of punishment, and leaders over many years have been silenced and imprisoned. We can understand how in that situation people felt so desperate that they turned to other methods, such as sabotage.[92]

Like Mandela two decades later, King was sensitive to the differences between the two movements that would make nonviolence more feasible and effective in the American case. But in the same speech he indicated a way that nonviolence could be brought to bear against apartheid. "Our responsibility presents us with a unique opportuntity," he told his British audience. "We can join in the one form of non-violent *action* that could bring freedom and justice to South Africa; the action which African leaders have appealed for in a massive movement for economic sanctions." Almost exactly one year after his London speech, King made another strong appeal for sanctions against South Africa in an address on behalf of the American Committee on Africa. "The international potential of nonviolence has never been employed," he said. "Non-violence has been practiced within national borders in India, the U.S. and in regions of Africa with spectacular success. The time has come fully to utilize nonviolence through a massive international boycott. . . ."[93]

King, who gave vigorous support to the sanctions movement for the remaining few years of his life, did not of course live to see the antiapartheid movement come to fruition without unleashing the violent revolution that so many thought would be necessary for the overthrow of official white supremacy. It is now possible to argue that the breakthrough that came with the release of Nelson Mandela and the unbanning of the ANC was as much, if not more, the result of international nonviolence as a fruit of the strategy of violent resistance inaugurated by the congress in the 1960s. The apartheid regime was not in fact decisively defeated on the battlefield or driven from power by a successful domestic insurrection. The armed struggle of the ANC served to remind the world that blacks were determined to be liberated from white oppression, but it was the moral disapproval of much of humanity that destroyed the morale and self-confidence of South Africa's ruling whites, and the increasingly effective economic sanctions that persuaded its business community and those in the government whom they influenced that apartheid had no future. Of course those sanctions would have been lighter and the disapprobation less sharp if the domestic resistance of the 1980s had not provoked the government into a final effort to use force to suppress dissent. But that domestic resistance was primarily a matter of withdrawing cooperation from the regime. Although violence occurred—and to some extent was encouraged by the ANC's call from exile to make the country "ungovernable"— the organized opposition within South Africa mounted a great domestic boycott to parallel the international one. The spirit of Gandhi, long since repudiated by the ANC, was alive and well in the nonviolent United Democratic Front, the domestic movement that rallied behind the ANC's goal of a nonracial democratic South Africa.[94]

In 1989, the Front defied government banning orders, and as the Mass Democratic Movement mounted the first disciplined nonviolent actions against specific segregation policies since 1960. Conspicuous as leaders of the marches and sit-ins that brought petty apartheid to an end even before the release of Nelson Mandela were clergymen such as Allan Boesak and Desmond Tutu both of whom had been greatly influenced by King and the church-based American Freedom Struggle—and the demonstrations they led featured the singing of African-American freedom songs. Nonviolence per se may not have been sufficient to liberate South Africa, but it is no longer possible to deny that it played a major role in bringing that nation to the brink of democracy. It would not be beyond the power of historical analogy to describe the successful anti-apartheid movement as Birmingham and Selma on a world scale.

· 7 ·

"Black Man You Are on Your Own": Black Power and Black Consciousness

Pan-Africanism in South Africa, 1944–1960

World War II awakened or strengthened independence movements in Asia and Africa and reduced the capacity of the Western imperial powers to suppress them. Unlike India, Indonesia, and Vietnam, where the war ended leaving strong nationalist parties or guerrilla armies ready to challenge European hegemony, the African colonies of England and France remained quiet and seemingly under control in 1945. But progress toward decolonization elsewhere encouraged African nationalists to act on the assumption that independence was a realistic aspiration. Except where there was substantial European settlement or other special conditions (as in Kenya, Rhodesia, Algeria, and the Portuguese colonies), these hopes were fulfilled by the early 1960s without the need for sustained guerrilla warfare. The emergence of new African nations helped inspire the American civil rights movement, as well as the anti-apartheid struggle in South Africa.

But the precise relationship of American and South African freedom struggles to the independence of black Africa and the Pan-African ideology that was associated with it was far from simple and straightforward. Were black Americans really Africans or had two or three centuries of exile and acculturation turned them into Americans? There could be no doubt about the Africanness of Bantu-speaking South Africans, but did the presence of substantial white and Indian settler minorities mean that a liberated South Africa could not be purely and simply African but would have to be a special kind of "multi-racial" or "nonracial" society? The 1950s saw a struggle within the African National Congress between the multi-racialists and the Africanists that ruptured the organization. In the late 1960s in the United States, advocates of Black Power and black nationalism challenged the integrationist ideology of the civil rights movement and gave new emphasis to African roots and identities. Shortly thereafter, the South African Black Consciousness movement borrowed some of the new African-American language of black pride and self-determination to question in a new way the ANC's insistent nonracialism. Both movements—Black Power and Black Consciousness—sought to encourage what Jean-Paul Sartre has called an "anti-racist racism"; they accepted the racial identity constructed for them by white oppressors and turned it against its creators by using it as a basis of solidarity and struggle against white domination.[1]

The postwar Pan-African movement was inaugurated at a conference in Manchester, England, in 1945. Unlike the earlier Pan-African conferences in which Du Bois and other African-Americans had played key roles, this one was dominated by representatives from Great Britain's African colonies, who anticipated that the winds of change in the British Empire would blow in favor of self-government. The only African-American in attendence was Du Bois himself, who was an observer rather than an official delegate. South Africa had two representatives, but they were chosen in an irregular and haphazard way; their main qualification was that they happened to be in England at the right time. The conference did pass resolutions denouncing the discriminatory policies of the Smuts government, but its deliberations attracted relatively little attention in South Africa. According to a leading historian of Pan-Africanism, the role of South Africa in the postwar movement was "peripheral" and "diminishing."[2]

Nevertheless, the election of Kwame Nkrumah as prime minister of the Gold Coast in 1951 and the subsequent struggle that led to the independence of Ghana in 1957 made black Americans and South Africans keenly aware of the rise of a free Africa. The ANC cabled congratulations to Nkrumah after his 1951 election victory and sent representatives to the Pan-African conference that the president of the new nation of Ghana hosted in 1958. Within the ANC a faction emerged that identified strongly with Nkrumah and his brand of Pan-Africanism. In the United States, the rise of Nkrumah did not have a dramatic effect on organizations and ideologies; but the hero of West African independence was cheered by a huge crowd in Harlem when he visited the United States in 1956, and interviewers seeking to gauge the impact of recent events in Africa on black Americans found a new sense of pride along with strong expectations that the regal image of Nkrumah would replace the negative stereotypes of blacks that were the stock-in-trade of white supremacists.[3]

The African-American organization that paid closest attention to Africa during the 1940s and early '50s was the Council on African Affairs, a small group of black radicals that sought to influence American policy on Africa. From its inception in 1937 the organization had taken a particular interest in South Africa; its principal founder was Max Yergan, who had spent seventeen years as a representative of the international YMCA in

South Africa. His South African experiences, which had included a significant behind-the-scenes involvement in African politics, had changed him from a Christian liberal into a Marxist-Leninist; from its founding therefore the Council on African Affairs was resolutely leftist and pro-Soviet. In 1939, Paul Robeson assumed the chairmanship of the council, and this celebrated singer, actor, and friend of the Soviet Union became the personification of the organization in the eyes of the public as well as a dominant force in its deliberations.[4]

In 1945, Du Bois invited Yergan and Robeson to participate in the Pan-African conference, but neither did so; and the council took no official notice of it, probably because the Communist party opposed the conference. (Its principal convenor, George Padmore, was known for his belief that Communism and Pan-Africanism were incompatible ideologies.) It was indicative of Du Bois' own growing sympathy for the Soviet side in the Cold War that he continued to be deeply involved in the affairs of the council. The rise of the militant anti-Communism and the witch-hunt atmosphere of the McCarthy era doomed the organization to a marginal position in African-American life. In 1948 Yergan underwent a strange conversion to fervent anti-Communism. This radical shift of allegiances of course split the organization, and Robeson and Du Bois emerged victorious in the ensuing struggle for control. The subsequent history of the council until its dissolution in 1955 was similar to that of other organizations that had been designated as "Communist fronts." It was in effect harassed out of existence and was replaced by the pacifist and interracial American Committee on Africa as the principal American expression of solidarity with African liberation. It is indicative of the weakness of black nationalism in the United States during the 1950s that there was no significant group of black intellectuals and activists who gave strong support to the non-Communist Pan-Africanism represented by Nkrumah and Padmore.[5]

The expression "Black Power," which would become the slogan for a black nationalist revival in the late '60s and early '70s, was first used in a widely noticed way in 1954 when it served as the title of black novelist Richard Wright's account of his trip to the Gold Coast to observe the nationalist movement. But Wright was the antithesis of a cultural Pan-Africanist who glorified African traditions. A confirmed modernist and rationalist, Wright deprecated the customs and values that Africans inherited from their precolonial past. He defended the independence movement as necessary to the modernization of Africa rather than as the revival of a glorious precolonial ethos, and his advice for Nkrumah and other emerging African leaders was "the militarization of African life" to root out the tribalism and superstition of "the fetish-ridden past." For Wright, who spoke for a segment of the African-American intellectual community, the emergence of "Black Power" in Africa or the United States was to be welcomed—but only if it was the fruit of the progressive scientific and technological spirit that had energized the West, and not the reflection of a romantic belief in the African soul.[6]

In South Africa, on the other hand, a mystical and philosophical Pan-Africanism was an indigenous growth as well as a way of thinking encouraged by the rise of independence movements elsewhere in Africa. Anton Lembede, the founder of the ANC Youth League in 1944 and its guiding spirit until his death in 1947 at the age of thirty-three, was a major exponent of the idea that Africanism represented an alternative to Europeanism in all of its aspects. For Lembede, the white man's civilization was based on materialism, rationalism, and individualism, whereas the true nature of the African was spiritualistic, intuitive, and communal. Rejecting Marxism as merely another Western materialistic philosophy, he called upon Africans to overthrow Western intellectual and economic domination by returning to what he conceived of as the essential outlook of their ancestors. Government would be based on the kind of localized consensus that had ruled traditional African societies, and the economy would be socialistic in the sense of community control of resources rather than as the fulfillment of a mechanistic Western ideology. Under Lembede's influence the Youth League initially adopted for its slogan the old Ethiopianist and Garveyite slogan of "Africa for the Africans"; the first affirmation in its 1944 creed was that "We believe in the divine destiny of nations." The nation in question was defined as all Africans "from the Mediterranean Sea in the North to the Indian and Atlantic Oceans in the South" who were expected to unify and "speak with one voice." It followed that "an African must lead Africans" because "no foreigner can . . . truly and genuinely interpret the African spirit which is unique and peculiar to Africans only." Lembede made it clear that the whites and Indians resident in South Africa were foreigners who should not presume to give leadership to the African liberation struggle.[7]

Lembede's notion that each nation has "its own divine mission" was influenced to some extent by his reading of European romantic nationalists, and at times he seemed to be propounding an Africanized version of the message that Afrikaner nationalists had derived from some of the same sources. But Lambede's cultural nationalism also had roots in the lived experience of Africans who had been deprived of ancestral lands and denied the right to follow traditional customs. It came ultimately from the same deeply rooted impulses that had given rise to Ethiopianism, Garveyism, and Africanist millennialism earlier in the century. His "imagined community" was not some specific historical nation that was seeking to regain its lost independence but rather black Africa as a whole. Africa, as "a black man's country," must be unified: "Out of its heterogeneous tribes," he wrote in 1946, "there must emerge a homogeneous nation. The basis of national unity is the nationalistic feeling of Africans, the feeling of being Africans irrespective of tribal connections, social status, educational attainments, or economic class."[8]

Within the African National Congress, however, Lembede's thoroughgoing cultural nationalism never became the predominant ideology. The active members of the Congress, including many of the more militant younger adherents who belonged to the Youth League during the late '40s, were too deeply embued with the cosmopolitanism and universalism that they had learned in mission schools, at Fort Hare Native College, or from their associations with white liberals and radicals to become wholehearted and consistent Africanists. The Youth League Manifesto of 1948 sought a middle ground between Lembede's Africanism and the universalist liberalism that had previously dominated the thinking of the ANC. Instead of calling for a definitive choice between Africa and Europe, it advocated "a policy of assimilating the best elements in European and other civilisations and cultures on the basis of what is good and durable in the African's own culture and civilisation." While affirming the Africanist principle that "Africa was, has been and still is the Blackman's continent," it went on to reject as "extreme and revolutionary" the kind of African nationalism that "centres around Marcus Garvey's slogan—'Africa for the Africans' "—and "the cry 'hurl the Whiteman to the sea' " in favor of a "moderate" version based on the realization that in South Africa "the different racial groups have come to stay." The goal therefore was not complete Africanization but rather "the abandonment of white domination" and "the inauguration of a free society where racial oppression and persecution will be outlawed." Lembede's influence persisted, however, in the manifesto's assertion that, as a practical matter, Africans would have to go it alone, for they "will be wasting their time and deflecting their forces if they look up to Europeans either for inspiration or help in their political struggle." Indians and Coloreds were also oppressed but in ways different from Africans; they should therefore form their own organizations which could then cooperate with the ANC "on common issues."[9]

The manifesto of 1948 was especially clear on one point—that there should be no cooperation with Communists. "There are certain groups," it noted pointedly, "which seek to impose on our struggle cut and dried formulae, which so far from clarifying the issues of our struggle, only serve to obscure the fundamental fact that we are oppressed not as a class, but as a people, as a Nation. Such wholesale importation of methods and tactics which might have succeeded in other countries, like Europe, where conditions were different, might harm the cause of our people's freedom. . . ." Commenting on the manifesto in a personal letter written in October 1949, Nelson Mandela, then national secretary of the Youth League, wrote that "The ground plan and cornerstone of our policy is African nationalism which is the exact antithesis of Communism." But after the Communist party of South Africa was banned in 1950 because of the danger that it was thought to pose for the apartheid regime, those Youth leaguers who had moved into the ANC leadership repudiated their earlier anti-Communism and cemented the alliance with the underground Communist party and its sympathizers—white and black—that would become a permanent ANC policy. Besides calculating that "the enemy of my enemy is my friend," emerging leaders such as Nelson Mandela, Walter Sisulu, and Oliver Tambo concluded that the United States, which was then supporting the South African government as a cold war ally, was now the principal international enemy of liberation movements, while the Soviet Union, whatever one thought of its form of government, was ready and able to aid in struggles against Western imperial domination. It was those former Youth Leaguers who retained a strong distrust of Communists (especially white ones) who became the Pan-Africanist dissidents of the 1950s.[10]

During the early 1950s, the ANC pursued a strategy of alliances with organizations representing the other South African racial groups. As we have already seen, the Defiance Campaign of 1952 was undertaken under the joint sponsorship of the ANC and the South African Indian Congress. In 1953, Professor Z. K. Matthews, a leading African intellectual and longtime ANC stalwart, proposed in a speech to the congress's Cape provincial conference that a "CONGRESS OF THE PEOPLE, representing all the people of this country irrespective of race or colour" be convened "to draw up a FREEDOM CHARTER for the DEMOCRATIC SOUTH AFRICA OF THE FUTURE." Shortly thereafter the Congress of Democrats, a small organization composed of white leftists, and the South African Coloured People's Organization were formed to join with the ANC and the SAIC in the "Congress Alliance." The Congress of the People, in which each of the congresses was equally represented, was held in June 1955, and out of it came the "Freedom Charter," the document that remains to this day the fundamental statement of the ANC's philosophy. Besides calling for political democracy and a partial socialization of the economy, it asserted in the first clause of the preamble that "South Africa belongs to all who live in it, black and white"—a clear repudiation of the Africanist view that South Africa was similar in all essential respects to the black nations that were struggling for independence elsewhere in the continent.[11]

The name given in the 1950s to the new doctrine of inter-group relations was "multi-racialism," and the Freedom Charter appeared to advocate a policy of ethnic federation rather than a straightforward form of majoritarian democracy. "ALL NATIONAL GROUPS SHALL HAVE EQUAL RIGHTS," the charter proclaimed, which meant "There shall be equal status in the bodies of state, in the courts and in the schools for all national groups and races." It was the multi-racial aspect of the charter that aroused the opposition of the minority that called itself "Pan-Africanist" and eventually broke away to form a separate organization. The Pan-Africanists objected strenuously to the equality of representation with the African majority that was accorded to the three smaller racial groups in the Congress Alliance and also apparently in the "democratic" government of the future. This denial of special claims of the Bantu-speaking Africans as the original inhabitants of most of South Africa and the overwhelming majority of the current population seemed to them perversely unfair, and they looked for a conspiratorial explanation. The villains of the piece, they concluded, were white Communists.

It did appear, in fact, that the Freedom Charter itself was written mainly by white members of the Congress of Democrats and was not even seen by ANC president Albert Lutuli before its presentation to the Congress of the People—although it was later formally approved without amendment by a special conference of the ANC. The socio-economic program of the charter—nationalization of banks and "monopoly industry," equal access to land, and the establishment of a welfare state—was in fact social-democratic rather than Communist in character; there was little in it that the British Labor party or the French Socialist party would have found objectionable. As advocates of a form of democratic socialism themselves, the Africanists were in no position to condemn the charter as the blueprint for a Communist South Africa. But it was their conviction that Communists concealed their true motives until they gained control of a movement or organization that could then be used to further the interests of the Soviet Union. Such an interpretation of the origins of the Freedom Charter could be found in an influential book published in 1956 by George Padmore, one of the leaders of the international Pan-African movement and a major influence on Pan-Africanist thinking in South Africa.[12]

It was, however, as *whites* as well as Communists or fellow travelers that the Congress of Democrats had allegedly conspired to highjack the African liberation struggle. Beneath universalist ideologies, the Pan-Africanists believed, were deeper loyalties determined by racial or ethnic origin. The Pan-Africanists also distrusted Indian leaders, some of whom were also Communists, for similar reasons; it was believed that they were maneuvering for ethnic advantage in a post-apartheid South Africa rather than simply enlisting in the struggle for majority rule. Many Africanists believed that whites and Indians were alien "settlers" in South Africa and not part of the nation that was seeking self-determination. (Coloreds, in light of their partially indigenous ancestry, might be included.) The harsh attacks on radical Europeans and Indians as untrustworthy outsiders in which some Pan-Africanist engaged led their opponents in the Congress Alliance to accuse them of racism.[13]

Robert Sobukwe, the most astute and articulate of Pan-Africanists, had a compelling and effective response to the charge that he and his followers were guilty of a black version of the racism that was the bane of South Africa. A man of undeniable humanity and integrity who was chosen to head the breakaway Pan-Africanist Congress in 1959, Sobukwe described the Pan-Africanists as the only true nonracialists or anti-racists because they alone denied the significance of the color distinctions that the regime had made the basis of apartheid. The Congress Alliance, he contended, had actually endorsed and perpetuated these white-imposed racial divisions by organizing itself as a federation of ethnically exclusive organizations and advocating special protections for minorities in a future South African constitution. Sobukwe did claim for the African majority the right to carry on its own struggle for national liberation, but his vision of the future was of a totally nonracial society in which groups currently defined by ethnicity or color would have no special political status. In the classic tradition of Western liberalism, he made individuals and not groups or classes the repository of rights, one of which was to be free of racial or ethnic discrimination.[14]

But there was an obvious weakness in Sobukwe's argument. The Pan-Africanists believed that the masses of Africans were acutely sensitive to the racial or national aspect of their oppression and could be mobilized most effectively for revolutionary action on the basis of an appeal to anti-white sentiments. But could racial feelings be turned off once they had been aroused and after they had fulfilled their function as a catalyst for action? Furthermore, as we have seen, the Pan-Africanists tended to believe that racial or national identities were more fundamental than identities based on class or mere membership in the human race. What reason, therefore, was there to expect that an African majority would not favor its own kind and discriminate against Indian and white minorities?

Differing views on the ontological meaning of race was at the heart of the philosophical quarrel between Charterists and Pan-Africanists. Supporters of the Congress Alliance, despite their willingness to make concessions to the de facto racial and ethnic divisions that existed in South African society, were reluctant to attribute any ultimate or deterministic significance to skin color or ancestry. If they were Marxists, they believed that identities based on class (or relation to the means of production) would prove in the long run to be more basic than those derived from race. In fact they tended to view a strong sense of racial identity as a form of "false consciousness" deliberately encouraged by the masters of capital. Liberals in the ANC drew primarily on Christian or humanistic conceptions of the potential solidarity of all humankind. On the surface the Pan-Africanists seemed to agree. Sobukwe denounced "the myth of race" and maintained that "Africanists take the view that there is only one race to which we all belong—the human race." Yet the conception of Pan-Africanism that the PAC had derived from Garvey, Du Bois, Lembede, Padmore, and Nkrumah was implicitly based on the notion that race was a profound and inescapable basis of personal loyalty and identity. What was it that potentially united all people of African ancestry throughout the world or merely in Africa itself if not race in some sense? Of course race did not have to be regarded as a purely biological or physical condition. In fact Pan-Africanist thought did not normally stress race in the biological sense and sometimes, as in the case of Sobukwe, denied it any significance except a symbolic one. What blackness symbolized, however, was the unique, irreducible historical experience of a subdivision of humanity, those people whose dark pigmentation had made them proper subjects, in the eyes of white Europeans, for enslavement and imperialist subjugation. If race was thus defined as the unavoidable legacy of racism, it was indeed difficult to deny its power.[15]

The Charterists and Pan-Africanists of the 1950s also differed on the historical and sociological meaning of South Africa. The Congress Alliance was premised on what Robert Sobukwe derided as "the fashionable doctrine of South African exceptionalism." According to this view of South African society, it differed in kind from the emerging nations in the rest of Africa, because of its substantial and permanent non-African minority populations. It might be justified for Tanzania, Ghana, or Nigeria to disregard its few European or Asian residents and conceive of its nationalism as a pure reflection of African culture and personality, but South Africa was a multi-racial and multi-cultural society, and its nationalism should derive from common political and economic concerns rather than from the cultural identity of a single group, albeit a majority. The Pan-Africanists, on the other hand, envisioned South Africa as essentially no different from other African societies. "Our contention," said Sobukwe at the inaugural convention of the PAC in 1959, "is that South Africa is an integral part of the indivisible whole that is Afrika. She cannot solve her problems in isolation from and with utter disregard of the rest of the continent." Since the aim of the PAC was the ultimate incorporation of South Africa into a United States of Africa, the existence of a white minority of less than one-sixth of the population in one corner of the continent would ultimately pale into insignificance.[16]

It would, however, be misleading to assume that a pure form of one or the other of these competing ideologies was wholeheartedly and unambiguously embraced by all the leaders and members of the rival organizations of 1959–60. A close examination of the rhetoric of both the ANC and the PAC reveals that each incorporated attitudes associated with the other, if only in a muted or covert fashion. The ambivalence of the ANC is evident in the fact that it remained a racially exclusive organization that did not in fact admit whites, Indians, or Coloreds directly into its ranks but required them to organize separately. An implict Africanism persisted among ANC supporters to the extent that they noted with satisfaction that an all-African organization was the flagship in the anti-apartheid fleet and would in the end have the strongest voice in deciding South Africa's future. PAC ambivalence was more obvious; it indulged freely in anti-white rhetoric while protesting loudly and strenuously against charges that its policy of racial mobilization meant that it had embraced a racially based nationalism that would deny meaningful citizenship in a future South African democracy to members of other races.

Perhaps the difference between the two organizations that had the greatest effect on the kind of support that they received from blacks was how militant and ready for bold action each was perceived as being. The top leadership of both organizations came from the same relatively well-educated middle class circles. But the PAC was led by less-experienced or -established figures who thought the ANC was insufficiently militant and action-oriented. They favored dramatic acts of confrontation and civil disobedience, which they hoped would lead to a massive, spontaneous uprising of Africans and the overthrow of white rule in a relatively short time. Younger people, the unemployed, and others who had relatively little to lose were readily caught up in this insurgent mood. The ANC, on the other hand, was more cautious and appeared to be hunkered down for a long and difficult struggle. Its leaders were more fearful of the physical force at the government's command and did not believe in mass spontaneity or the possibility of a quick victory over white supremacy. Africans who were older, had steady jobs, and were not prepared to risk everything in challenging the government before they had a realistic chance of succeeding were likely to be ANC supporters. In addition, the ANC, partly because of its ties to the Communist party with its longstanding commitment to organizing unions among blacks, had stronger support from industrial workers. Its affiliation with the South African Congress of Trade Unions gave it an access to the shop floor that the PAC could not match. For their part, organized black workers were more inclined to strike for immediate gains than to engage in a mass insurrection. It is not surprising therefore that it was the PAC that confronted the regime at Sharpeville and dragged the ANC into a make-or-break confrontation with the government that it might have preferred to avoid. The result, as we have seen, was the banning of both organizations.[17]

After 1960, while the relatively cool-headed Robert Sobukwe was languishing in prison, some of the PAC's more reckless adherents carried its policy of militant activism to another stage by attempting to strike at white rule through terrorist acts. An affiliated underground organization known as *Poqo* launched a series of violent attacks on randomly selected whites and on African collaborators with the regime; but a determined government response and poor security within PAC ranks soon led to the arrest of most of its members. The difference between the more controlled and less lethal violence of the ANC's military arm, *Umkonto We Sizwe*, and the out-and-out terrorism of *Poqo* highlighted the contrast between the ANC's political astuteness and ideological eclecticism and the PAC's rashness and rigidity.[18]

The Rise of Black Power in the United States

If we conceive of black nationalism in the way that George Padmore, Robert Sobukwe, and Kwame Nkrumah did—as an idealist Pan-Africanism that is neutral in the cold war and carries on a struggle against imperialism that is independent of the one sponsored by the Soviet Union—we would be hard pressed to find much of it in the United States in the 1950s and early '60s. Nkrumah may have inspired black pride, but he did not find many close students and followers of his philosophy. George Padmore's seminal book of 1956 arguing for the incompatibility of Communism and Pan-Africanism was not even published in the United States until 1971.

Du Bois and Robeson, for the most part, ignored Pan-Africanist thought that was not also pro-Soviet, and their preeminence among black radicals discouraged debates over the nature of black liberation similar to the ones taking place in South Africa. But black nationalism in a special American sense was undergoing a revival among some of the least articulate and most desperate segments of African-American society—the inhabitants of northern urban ghettos. This reawakening would sew some of the seeds for the multi-faceted resurgence of African-American nationalism that took place in the late 1960s.

Before the 1950s, the desire for a separate African-American nationhood was often—as in the case of the Garvey movement—closely associated with hopes for the redemption or liberation of Africa and the advocacy of at least some black American emigration to the mother continent. But at times the desire for independent nationhood, or at least its cultural and psychological equivalent, was more loosely tied to Africa and envisioned autonomy as being achieved within the existing territory of the United States or elsewhere in the Western Hemisphere. Disillusionment with the United States and the prospects that it offered for the equality and dignity of black people, rather than strong attraction to Africa, was the common thread running through the entire history of African-American separatism. In 1947, W. E. B. Du Bois cogently described the essential context of American black nationalism in all periods. "The so-called Negro group . . . while it is in no sense absolutely set off physically from its fellow Americans, has nevertheless a strong, hereditary cultural unity born of slavery, common suffering, prolonged proscription, and curtailment of political and civil rights. . . . Prolonged policies of segregation and discrimination have involuntarily welded the mass almost into a nation within a nation." What is striking about this description of the essence of African-American nationalism—coming as it did from the father of Pan-Africanism himself—is that it makes no direct mention of Africa. It also suggests how hopes of ending segregation and discrimination could have led many black leaders to identify with a reformed, racially egalitarian America. (Du Bois, however, would have insisted that this commitment to the American Dream should not require a loss of national or ethnic consciousness.) Belief that this project was unrealizable because of the strength of white racism, on the other hand, has led to separatism—the desire for secession from the United States, if not in a literal political and territorial sense, at least in a cultural and psychological one.[19]

In the late 1950s Americans began to notice a new manifestation of this broader nationalist impulse—the growing and increasingly visible Black Muslim movement. The Nation of Islam was founded in Detroit in 1930 by a shadowy figure who went under various names, one of which was W. D. Fard. Fard claimed to be the reincarnation of Noble Drew Ali, the recently deceased leader of an earlier Islamic cult called Moorish Science that had established several temples in northern cities between 1913 and

1925. After Fard mysteriously disappeared, Elijah Poole, who took the name of Elijah Muhammed, became the leader of the cult and the "messenger" sent by Allah to redeem African-Americans from Christianity and white domination. The Nation grew slowly during the 1930s and actually declined during the war years when Elijah Muhammed was imprisoned for draft resistance. But beginning in 1945 the movement began to grow steadily and during the late 1950s this growth accelerated markedly. The number of temples doubled from 15 in 1955 to 30 in 1959, giving the Nation a presence in 28 cities in 15 states. In that year a television documentary viewed by millions of Americans represented the nation as a black "hate group."[20]

Elijah Muhammed preached that blacks were the original human beings created in the image of God, whereas whites were "devils" who had been created later by a mad scientist and given power to rule the world for 6000 years—an era that would soon come to an end. Islam was the black man's religion (Elijah Muhammed claimed that "so-called Negroes" were really "Asiatics" and not racially distinct from Arabs), and Christianity was the white man's faith that had been taught to blacks in order to keep them subservient and exploitable. Divine retribution would, in the near future, be visited on the white devils, and blacks would be restored to their rightful place as masters of the world. In expectation of this apocalypse, blacks should separate themselves from whites and endeavor to establish an independent nation of their own, either by emigrating to Africa or by gaining sovereignty over a portion of the United States (usually described as several southern states). It reflected the vagueness of the Nation's sense of Africa and its meaning for black Americans that it was the prospect of a separate nation within North America that was most often proclaimed as the movement's political objective. It seems likely, however, that the minds of the faithful were more apt to dwell on the prospect of divine intervention to come than on strategies for the achievement of a separate nationhood in the here and now. According to E. U. Essien-Udom's classic study of 1962, "The Nation of Islam is not a political movement. Although black nationalism is ideally a separatist type of political ideology, the Nation of Islam is in fact apolitical. It is also nonrevolutionary."[21]

The Nation, like other successful messianic or millenarian religious movements, created a separate world for its converts that isolated and protected them from the pain of confronting the world outside. For the mostly lower-class urban blacks who were attracted to the movement, it offered relief from poverty, desperation, painful feelings of worthlessness or inferiority, and the physical dangers of drug addiction or criminality. Significantly, its greatest appeal was to people in northern ghettos who would gain little from the achievement of the civil rights movement's goal of ending legal segregation in the southern states. Successful in turning convicts and drug addicts into sober, law-abiding citizens, the Nation posed no direct danger to the American racial or economic status quo. The anger of its adherents against whites was not usually channeled into confrontational action but found its outlet mainly in ritualistic denunciations and eschatological fantasies. The news reports suggesting that the Muslims were about to go to war against white America were not the result of a real threat to the social order; they stemmed rather from the guilt and anxiety that the civil rights movement was producing among whites in the late '50s and early '60s. The usual subtext of reports on "the hate that hate produced" was that something needed to be done quickly to address legitimate African-American grievances or a race war might occur. The civil rights movement capitalized on this fear by offering itself as an antidote to the spread of the kind of violence-prone black racism that the Nation allegedly represented.

But the angry racial separatism of the Muslims had the potentiality of being redirected from the realm of millenarian expectation and moral rehabilitation (where Elijah Muhammed tried to keep it) into the sphere of political confrontation and conflict that the nonviolent civil rights movement had created. The pioneer of this refocusing of nationalist sentiments was of course Malcolm X, born Malcolm Little. His celebrated autobiography—one of the great classics of African-American literature—tells a story too well known to require repeating beyond simply recalling that he had risen from a life of petty crime and penal servitude to become the charismatic minister of the Nation's New York temple and that it was mainly as a result of his efforts that the Nation experienced its phenomenal growth during the late '50s and early '60s. A brilliant orator and commanding personality, he brought Elijah Muhammed's message concerning the sins of white America and the proximity of divine retribution to a black audience that went far beyond the ranks of potential converts to a religious cult that demanded a total commitment and strict discipline from its members. When Malcolm spoke, listeners found that they did not need to accept the literal truth of his fantastic account of human history but could appreciate it instead as appropriate allegorical or symbolic rendering of the black experience.[22]

Malcolm's resignation from the Nation of Islam in 1964 stemmed in part from the tension in his relationship with Elijah Muhammed that developed when Malcolm discovered that his leader practiced the marital infidelity that he had strictly forbidden to his followers. Also threatening his position in the movement was Elijah Muhammed's growing jealousy at the fact that Malcolm's fame was beginning to exceed his own. But added to these personal antagonisms was the issue of whether Muslims should engage in political action. Malcolm was impressed by the activism of the civil rights movement, although he noisily repudiated its goal of integration; in 1962 and '63, he attempted on several occasions to get permission from the Messenger of Allah to participate in demonstrations or other actions aimed at increasing black employment opportunities in New York and New Jersey. Permission was invariably denied, but Malcolm and the members of the New York temple sometimes violated these instructions and surreptitiously lent their support to the battle for economic opportunity in the urban North. Malcolm hovered on the fringes of the March on Washington of 1963, expressing his disdain for what he called "the farce on Washington," but nevertheless displaying an inordinate interest in something that Elijah Muhammed simply ignored. The occasion for the suspension from office that ultimately led to Malcolm's resignation was his comment on the assassination of John F. Kennedy—the famous "chickens have come home to roost" remark—implying that white America was being violently punished for its violence against African-Americans and people of color throughout the world. Although this comment was in accord with Elijah Mohammed's prophecy of divine retribution against the white devils, it was too pointed, topical, and controversial to suit the Messenger of Allah.[23]

In the period of slighty more than a year between his break with Elijah Muhammed and his assassination, Malcolm made a pilgrimage to Mecca that led to his endorsement of an orthodox, non-racialistic form of Islam, paid a visit to Africa where he found himself hailed as an authentic representative of black America, founded a nonsectarian political organization—the Organization of Afro-American Unity—that attempted with limited success to bring black Americans together on a platform of black nationalism, and gained credibity among black activists for his criticisms of the civil rights movement's commitment to nonviolence.

His statements during this period lacked perfect clarity and consistency; for example he vacillated between rhetoric that appeared to justify an anti–imperialist ghetto uprising and a more moderate emphasis on community organizing and use of the ballot to improve the lot of African-Americans within the existing structure of American society. He repudiated the Muslim claim that whites were "devils," but he lambasted white liberals as false friends of blacks and denied them a place in the black freedom struggle. He flirted with social and economic radicalism but continued to view race rather than class as the crucial division in American society. Exploring a variety of strategies for fighting white supremacy in the United States, he failed in the months before his assassination to settle on a definite program of action. His famous call for justice "by *any means necessary*" did not specify which means were likely to prove most fruitful. But at least it ruled out a commitment to nonviolent protest as the only acceptable method.

Nonviolence without even the sanctioning of self-defense was in his view a betrayal of black manhood and a concession to white ideas of black submissiveness and inferiority. He also regarded integration as a surrender to white supremacy; for its aim of total assimilation into white society implied that blacks had nothing worth preserving. His affirmation of a positive black culture and identity that must be respected, preserved, and given some form of political expression put him in the mainstream of black nationalist thought extending back to Marcus Garvey, Bishop Turner, and Martin Delany —and it was the main source of his growing popularity among African-Americans at a time when the dream of integration was beginning to fade. But he was unable to resolve the practical question of how the struggle for black self-determination should be conducted.[24]

The black writer and activist Julius Lester wrote in 1968 that "More than any other person, Malcolm X was responsible for the growing consciousness and new militancy of black people." Indeed the three years since Malcolm's death had seen his basic ideas become common currency among black intellectuals and activists. Most of the rhetoric and policies of the "Black Power" movement were anticipated by Malcolm, especially during the last year of his life. His basic message was increasingly endorsed because of changing circumstances that made his ideas even more persuasive in the period immediately after his assassination than they had been while he was still alive.[25]

A major turning point was the bloody and devastating riot in the Watts district of Los Angeles in August 1965. Coming only a few days after the civil rights movement had reached its legislative climax with the passage of the Voting Rights Act, it revealed that the end of legal segregation and racially discriminatory voting restrictions in the southern states did not address the main problems of poor, urban blacks, who suffered less from a denial of their civil rights than from unemployment, inadequate housing, and limited educational opportunities. Fully enfranchised, they often did not bother to vote because none of the parties or candidates addressed their problems. The residential segregation they suffered from was sustained by a bewildering complex of laws, regulations, and institutional practices that were ostensibly impartial or color-blind but had the effect of disadvantaging blacks more than whites. Martin Luther King, Jr., attempted in Chicago in 1966 to use nonviolent direct action to bring about open housing and equal employment opportunities, but he could not overcome the combined opposition of the political machine, the business community, and the working-class or lower middle-class white "ethnics" who had few qualms about using violence to prevent black penetration of their neighborhoods. As riots broke out in other cities in the summers of the late '60s and as a white "backlash" against any proposals to help blacks that went beyond the narrowest conception of legal and political equality became a decisive force in American politics, it became self-evident that white racism—increasingly defined by blacks as a structure of social relations rather than as an ideology or attitude—was too deep and pervasive to be overcome by civil rights laws that addressed the symbols of black inequality rather than the underlying causes. What then was to be done? For disciples of Malcolm X, the answer was that blacks should give up the hopeless and demeaning struggle for integration and seize the power to determine the destiny of their own ethnic community "by any means necessary."

The most compelling alternative to following the example of Malcolm X during the two or three years after his death was to focus on poverty rather than racism per se as the problem to be addressed. Since African-

Americans were a minority of all impoverished Americans, this redefinition of the problem opened the possibility of coalitions with other disadvantaged groups, such as Indians, Mexican-Americans, Appalachian whites, and segments of the white urban working class. It was also an approach that seemed for a time to have some support from the federal government. In his Howard University speech of June 1965, President Lyndon Johnson announced that the logical next step in the pursuit of racial equality was to go beyond equal rights and begin to work for the equalization of life chances by waging a "war on poverty." Veteran black leaders such as Bayard Rustin and A. Philip Randolph proposed a New Deal-type coalition between blacks, white liberals, and organized labor to push for vastly increased government expenditures for social programs designed to provide all Americans, but especially blacks, with a decent standard of living.

But the war on poverty and the politics of interracial social democracy were casualties of the Vietnam War and the rightward drift of American politics leading to the strong showing of George Wallace and the election of Richard Nixon in 1968. Martin Luther King, Jr., shared the view that the economics of poverty and privilege had to be the new focus of the freedom struggle, but he also appreciated the depths of American racism and lacked the faith of Rustin and Randolph in the willingness of white liberals and labor unions to support a redistribution of wealth that would primarily benefit blacks. Unlike them, he also realized that the Vietnam War was a disaster for those hoping for racial justice in the United States and that stopping it was a precondition for progressive domestic action. His last national campaign, which he did not live to see come to fruition, was an interracial poor people's march on Washington. Unlike most other black leaders of the late '60s —whether militant and revolutionary or moderate and reformist—King never surrendered his belief that nonviolent direct action could play a major role in the struggle for a just and decent society. But now it was the poor of all races and not the African-American community as such that he hoped to lead in nonviolent protest against economic inequality.[26]

It was the reaction of young black civil rights activists in the South against two of King's cardinal principles—interracialism and nonviolence—that provided the immediate context for the emergence of the Black Power slogan. For many of them, the failure of the Democratic party in its 1964 convention to seat the insurgent black delegation from Mississippi tarnished the image of mainstream white liberalism beyond repair. In both SNCC and CORE, the spirit of "black and white together" that had characterized both organizations before 1963 had given way by 1965 to a growing feeling that the presence of whites in the movement was inhibiting to the growth of black initiative. Even those whites who volunteered for hazardous duty in the South were criticized for their reluctance to accept black leadership and their inablity to relate successfully to black communities. This increasing distrust of left-leaning whites did not derive so much from relations with veterans of early civil rights campaigns like Bob Zellner of SNCC or Jim Peck of CORE as from the tension between black organizers and the hundreds of inexperienced volunteers who had flooded into Mississippi during the Freedom Summer of 1964. White students from the Ivy League and rural southern black activists did not in fact have much in common and were bound to misunderstand each other a good deal of the time. By 1966 racial exclusiveness was the basic policy of both SNCC and CORE. Even stronger emotions surrounded the issue of nonviolence versus self-defense. The brutal beatings and killings of civil rights workers who had followed King's rules for nonviolent engagement and whose pleas for federal protection had gone unanswered had created a deep reservoir of frustration and anger.[27]

Preparations for a civil rights march in Mississippi in June 1966 —to protest the shooting of solitary marcher James Meredith a few days earlier—brought into the open the long-simmering conflicts between King and SCLC, and SNCC, now led by Stokely Carmichael. Carmichael, a charismatic orator and agitator, had been born in the West Indies, grew up in New York City, where he attended an integrated elite high school, and emerged as a leader of militant students at Howard University in the early stages of the civil rights movement. The immediate issues in Mississippi were whether whites should be allowed to participate in the march and whether a black self-defense organization, the Deacons for Defense, should provide armed protection. The compromise hammered out authorized the inclusion of both the Deacons and white sympathizers, although relatively few whites actually showed up. On the march itself a rhetorical struggle developed between King's gospel of racial reconciliation and Carmichael's stress on polarization and conflict. Finally in Greenwood on June 16, Carmichael, fresh from being held by the police, announced that he was fed up with going to jail and tired of asking whites for freedom. "What we gonna start saying we want now is 'black power.' " He then shouted "Black Power" several times and the audience shouted it back. The context reveals that the original implications of Black Power were self-defense against racist violence and an unwillingness to continue petitioning whites for equality. From now on, blacks would confront power with power rather than offer love in return for hate. According to one Black Power advocate: "What King apparently did not foresee was that his strategy of *confrontation* would also unveil white barbarism to such a degree that young blacks would reject his philosophy of nonviolence."[28]

The panic over the Black Power slogan in the white press in 1966 was due primarily to its association with violence, which made it seem part of the same spirit that was erupting in the ghettos. But, initially at least, the only violence that was being sanctioned was self-defense against racist assaults. The secondary association with racial exclusiveness was particularly shocking to white liberals who had identified strongly with the ideal of integration. Charges of black racism compelled African-Americans to make the argument, originally put forth by Malcolm X, that blacks could not be racist because they lacked the power and inclination to dominate whites the way that whites continued to dominate blacks. Here differing definitions of racism—for liberal whites it was a prejudiced attitude and for Black Power advocates it was a hierarchical social order—made communication difficult. But the essence of Black Power was neither violence nor the exclusion of whites; it was rather self-determination for black people. According to Julius Lester, it meant simply that "Black people would control their own lives, destinies, communities. They would no longer allow white people to call them ugly." Blacks were tired of having whites define who they were and what they might become, especially since white "friends of the Negro" failed to deliver on their promises of racial justice and at times used their egalitarian rhetoric to cover up substantive inequalities.[29]

During the racial polarization that took place in the years between 1966 and 1968, liberal whites tended to withdraw their active support from the struggle for racial equality, either because they believed that the goal already had been achieved or because they saw no place for themselves in the redefined freedom struggle. At the same time, blacks from a variety of ideological backgrounds were endorsing Black Power in the basic sense of community control and self-determination. Shortly after the events in Mississippi, a prominent group of black clergymen took out an ad in the *New York Times* endorsing the idea that blacks must develop "group power," because they had been oppressed as a group and not as individuals and had as much right as other American racial or ethnic groups to unify and exercise power on behalf of their own community. In 1967, a national Black Power conference was held at which a range of black organizations, including the NAACP and the National Urban League, were represented. Its principal convener, the Reverend Nathan Wright, Jr., described the purpose of Black Power as going beyond civil rights and getting black people to address themselves to "the far more basic business of the development by black people for the growth in self-sufficiency and self-respect of black people."[30]

These early formulations of the Black Power program did not directly challenge the status quo of American society. They did not in fact sanction either a total and permanent separation of the races or revolutionary action to liberate blacks from oppression. They merely substituted the idea of corporate or group integration for the individualist version that had previously prevailed. According to Nathan Wright: "The Thrust of Black Power is toward freeing the latent power of Negroes to enrich the life of the whole nation." What blacks were doing, he argued, was following the example of other ethnic groups: "The basic American tradition is for each rising ethnic group to devise and execute its own plan for economic, political, and civic freedom and development. So it must be with the Black people of our land." Individualist integration, according to Wright, had not been a goal of other groups, and it need not be for blacks.[31]

To be sure, Wright was one of the least militant of the major Black Power advocates of 1966 and '67; he was essentially a conservative whose thinking recalled at times the accommodationist "self-help" tradition of Booker T. Washington. But those who used a more confrontational rhetoric often ended up advocating a reformist ethnic pluralism similar to Wright's. According to the book that in 1967 was taken as the definitive statement of the new racial philosophy, *Black Power* by Stokely Carmichael and Charles V. Hamilton, "The Concept of Black Power rests on a fundamental premise: *Before a group can enter the open society, it must first close ranks.*" The aim was "bargaining strength in a pluralistic society," and the model to be followed was the way that white ethnic groups such as the Jews, Irish, and Italians had been able to exert political power by voting as a bloc.[32]

But another argument in the book had more radical implications —the analogy made in the first chapter between the situation of African-Americans and that of colonized peoples of Africa and Asia. Here the authors likened the internal form of colonialism that characterized black-white relations in the United States to the oppressive system of white domination that prevailed in South Africa and Rhodesia. If, in fact, "black people in this country form a colony, and it is not in the interest of the colonial power to liberate them," what reason was there to expect that the mobilization of blacks as a pressure group within the American political and economic system would result in their incorporation on a basis of equality? In his speeches and writings of 1966 and '67, Stokely Carmichael—the chairman of SNCC and the most visible of those who explained Black Power—was in the process of shifting his allegiance from the reformist model of ethnic mobilization in a pluralist society to a revolutionary model of national liberation from colonialism. "Traditionally for each new ethnic group," he had told the readers of the *Massachusetts Review* in September 1966, "the route to social and political integration in America's pluralistic society has been through the organization of their own institutions with which to represent their communal needs within the larger society. This is simply what the advocates of Black Power are saying." But in a speech in London in July 1967, he sounded a different note:

> Black Power to us means that black people see themselves as part of a new force, sometimes called the Third World: that we see our struggle as closely related to liberation struggles around the world. We must hook up with these struggles. We must, for example, ask ourselves: when black people in Africa begin to storm Johannesburg, what will be the reaction of the United States? . . . Black people in the United States have the responsibility to oppose, certainly to neutralize, white America's efforts.[33]

In London Carmichael stopped short of calling for an African-American insurrection in support of the international anti–imperialist struggle. By 1968, however, he was openly advocating revolution and paying homage to Frantz Fanon as the prophet of decolonization through violence. But dissension quickly developed within the radical wing of the Black Power movement between those like Carmichael who believed that blacks were victimized primarily as a race and must therefore endeavor to separate themselves from whites and deepen their connections to the African motherland and those who believed that they also were oppressed as a class and might therefore establish alliances with other potentially revolutionary segments of American society.[34]

The most conspicuous exponents of the latter position were the supporters of the Black Panther party, initially established in 1966 in Oakland, California, and by 1968, according to historian Manning Marable, "the most revolutionary national organization in the United States." When Stokely Carmichael and other militants from SNCC attempted to combine forces with the Black Panthers the differences of opinion on the nature of black oppression came into the open. Carmichael resigned as prime minister of the Panthers in 1969 because of the ties the party had established with white leftists. Those who had begun as Black Power radicals increasingly divided into two warring factions—those who stressed racial separatism and cultural nationalism and those who, following the Black Panthers, moved toward the Marxist conception of an anti-capitalist revolution, but with the provisos that the revolution in the United States would be led by blacks from the ghettos rather than by the predominantly white industrial working class and that—in the international struggle—people of color rather than the most advanced segments of the proletariat would be in the vanguard. Huey Newton, the leader of the Panthers, derided those who looked to African culture as the basis of a separatist identity as "pork chop nationalists," and in Los Angeles the Panthers engaged in violent skirmishes with members of an extreme separatist and cultural nationalist group (called simply US) led by Ron Karenga. Well into the 1970s there were bitter quarrels among black radicals between separatist nationalists—who tended to become more inward looking and less confrontational as time went on—and revolutionary nationalists, influenced by Marxist-Leninism, who bore the brunt of official repression and ended up as a small group of theorists rather than as actual instigators of black rebellion.[35]

But many of those who invoked the Black Power slogan in the late 1960s and early '70s never embraced a radical black nationalism of either variety and continued to stress the reformist ethnic pluralism that had been the original meaning of the slogan. Besides the radical versions, there were conservative and liberal interpretations of what proved to be an extremely elastic conception. These formulations eventually forced many of the radicals to disown the slogan or to see it as merely representing a stage on the way to a fully developed nationalist consciousness. Economically conservative black leaders emphasized black self-help and entrepreneurship, virtually replicating the pro-capitalist "bootstrap" philosophy of Booker T. Washington. Such formulations even made it possible for Richard Nixon and the Republican party to embrace Black Power as black capitalism in the election of 1968. Black politicians in the Democratic party generally defined the concept as a mobilization of black voters behind stronger civil rights legislation and liberal reform. Perhaps the most articulate and thoughtful of those who defined Black Power in this way was Shirley Chisholm of New York, the first black woman to serve in Congress and in 1972 the first African-American to mount a serious campaign for the presidential nomination of one of the major parties.[36]

As the radical '60s gave way to the relatively conservative '70s, it became clear that Black Power had made a significant difference in the attitudes of black America. Especially evident was a significant increase in racial pride and self-esteem. The slogan "Black is beautiful" summed up the positive affirmation of black identity that had replaced the widespread sense of ugliness and inferiority that psychologists in the 1950s had found to be widespread among blacks. There was also an increasing willingness to identify with African culture; African-Americans in the late '60s and early '70s wore African clothes, adopted African hairstyles, and began to celebrate African holidays. A coherent African-American cultural ethnicity was in the process of being constructed out of a combination of African and specifically African-American traditions. But in political and social terms what had triumphed was a validation of black ethnic solidarity and action within the context of a liberal pluralist society and not the radical alienation from the American political and social system that had characterized the black nationalism of a Stokely Carmichael, a Huey Newton, a James Foreman, or an Imamu Baraka.[37]

By killing the dream of integration into a color-blind society, Black Power delivered a body blow to the utopian view of America as a land where a new kind of citizen would emerge free of inherited ethnic or racial associations—the essential human being existing outside of history and culture. But as other groups imitated the black example of identity politics and the assertion of group interest as the highest good, giving rise to the notion that rather than a "melting pot," the nation was a "mosaic" of fixed and permanently distinguishable parts, it became an open question whether an inclusive American identity and common civic consciousness was still possible. In the 1970s and '80s the broadly humanistic and cosmopolitan vision of America put forth by black integrationist leaders such as Martin Luther King, Jr., A. Philip Randolph, and Bayard Rustin in the 1950s and '60s showed few signs of life. The demise of the dream of "black and white together" was historically explicable and arguably the product of a more realistic assessment of national capacities, but it was nevertheless a tragic failure of American civic culture to realize its highest aspirations.

Black Consciousness in South Africa

The Black Consciousness movement of the 1970s appears at first glance to have been the most obvious case of imitating an American movement in the entire history of black protest in South Africa. But close examination of the circumstances of its growth and the content of its ideology shows that the African-American influences were less important than local conditions and indigenous currents of thought. The reading of Stokely Carmichael, Eldridge Cleaver, Malcolm X, James Cone, and other American Black Power advocates was clearly a stimulus, but the adoption of African-American concepts and slogans was selective rather than wholesale, and the ideas that were appropriated were often reinterpreted to fit South African conditions.

Black Consciousness rose to fill the vacuum created by the banning of the ANC and PAC in 1960, but only after a hiatus of nearly a decade that saw little organized and visible political activity among Africans. The focus of attention in the early '60s was on the futile efforts of the outlawed organizations to maintain a presence within the country through sabotage or other acts of violence against the regime. The repression that succeeded in making the Congresses virtually invisible within South Africa after 1964 meant that they existed thereafter mainly as émigré organizations seeking to organize guerrilla forces in neighboring African states and to rally international support against the apartheid regime. Inside South Africa, the voices raised in public against the government's racist policies during the mid to late 1960s tended to be those of white dissidents. The ANC's white auxiliary, the Congress of Democrats, was banned under the Suppression of Communism Act in 1962, and most of the white radicals who had been conspicuously associated with it were placed under personal banning orders throughout the decade. Consequently, white dissent came primarily from self-styled liberals who favored eliminating racial discrimination and thereby transforming South Africa into a Western-type democracy (rather than into a socialist society of either the Marxist or Africanist type).[38]

The clearest organizational manifestation of this impulse was the Liberal party, founded in 1953 under the leadership of the novelist Alan Paton, among others. The Liberal party welcomed black members and was the only interracial party in South Africa until 1968 when it disbanded rather than comply with a new law prohibiting racially mixed political organizations. The principal plank in its platform after 1954 was universal suffrage, but there was virtually no support among white South Africans for creating a black-majority electorate, and the Liberals failed to elect a single member to parliament. At the time of Sharpeville, a few young Liberals gave up on constitutional reform and endorsed direct action. Liberals in Cape Town cooperated with the Pan-Africanist Congress in organizing protest demonstrations immediately after the massacre, and during the early '60s, the sabotage campaign of *Umkhoto we Sizwe* was supplemented by that of the African Resistance Movement, a secret organization of young white Liberals and others committed to the violent overthrow of the Afrikaner Nationalist government. The only white ever executed for resistance to apartheid was the Liberal saboteur Frederick

John Harris, who was put to death in 1964 after a bomb he had planted killed one person. Another militant Liberal, Patrick Duncan, resigned from the party in 1963 in protest against its official policy of nonviolence and, after fleeing abroad, was accepted as the first white member of the PAC. The odd symbiosis between the radical wing of the Liberal party and the Pan-Africanists was due primarily to a shared anti-Communism and a belief that the ANC was under the influence of the Soviet Union.[39]

Electoral failure, internal dissension, and the banning of its leaders put the Liberal party on the path to oblivion even before its forced dissolution in 1968. Besides being hobbled and ultimately destroyed by repression, it simply failed to attract enough white support to become a force in European politics. It actually fared better among Africans, who eventually constituted a majority of the membership, but their numbers remained insufficient to influence the liberation movement. (Its most conspicuous black supporter was Jordan Ngubane, a prominent intellectual and founder of the Youth League who had never retreated from the league's original anti-Communism.) The liberal ideology nevertheless took root on the campuses of the English-speaking white universities and came to predominate in the principal political organization on these campuses, the National Union of South African Students (NUSAS). Like the Liberal party, NUSAS was opposed to racial segregation and sought to involve blacks in its own activities. Although the organization was not permitted to organize on the campuses of the "tribal colleges" established under apartheid, it recruited among the small and diminishing number of Africans allowed to attend the predominantly white universities through loopholes in the separate education laws and also solicited representatives from the African colleges to attend its conventions as guests or observers. After some of its members were implicated in the African Resistance Movement in 1964, NUSAS became less militant in its opposition to the regime and devoted itself thereafter mainly to fighting against state interference with academic freedom. But it retained its commitment to a racially integrated society and its own policy of interracial membership stood as symbolic protest against the pattern of total separation that the government was imposing with increasing rigor.[40]

At the 1967 annual NUSAS conference at Rhodes University in Grahamstown, the black delegates were forced by the host institution to eat separately from whites and to occupy separate living quarters far from the conference venue. Among those subjected to this treatment was Steve Biko, a student at a medical school for nonwhites established under the auspices of the University of Natal in Durban. At the July 1968 conference Biko provoked a searching discussion of whether there was any point in Africans continuing their affiliation with NUSAS in the light of their minority status and second-class treatment. He pressed the issue again at a meeting of the University Christian Movement (UCM), another interracial organization that also met during the winter vacation period of 1968. The UCM, which was allowed to organize formally on black campuses because of its seemingly nonpolitical religious charater, had more black members than white, making it a better springboard for independent black political action than NUSAS. At its meetings, Biko proposed the establishment of an all-black student movement to supplement NUSAS. A year later, the South African Student Organization (SASO), with Biko as its first president, was formally established at a conference held at the African college at Turfloop in the northern Transvaal. At first SASO operated under the umbrella of NUSAS, which it continued to recognize as the "national organization" of students. But in 1970 it severed all ties with the parent body and endeavored to extend the influence of its separatist philosophy—summed up in the slogan "Black man, you are on your own"— beyond black students and into the larger African community. In 1972, the Black People's Convention was founded as a coalition of African organizations committed to the ideology that was now being called "Black Consciousness." Another wing of the movement and perhaps its most active and creative component in the early 1970s was Black Community Programmes (BCP), the coordinating body for a variety of local self-help initiatives, such as community medical clinics and home industries for the unemployed poor. These were run by blacks for blacks, but were financed mainly by anti-apartheid church groups both within the country and abroad.[41]

It was initially difficult for both the government and the international anti-apartheid movement to determine if Black Consciousness represented resistance or accommodation to white supremacy. BC broke sharply with both the ANC's doctrine of multi-racialism and the PAC's confrontational militancy. The emphasis on racial separatism and self-help rather than direct resistance against apartheid beguiled the authorities into treating the movement with kid gloves for about two years in the hope that it could be enlisted behind the government's plans for separate political development. Reformist ethnic politicians, such as Chief Gatsha Buthelezi of the Zulus, who were developing a strategy for using the "homeland" governments as agencies for reforming apartheid from within (a policy that the ANC would eventually condemn as "collaboration"), for a time had high hopes that BC could be harnessed to this project. The ANC, observing the emergence of the new ideological tendency from outside the country, saw no reason to applaud a movement that seemed to be playing into the hands of segregationists. The decisive action that established the anti-apartheid bona fides of Black Consciousness was its rejection in 1972 of governmental schemes for separate development. It did not endorse violence or revolution, but its repudiation of limited black autonomy through territorial separation strongly implied that all of South Africa, rather than the small portions assigned to Africans by the government, was the birthright of blacks. In Biko's own words, "we black people should all the time keep in mind that South Africa is our country and that all of it belongs to us."[42]

This assertion of exclusive black ownership was probably a conscious repudiation of the Freedom Charter's famous opening line: "South Africa belongs to all who live in it, black and white." In some respects, Black Consciousness was a revival of the anti-charterist ideas of the Pan-Africanist Congress. Biko himself came from a family with PAC connections; he had an older brother who was arrested and held without trial for suspected involvement in *Poqo*, the PAC's terrorist offshoot. In 1972, Biko's paper at a Cape Town conference on "Student Perspectives on South Africa" explicitly traced the BC ideology back to a "group of young men [in the 1950s] who were beginning to 'grasp the notion of their peculiar uniqueness' and who were eager to define who they were and what." These forerunners of BC opposed "the ease with which the leadership [of the ANC] accepted coalitions with organizations other than those run by blacks. The 'People's Charter' adopted in Kliptown in 1955 was evidence of this." In Biko's view, therefore, the Africanists of the 1950s—those who opposed the Freedom Charter and eventually broke away to form the PAC—produced "the first real signs that the blacks in South Africa were beginning to realize the need to go it alone and to evolve a philosophy based on, and directed by, blacks. In other words, Black Consciousness was slowly manifesting itself." A reluctance to cooperate with white liberals and radicals on the grounds that all whites were beneficiaries of the system of oppression and could not be trusted to act on behalf of the black community was an attitude common to Pan-Africanism, Black Consciousness, and Black Power.[43]

But Black Consciousness departed in some significant respects from Pan-Africanism, most obviously in its definition of "black." For Biko and his colleagues, all those previously described in negative terms as "nonwhites"—Indians and Coloureds as well as indigenous Africans—were to be considered "black" so long as they identified with the struggle against racial oppression. Blackness then became less a matter of ethnic ancestry than of a raised consciousness. On the other hand, all people of African ancestry and pigmentation were not automatically black; those who accepted white domination and cooperated with their oppressors continued to deserve the appellation "nonwhite."[44]

This repudiation of a strictly genetic view of blackness paralleled a subtle and little noticed difference between African-American nationalism of the 1960s and the earlier varieties associated with Edward Blyden, Alexander Crummell, and Marcus Garvey. As we have seen, these forerunners were men of dark complexions who distrusted mulattoes and at times openly disparaged them. But in the 1960s, the foremost champion of blackness could be the light-skinned and red-haired Malcolm X. Obviously no strict genetic test was being applied, and discussion of the historical significance of skin-color variations among African-Americans became virtually taboo. The implicit message was that one was as black as one felt, and that people of African ancestry who retained the integrationist view that white culture was superior to black culture continued to be "Negroes" rather than "blacks," however dark complexioned they happened to be. Even people who were of mostly white ancestry and appearance could be as black as any other African-American provided that they were part of the group that had been historically classified and treated as such (in accordance with the extraordinary American custom of considering anyone of known black ancestry to be black) and provided also that they currently identified themselves with the struggle for black liberation and self-determination. Whether or not the new American affirmation of a non-genetic blackness influenced the racial thinking of Black Consciousness, there can be no doubt that both movements innovated significantly in making race consciousness more a matter of existential choice and political awareness than of biological determination.[45]

Another way that Black Consciousness departed from the Pan-Africanist precedent and drew closer to American black nationalism of the 1960s was in its emphasis on psychological rehabilitation as a precondition for political resistance. The Pan-Africanists had believed that the masses were fully conscious of the injustices perpetrated upon them because of their race and that the anger that they naturally felt made them ready at any time for a massive uprising against white domination. Leaders needed only to provide the spark in the form of some dramatic act of confrontation and provocation. But the fact that Sharpeville had led to massive repression and political quiescence rather than to a general uprising of Africans had cast doubt on their belief in revolutionary spontaneity. For the advocates of Black Consciousness, the state of affairs in post-Sharpeville South Africa revealed that blacks were held in subjugation not merely by force but by their own sense of impotence and inferiority. Consequently the primary task of their movement was to "conscientize" black people, which meant giving them a sense of pride or a belief in their own strength and worthiness. Only in this way could the psychologically debilitating effects of white domination be overcome. BC's rejection of alliances with white liberals and radicals was based on a conviction that the whites in such relationships tended to assume authority and behave paternalistically, thus preventing blacks from overcoming their inferiority feelings. Malcolm X's distinction between forced segregation and voluntary "separation" was central to the South African Black Consciousness movement, and the slogan "Black is beautiful" had as much resonance for its adherents as it did for American Black Power advocates.[46]

But the idealist view that consciousness precedes praxis was more clearly and insistently affirmed in South Africa than in the United States—at least in the early and classic formulations of the Black Consciousness philosophy. The very difference in the names generally assigned to the two movements suggests a muted philosophical difference. In the United States, the growth of black pride and a positive sense of identity was not divorced conceptually in most formulations from the actual exercise of black power. Awareness of a positive black identity was indeed a precondition for community organization and the application of political pressure, but consciousness was expected to be translated quickly into forceful action, and the exercise of power in turn was supposed to be essential for the full development of consciousness.

The most obvious reason that consciousness was divorced more sharply from power in early Black Consciousness thinking was the significant difference in the political rights and economic conditions of blacks in the United States and South Africa. Only people who could vote could plan to exercise power at the polls, and it was futile to think about a separate black economy if blacks had virtually no resources they could mobilize. Furthermore, the South African government was willing to tolerate the public expression of BC ideology only so long as it remained convinced that the movement was a purely intellectual and cultural one that was not actually proposing any kind of political resistance. After the Black People's Convention was established in December 1971 as a broad-based "political" expression of Black Consciousness (an initiative that Steve Biko initially opposed because it would expose the movement to persecution), the government concluded that the rhetoric of "conscientization" was a cover for seditious action and that BC ideology could act as a stimulus to acts of defiance and insurrection. Eight BC leaders, including Biko himself, were banned in early 1973. The following year the leading BC activists in Durban were arrested for treason after they defied a government ban on holding a rally to celebrate the victory of FRELIMO over the Portuguese in Mozambique.[47]

But the idea that consciousness was itself a kind of power had an intellectual basis as well as a tactical one. One of the features of Black Consciousness that distinguishes it from the mainstream of the African-American movement was the extent to which religious beliefs and associations shaped its ideology and mode of operation. Virtually all of its leaders were practicing Christians with affiliations to one or another of the mainstream interracial churches, and the movement's institutional origins were as much in churches and religious associations as in in student organizations. It will be recalled that the United Christian Movement was the cradle of SASO and that the principal white patronage and financial support that the movement received was from church groups both in South Africa and abroad. Of particular importance in aiding and abetting the movement was the resolutely anti-apartheid Christian Institute, which was led by radical white clergymen. Anti-racist white ministers such as the Methodist Basil Moore, the renegade Dutch Reformed predikant Beyers Naude, and Anglican priest Aelred Stubbs were strong supporters and major facilitators of the Black Consciousness movement. The ban on cooperation with white liberals did not extend to fearless men of God who saw BC as a religious movement designed to purge the Church of the sin of white supremacy. A large number of black clergymen, mostly within the "historic" churches originally established by European missionaries, became prominent advocates of Black Consciousness; they predominated in the leadership of the "adult" wing of the movement—the Black People's Convention.

To some extent, to be sure, association with religion was a matter of convenience for the student activists who remained in the forefront of the movement. The one place in the late '60s and early '70s where blacks could express themselves with some freedom was within the churches or in associations that enjoyed religious sponsorship. But it would be cynical to leave it at that and ignore genuine religious convictions of a Steve Biko and a Barney Pityana, to say nothing of the religious basis of the philosophy itself. The belief that a new consciousness could transform physical reality, or that spiritual truth could overcome vast differences in power, was a profoundly religious one. It assumed that God was on the side of the most downtrodden and despised portions of humanity, and that once the sufferers realized that they were the chosen of God the end of their agonies would be in sight.[48]

The aspect of the American Black Power movement that had the most more direct and significant impact in South Africa was an intellectual tendency that is usually viewed as peripheral to the mainstream development of American black nationalism—the effort of clergymen and religious thinkers to formulate a Black Theology. Beginning with the 1966 defense of Black Power by a distinguished group of African-American ministers and emerging full-blown with the publication of James Cone's seminal *Black Theology* in 1969, Black Theology caused much controversy within religious circles but attracted relatively little attention outside of them. That the black nationalist revival of the 1960s began with Malcolm X's condemnation of Christianity as a white man's religion and was stimulated by a negative reaction to Martin Luther King's Christian nonviolence made Christian theology seem like an unpromising source of Black Power sentiments. Furthermore, none of the more radical exponents of Black Power who attracted the attention of the press in the late '60s—Stokely Carmichael, Rap Brown, James Foreman, T. Huey Newton, or Eldridge Cleaver—manifested a positive view of Christianity.

But a small number of black ministers and theologians went to work in the late '60s and early '70s reformulating Christian doctrine in the light of the Black Power revolt and the resurgence of nationalist and separatist ideas in the black community. The most radical among them was Albert Cleage, pastor of the Church of the Black Madonna in Detroit, who attempted to outdo Elijah Muhammed as a theorist of black supremacy by arguing that the Jews of the Bible were black people, that Christ was a black messiah who had come to save his chosen people, and that contemporary white Christian and Jews had usurped a heritage and promise of divine favor that properly belonged to blacks alone. But Cleage was too radical and heterodox for most of the seekers of a distinctive African-American theological vision. When South Africans sought inspiration for a black theology of their own, they found it primarily in the writings of James Cone, a prolific author of serious theological works who was appointed a professor at Union Theological Seminary in New York after the publication of his first book in 1969. Cone was not merely a distant intellectual stimulus; he established direct connections with black theologians in South Africa, contributing papers to their symposia and commenting on their work.[49]

Cone and the theologians of Black Consciousness in South Africa agreed that white missionaries had preached a form of Christianity that had helped to sustain racist and colonialist oppression. It had not only helped to justify slavery and imperialism but had also taught black converts that their own cultural traditions were worthless and that resistance against white domination was sinful. But this was not the fault of the Gospel itself; it had resulted rather from an interpretation of it that served the selfish interests and sinful appetites of Europeans. Blacks had the right and the need to interpret the Christian religion in the light of their own situation as an oppressed people. Passages in the New Testament that presented Jesus as the champion of the poor and oppressed were the basis for a theology of liberation. Christ himself was black, if not literally at least in the sense that blackness had come to symbolize the state of being oppressed that He had been sent to overcome. In Cone's formulation of 1970, "blackness is an ontological symbol and a visible reality which best describes what oppression means in America." For the South African black theologians, it stood for oppression in their country in an even more obvious sense. To affirm blackness as a positive identity in either society was to be freed in spirit and committed to a struggle for liberation from physical oppression. As the South African theologian Manas Buthelezi put it: "As long as somebody says to you, You are black, you are black,' blackness as a concept remains a symbol of oppression and something that conjures up feelings of inferiority. But when the black man himself says, 'I am black, I am black', blackness assumes a different meaning altogether. It then becomes a symbol of liberation and self-articulation."[50]

It would be wrong to suppose, however, that those in South Africa who were stimulated by Cone to pursue the project of creating a distinctive black theology ended up in total agreement with his forthright apology for Black Power. In the end Cone was too extreme in his separatist rejection of whites to meet the needs of most African churchmen, who, despite their endorsement of Black Consciousness, remained in denominations that had both white and black communicants. Once again, the locus of religious involvement in African political activity was the "historic" or mission churches rather than in the African independent churches. What Black Consciousness implied for these churches was that their black majorities should have a proportionate role in the leadership or hierarchy. It could not mean that these churches should be exclusively black or that black clergymen who condemned apartheid were ready to stop cooperating with the small number of white ministers and priests who were doing the same. The proponents of Black Theology showed an interest in the independent churches and expected to learn something from them about how traditional African religion could be reconciled with Christianity, but they found it difficult to make common cause with the leaders of these normally nonpolitical and otherworldly sects. For institutional as well as intellectual reasons, therefore, they could not fully embrace Cone's thoroughly ethnocentric version of the Christian faith.[51]

When Cone denied the universality of Christ's offer of salvation, arguing that "Jesus is not for all, but for the oppressed, the poor and unwanted of society, and against oppressors," he was coming too close for the comfort of the South Africans to saying that whites were beyond redemption. According to Lutheran Bishop Manas Buthelezi, the Gospel dictated a very different attitude: "It is now time to evangelize and humanize the white man," he wrote in 1973, thus reversing the original mission relationship without sacrificing the basic Christian idea of a universal salvation. In his early writings, Cone often expressed a categorical hostility to whites that seemed to make reconciliation with them impossible. He also flirted with what more orthodox Christians could only view as heresy when he affirmed that "Black Power is not only consistent with the gospel of Jesus Christ, but . . . it is the gospel of Jesus Christ." For Black Consciousness theologians such as Buthelezi and the Colored Dutch Reformed minister Allan Boesak, reconciliation of the races could not be achieved without black liberation but it remained the ultimate goal of genuine Christians; for them the Gospel transcended human ideologies and could never be reduced to a finite political meaning. Boesak, in an important book of 1977 endorsed Black Power as "the power to be," but rejected the tendency in American Black Theology toward "a complete identification [of the Gospel] with Black Power's political program (in all its expressions)." As critics of Afrikaner nationalism with its idolatry of the *volk*, South African black theologians were on guard against making national feeling and religious faith synonymous.[52]

The differences between the versions of Black Theology promulgated in the two countries support the more general conclusion that the Black Consciousness movement was influenced by the American Black Power philosophy but did not slavishly imitate it. The most obvious borrowings can be found in early SASO documents. The Policy Manifesto of 1970 featured the free appropriation of current African-American ideas and slogans, suggesting strongly that there were significant similarities in the situation of black people in the two societies and comparable solutions to their problems. Repeating, almost verbatim, a famous phrase from Carmichael and Hamiltion's *Black Power*, the manifesto accepted "the premise that before the black people should join the open society, they should first close their ranks, to form themselves into a solid group to oppose the definite racism that is meted out by the white society, to work out their direction clearly and bargain from a position of strength." The manifesto repudiated "integration" if it meant "the assimilation of blacks into an already established set of norms drawn up and motivated by white society," but went on to endorse an integration based on "the proportionate contribution to the joint culture of the society by all constituent groups."

In this formulation, Black Consciousness was not yet a revolutionary black nationalism but rather a reformist pluralism similar to the moderate or mainstream version of Black Power. Difficult to determine is whether this seemingly unrealistic conception of what was possible in South Africa without violent confrontation reflected the honest beliefs of Biko and the founders of SASO or was, on the contrary, an expedient cover for the political organization of blacks under the eyes of a government intensely fearful of the revolutionary potential of the African majority.[53]

Historian Gail Gerhart has uncovered an internal SASO document describing a 1970 seminar discussion on the applicability of the Carmichael and Hamilton injunction that "before entering the open society we must close ranks," which shows an appreciation of the differing contexts of Black Power and Black Consciousness. "This group," the document reports, "made the observation that an open society in this country can only be created by blacks, and that for as long as whites are in power, they shall seek to make it closed in one way or the other. We then defined what we meant by an open society. . . . The group ended up by stating that the original statement should read 'before *creating* the open society we should first close our ranks.' "[54]

In his 1971 paper at the conference on Student Perspectives on South Africa, Biko discussed at some length the relationship of Black Consciousness to Black Power and argued that the influence of the latter on the former had been exaggerated. A more important impetus, he claimed, was "the attainment of independence by so many African states in so short a time. . . . The fact that American terminology has often been used to express our thoughts is merely because all new ideas seem to get extensive publicity in the United States." Five years later, when the government brought some Black Consciousness leaders to trial for celebrating the victory of FRELIMO in Mozambique, Biko was called to the stand and questioned closely about the origins of the movement. In answer to a question on the relationship to Black Power, he made a fundamental distinction between the two manifestations of black self-determination:

> I think the end result of Black Power is fundamentally different from the goal of Black Consciousness in this country, that is, Black Power . . . is the preparation of a group for participation in an already established society, a society which is essentially a majority society, and Black Power therefore in the States operates as a minority philosophy. Like you have Jewish power, Italian power, Irish power and so on in the United States. The Black People are merely saying that it is high time that they are not used as pawns by the other pressure groups operating in American society. They must form themselves into a definite pressure group, because there were common problems with black people in the United States but essentially they accept that they are a minority group there and when they speak of bargaining . . . they are talking within the American context of the ballot box. They want to put up the kind of candidates they like and be able to support them using black votes. . . . And

> the other thing which causes that: The American Black man is, essentially you know, he is accepted; he is truly American in many ways. You know he has lived there for a long time. All he is saying is that "Man I am an American, but you are not allowing me to live like an American here in America." He has roots with Africa . . . but he does not reject his Americanness.[55]

Biko's understanding of Black Power, which was probably derived mainly from Carmichael and Hamilton's book, was actually a fairly accurate perception of the lasting or residual meaning of the concept that had survived the suppression and decline of the more radical black movements. But in distinguishing between the operation of Black Power within the context of a benign American ethnic pluralism and the implied claims of Black Consciousness as a "majority philosophy" in an undemocratic South Africa, he shied away from drawing the conclusion to which his logic pointed—that reform, persuasion, and peaceful pressure, which might work in the United States, had little chance in South Africa. Elsewhere in his testimony, he explicitly denied that Black Consciousness would lead to a black revolution. In a somewhat tortuous interpretation of the practical meaning of a statement of the Black People's Convention that blacks needed to form a "power bloc" for the purpose of bargaining on the basis of strength with the white community, he conceded that blacks were not yet strong enough to make radical demands and that it might take "over 20 years of dialogue between blacks and whites" to achieve real success. Eschewing "armed struggle" or even "confrontational methods" leading to civil disobedience, he affirmed that "our operation is basically that of bargaining and there is no alternative to it. It is based mainly on the fact that we believe we have interpreted history correctly, that the white man anyway is going to have to accept the inevitable."[56]

Biko seemed to be saying that the methods that would work for a minority in the context of democratic pluralism in the United States would also work for a disfranchised majority in South Africa. But a more realistic assessment of the situation might have suggested that a racist minority could not be persuaded to cede power to a black majority without a fight and that it would do everything in its power to prevent the majority from gaining the kind of leverage that Biko predicted that it would gradually and peacefully acquire. Biko may well have been masking his real views in an effort to keep his movement alive and within the law. But, if we take him at his word, there is a considerable gap between his analysis of the situation and how it differed from the American and the kind of action that he was publicly prepared to contemplate.

Biko's advocacy of organization, self-help, and "conscientization" —with no clearly specified program for political resistance—resembled the stance of cultural nationalists in the United States. Like them he seemed to be saying that for the time being black people should devote themselves mainly to building their self-esteem. Just as the African-American cultural nationalists were criticized by the Black Panthers and other revolutionary nationalists for their lack of political militancy and failure to address the class basis of racial oppression, so Biko was criticized by the ANC, the South African Communist party, and assorted freelance Marxists for his idealist conception of the power of consciousness and his failure to link up with the struggle of black workers for economic justice. In South Africa in the 1970s there was in fact a significant development of extralegal black trade unions and, beginning in Durban in 1973, a number of major strikes, some of which were partially successful. The Black Consciousness movement made some attempt to promote unions which shared its viewpoint, but most labor activism was explicitly or implicitly based on a Marxian-type class ideology rather than on the BC view that race was the fundamental reality. Young white radicals, frozen out of the Black Consciousness organizations, secretly affirmed their allegiance to the ANC with its policy of nonracialism—which since 1969 had meant that whites could actually join the organization—and worked with some effectiveness to aid black workers to establish unions. Black Consciousness groups on the other hand remained overwhelmingly middle-class in membership and established only weak connections with the black labor movement—the sphere of activity that produced the most significant gains in real power for Africans during the 1970s.[57]

But the Black Consciousness movement was not without political consequences. The circulation of its ideas beyond the colleges and universities to the high school students of Soweto helped to set off the revolt of June 1976. The brutal suppression of student protests against government efforts to require African students to have instruction in the hated oppressor language of Afrikaans touched off student strikes and riots throughout the country and plunged South Africa into its most serious domestic crisis since Sharpeville. The organization that called the demonstration of June 16—upon which the police fired with bloody proficiency—was the Soweto Students Representative Council, which had been founded by the local members of the South African Students' Movement, a national organizational of black high school students inspired by SASO and clearly under BC influence. The government had no doubt about who was ultimately responsible, and it proceeded to ban all of the Black Consciousness groups. The arrest, torture, and murder of Steve Biko in 1977 climaxed the massive effort to stamp out the movement with which he was identified.[58]

Unlike what happened after Sharpeville, the repression that followed Soweto did not lead to a long period of political inactivity and apparent black resignation in the face of overwhelming white power. Soweto in fact is now recognized as a turning point in recent South African history; as a result of the impression it conveyed to the world of the utter viciousness of the white regime, it energized and empowered the international antiapartheid movement. Less easy to calculate but nevertheless undeniable was the effect on black South Africans of the fact that their children had been willing to risk their lives by defying the regime on an issue that involved black pride and cultural identity. The adult Black Consciousness organizations did not plan or direct the uprising, but they could take some credit for instilling the mood of black self-assertion that produced it.

Nevertheless, the historical fate of the Black Consciousness ideology after 1977 defied predictions made at the time that its way of thinking would soon predominate in the black resistance movement, eclipsing the nonracial nationalism of the ANC. An estimated 60 percent of the student rebels who fled the country after the uprising were recruited into the guerrilla army that the ANC was organizing in friendly African states. The Pan-Africanist Congress, which as we have seen was ideologically closer to Black Consciousness, was in no position to receive them in large numbers. Disabled by factionalism and incompetent leadership and without the reliable supply of arms that the ANC received from the Soviet Union and the nonmilitary help it obtained from Western supporters—the PAC was virtually defunct by the late '70s. While young recruits in the guerrilla army camps were being indoctrinated in the ANC view of the world by veteran émigrés, the Black Consciousness activists who were arrested and sent to Robben Island were being reeducated by Nelson Mandela, Walther Sisulu, Gavin Mbeki, and other ANC leaders who had been incarcerated there since the early 1960s. Hence it was the ANC and not organizations that tried to carry on in the Black Consciousness tradition that derived the most benefit from the "conscientization" of blacks that was occurring in the late 1970s.[59]

Within South Africa, many of the staunchest supporters of Black Consciousness reorganized in 1978 to form the Azanian People's Organization (AZAPO). But from the outset, the new group revealed that it was deviating in significant ways from the original BC ideology. Like those Black Power advocates in the United States who had first defined their oppression in exclusively racial terms and then had come to appreciate the class dimension of black inequality, AZAPO defined apartheid as not merely a form of racial oppression but also as a form of class domination. What needed to be overthrown was "racial capitalism," and the agent of revolution would not be blacks as a national or racial group but the black working class as the victim of *both* race and class oppression. Whites, including white workers (who were not true proletarians but rather part of the exploiting group) would have no part in the liberation struggle. Hence the main thread of continuity between the PAC, Black Consciousness, and AZAPO was the exclusion of whites from the resistance movement and by implication from charter membership in the post-apartheid South African nation.[60]

It is beyond the scope of this study to provide a detailed account and full analysis of the events of the tumultuous 1980s in South Africa. But a brief summary of the general pattern of black resistance and ideological development can convey a sense of the fate of Black Consciousness. The main source of domestic resistance to the apartheid regime beginning in 1983 was the United Democratic Front (UDF), a coalition of organizations African, Colored, Indian, *and* white—that was originally established to protest against the constitutional changes that the government was proposing in order to give a limited form of political representation to Indians and Coloreds, but not to Africans. Indicative of the new interracialism was the fact that NUSAS, the predominantly white student organization from which SASO had seceded in 1970, was among the affiliating groups that founded the UDF, and one of its former presidents became a member of the UDF'S national executive committee.[61]

The new federation quickly identified itself with the Freedom Charter and, becoming bolder, with the ANC itself. One impetus for making this connection was the fact that the ANC had grown in strength and visibility since the time before Soweto when it seemed to be merely an exile group with virtually no visible presence within the country. Its forces augmented by refugees from the Soweto uprising, the ANC was able to carry out a number of spectacular acts of sabotage within South Africa in the late '70s and early '80s. Since the rival PAC remained considerably smaller and less active, the conviction grew within the black communities of South Africa that the main source of resistance against apartheid was the ANC and that its camp was the place to be if one wanted results. Embracing the Freedom Charter meant welcoming all racial groups, including whites, into the movement and setting as the goal of the struggle a racially inclusive democratic South Africa rather than a state that gave official priority to African interests and cultural values. The opposition to the Charterists, as they were now called, came from AZAPO and the Black Consciousness alternative to the UDF as a confederation of community groups, the National Forum. But it was clearly the UDF that won the support of most blacks and that took the lead in the wave of boycotts, strikes, and demonstrations that characterized the mid 1980s and created the last great crisis of apartheid. The rise of the Congress of South African Trade Unions (COSATU), a labor federation closely allied to the UDF and the ANC, made possible a coordination of political and industrial action against the regime that went far beyond any earlier black challenge to apartheid and could be held in check only by an unprecedented (and internationally unacceptable) level of repression.[62]

Did the Black Consciousness movement and the closely related tradition of Pan-Africanism therefore simply shrivel up and die except in the thinking of a minority that was relegated to the periphery of the struggle? Some former advocates of BC who now joined the Charterist movement maintained that their previous persuasion had served its historical function by increasing black self-confidence and willingness to challenge white supremacy but that its ethnocentric racial exclusionism had outlived its usefulness. Since blacks were clearly in charge of the movement and white supporters were deferring to their leadership, the old problem of white paternalism and black deference no longer seemed to exist. Longstanding fears of "alien" Communist domination of the liberation struggle receded in the '80s as the Soviet Union withdrew from involvement in African conflicts and as the cold war itself began to wind down. At the same time, the Communists and Marxists of all races who continued to be influential in the organization could be counted upon to fight for a nonracialism that was compatible with their basic belief that consciousness of class and not of race was the key to revolutionary change.[63]

It can also be argued, however, that the mainstream of the liberation movement had absorbed part of the Black Consciousness philosophy and that an unacknowledged accommodation between the two protest traditions had become possible under the ANC banner. The "multi-racialism" that the ANC had espoused in the 1950s—the policy of separate racial or ethnic congresses and the commitment to write special protections for non-African groups into the constitution of a post-apartheid South Africa—had quietly given way to a consistent "non-racialism" that dictated straight majority rule on the basis of "one person, one vote." On the surface, this inclusive South African nationalism was the antithesis of a Pan-Africanist or self-consciously *black* nationalism. It meant, in Nelson Mandela's words to the United States Congress in 1990, "that this complex South African society, which has known nothing but racism for three centuries, should be tranformed into an oasis of good race relations, where the black shall to the white be sister and brother, a fellow South African, an equal human being—both citizens of the world."[64]

There is no reason to doubt Mandela's commitment to a South African equivalent of Martin Luther King's "beloved community," but it might be worth recalling that a post-apartheid constitutional color-blindness was precisely what Robert Sobukwe and the Pan-Africanist Congress had advocated in the late 1950s. Of course a major difference remained: the ANC welcomed all racial groups into the struggle, whereas Africanists and BC advocates would exclude whites until after liberation when they would be offered equal citizenship if they could prove their loyalty to the black nation. The nonracial membership policy of the ANC began with admission of whites to *Umkonto We Sizwe*, the armed affiliate of the ANC in 1961, was extended in 1969 to allow "white revolutionaries" to be members of the parent organization itself, and was completed in 1985 when whites, Indians, and Coloreds were made eligible for election to the executive committee. But the demographic facts of life meant that Africans were the overwhelming majority within the movement and would also predominate in a free democratic South Africa. What practical purpose would have been served if Africans had emphasized their cultural differences from other racial or ethnic groups or claimed special prerogatives as the true natives of the country?[65]

The immediate task, after all, was liberation, not from all association with whites or from Western cultural influences, but from the political, economic, and social tyranny of a white supremacist oligarchy. A South Africa based on "one person, one vote" and straight majority rule would have whatever cultural identity the majority wished to affirm. By some such reasoning, it would have been possible for those whose true sentiments partook of Pan-Africanism or orthodox African nationalism to coexist in the ANC with those whose nonracialism came from ideological conviction—Christian, Marxist, or liberal democratic. Just as the question of whether South Africa would become a welfare state based on capitalist relations of production or a genuinely socialist society after liberation could be postponed until after white rule was overthrown, so the question of a national culture for South Africa could be deferred until the people were able to decide it for themselves. What may have been crucial to the ANC's victory in the struggle for popular support was less the specific character of its ideology than the fact that its primary commitment to political democracy meant that it could accommodate a greater range of ideological styles or preferences than its more rigid and intolerant rivals were able to do.

Comparing Black Power and Black Consciousness

Black Power and Black Consciousness had a great deal in common, beyond the sharing of slogans like "Black is beautiful" and "Before a group can enter [or create] the open society, it must first close ranks." Perhaps the most durable contribution of both was to instill in many black people a new sense of self-worth and competence that made traditional patterns of racial deference impossible to maintain. The rejection of white leadership and significant participation in the freedom struggle that the two movements shared had more lasting effects in the United States, but the contrast must be qualified by the acknowledgment that a minority has reason to feel more anxious about its ability to determine its own destiny than a majority; it can much more easily find itself the instrument of some other interest than its own. Clearly the ideal of total assimilation into a middle-class society and culture that reflected only European or Euro-American values and historical experiences was now recognized as a confession of cultural inferiority and was no longer an acceptable ambition for blacks in either society. Those in the United States who had been lured by the image of a melting pot of races and nationalities and those in South Africa who had been persuaded by missionaries that Africans could be reborn as white Christians with dark skins had learned that proposing to whiten black people—literally or figuratively—was a genteel way of advocating genocide.

On a more practical level, the emphasis on community organization and self-help that was common to both movements had empowering consequences. In South Africa, the communal resistance of the '80s built to some extent on the community organizing of the '70s, much of which was associated with the Black Consciousness movement. In the United States, the election of African-Americans in substantial and increasing numbers to federal, state, and local offices was the result not simply of voting rights legislation but also of Black Power's call for mobilizing the vote behind black candidates and causes. In 1986 American Black Power asserted itself on behalf of South African liberation when the political clout of

African-Americans was instrumental in getting legislation applying strong sanctions against South Africa passed by Congress over a presidential veto. On balance, therefore, both movements had healthy and liberating consequences.[66]

But the movements were far from identical, which is scarcely surprising given the fact that the contexts in which they operated were in some ways radically different. The American movement was more diverse and variegated. In a strict sense it was not a single movement at all but several related tendencies of thought and action, ranging from Republican-sponsored Black Capitalism to a few attempts at terrorism by tiny urban guerrilla groups. Between the fringes, the movement divided into ethnic pluralists, separatist nationalists, and revolutionary nationalists. The pluralists were likely to believe that mobilizing blacks as a pressure group could reform America's liberal capitalist system; the separatists wanted to secede from it culturally and, if possible, physically; and the revolutionists envisioned blacks leading an uprising of oppressed peoples and classes to overthrow it. The most militant debated among themselves the importance of a distinctive black culture in group mobilization. For some cultural autonomy was crucial, almost an end in itself; for others it was a diversion from the politics of making a revolution against American capitalism and imperialism (which would include making appropriate alliances with other oppressed peoples). The Black Consciousness movement, by contrast, was relatively unified in policy and leadership. It was not entirely monolithic; differences were developing even before the Soweto crisis between those who considered the oppression of blacks purely a matter of race and those who were beginning to perceive that apartheid also had a profound class dimension. But there were no dramatic schisms or major public disagreements within the movement before its suppression in 1977.[67]

This difference reflected the contrast between protest in a liberal democracy, with constitutional protection of civil liberties, and in a state that permitted some freedom of speech to its white citizens but tried to maintain totalitarian control over black expression. The contrast was not absolute; SNCC, the Black Panthers, and other militant groups were victims of FBI and police harassement, "dirty tricks," and even murderous attacks. But these assaults did not occur until after they had worked out and promulgated their basic ideas and programs in relative freedom. The chance to write and speak freely invited a diversity of views about how best to respond to the post-civil rights predicament of blacks and provided ample opportunity for ideological and tactical disagreements. In South Africa, Black Consciousness adherents knew from the beginning that advocacy of violence or even militant nonviolence would lead to immediate proscription. The movement had to walk a tightrope between accommodation to the regime and revolutionary assertion; this balancing act limited the scope of discourse and action. Part of the explanation for the fact that Black Consciousness relied so heavily on churches and church-sponsored organizations as a vehicle for its message was that religious expression was less closely monitored than other forms. In the United States, the more charismatic or notorious Black Power advocates had many forums; they were interviewed on television and radio, wrote articles for prominent liberal journals, had their utterances reported (sometimes accurately) in daily newspapers, and published their books with major commercial publishers.

Steve Biko put his finger on the basic difference between the situations faced by the two movements. One embodied—in its most characteristic and durable expressions—the desire of a minority to be included, *but on its own terms*, within a society that it could never dominate. The other reflected the ambition of a majority to rule in its native land. This difference seems so fundamental that the degree of similarity that our inquiry has revealed may seem surprising. But numbers are not the whole story. Blacks in South Africa were even more of a minority from the standpoint of how much power they were officially allowed to exercise than African-Americans. But their potential power was of course much greater. The sense of that potential power, however long it might take to be realized, may be part of the reason why representative expressions of black protest in South Africa since the 1960s have generally been delivered in a calmer and less angry tone of voice than the equivalent expressions of African-American grievance. Black Americans have been haunted by the inherent limits of their potential power as a minority community and have at times found it very difficult to conceive of true self-determination ever occurring. The persistence of white racism and black disadvantage has aroused anger, but the difficulty of finding a credible strategy to achieve liberation has bred frustration, if not despair. The fear that the most determined and courageous resistance to white supremacy might prove unavailing helps to account for the affirmation in African-American writing of what Huey Newton called "revolutionary suicide," the willingness to die bravely and defiantly in a struggle that has no guarantee of success. The most famous expression of this sentiment is Claude McKay's poem of 1919, "If We Must Die":

> If we must die, O let us nobly die,
> So that our precious blood may not be shed
> In vain; then even the monsters we defy
> Shall be constrained to honor us though dead.[68]

From a more conventional and pragmatic point of view, however, Black Power was a greater success than Black Consciousness. Its pluralist version, especially in its political manifestation, clearly increased the ability of blacks to advance their own interests and defend themselves against racism. Black Consciousness, by contrast, failed to exert sufficient pressure to make apartheid unworkable and was superseded by a movement that played down BC's message of black pride and solidarity. Black Consciousness failed in practical terms because the white minority government of the 1970s was unwilling to allow blacks to acquire the kind of bargaining power that might actually bring genuine reform and had the strength and ruthlessness to prevent it. BC ideology was eclipsed by Charterism, not only because of the strategic advantages of the ANC that have already been described, but also because the international pressure that the liberation struggle needed to help make the government receptive to basic change could not readily be brought to bear on behalf of a movement that seemed to be espousing black chauvinism. An inestimable advantage that the ANC possessed in its competition with the PAC and Black Consciousness groups for international support was that its official ideology transcended race in the name of a common humanity.

The success of the ANC of the 1980s in appealing to "the conscience of the world," which included the United States and the nations of Western Europe as the prime sources of economic leverage over South Africa, raises the difficult and sensitive issue of whether African-American leaders and intellectuals since the late '60s have hurt the cause of black liberation by generally assuming that there was no such thing as a white American conscience to be appealed to and that the only path to racial justice is through an ethnocentric exercise of power and pressure. The question becomes inescapable if we contrast the relatively self-contained African-American struggle—with its reluctance since the late '60s to cooperate closely with sympathetic members of other races—with the way that the main liberation movement in South Africa has opened its doors to the full participation of all of South Africa's racial groups. Of course the significant difference in the nature of the two movements must be acknowledged. The African National Congress has been the government-in-waiting of a multi-racial nation with a black majority, while the NAACP and the Congressional Black Caucus are lobbying or exerting political pressure on behalf of a minority. Clearly there is a need for the straightforward representation of blacks as a group with particular needs and interests. But why is there no broad-based, anti-racist political movement in the United States, a multi-cultural alliance of all people who believe in racial equality and multi-cultural democracy? There are such anti-racist organizations in an increasingly multi-racial Europe—SOS RACISME in France is one example. Perhaps the level of distrust among racial groups in the United States has reached the point that no one group would be able to take the initiative because they would be viewed by the others as pursuing an agenda of their own. If white progressives took the lead, the fear of white paternalism and cultural hegemony would undoubtedly inhibit the involvement of people of color, especially blacks. The cultural left's image of America as a racial and ethnic mosaic seems to encourage groups to firm up their boundaries and perfect their interior designs rather than seek a way to make each fragment part of a larger pattern that reveals a vision of liberty and justice for all —that, in other words, reconstructs fundamental American values until they provide a clear mandate for racial and cultural democracy.

Despite all the contextual differences between the struggles against racism in the United States and South Africa, there may still be something Americans can learn from the nonracialism of the African National Congress. What made a nonracial vision possible was the heroic devotion to it of ANC leaders such as Nelson Mandela, who did not allow twenty-seven years of imprisonment to embitter him against whites as a race, however much he hated the apartheid system. Perhaps if Martin Luther King, Jr., had lived for his full span of years he could have kept his image of an interracial "beloved community" alive in the United States although, as we have seen, it was already badly tarnished at the time of his death. In any case, it scarcely behooves white Americans to expect forgiveness and salvation at the hands of a black messiah who will lead us all into the millennium. But there was something else that made nonracialism possible in South Africa —the heroic contribution of some whites to the cause of black liberation. At the founding meeting of the United Democratic Front in 1983, Allan Boesak made the case for including whites within the movement:

> We must remember that apartheid does not have the support of all the whites. There are some who have struggled with us, who have gone to jail, who have been tortured and banned. There have been whites who died in the struggle for justice. We must therefore not allow our anger over apartheid to become the basis for a blind hatred of all whites. . . . Let us, even now, seek to lay the foundations for reconciliation between whites and blacks in this country by working together, praying together, struggling for justice.[69]

The white freedom fighters to whom Boesak referred were relatively few in number and many of them espoused a Marxist-Leninist ideology that no longer seems viable. But just as John Brown, William Lloyd Garrison, and Albion W. Tourgée were heroes for the black militants who launched the modern freedom struggle when they founded the Niagara Movement of 1905, so white radicals such as Bram Fischer, Ruth First, Neil Aggett, and Joe Slovo have been heroes for black liberationists in South Africa. The strike of 100,000 black workers to protest the death in detention of white trade unionist Neil Aggett in 1982, and the huge black turnout for his funeral, reveal that there was no color qualification for anti-apartheid martyrdom. Reports of militant African youth in the townships during the '80s chanting the name of Joe Slovo to defy the police was a graphic reflection of how a white radical could come to personify the liberation struggle. At his trial in 1966, Bram Fischer, scion of a distinguished Afrikaner family and the lawyer who had defended Nelson Mandela two years earlier, justified his support of the ANC and the Communist party as contributing to the survival of whites in a liberated South Africa. Pan-Africanists later quoted his testimony to show that white nonracialists had ulterior motives for supporting the ANC, but Fischer's higher form of loyalty to his own people, his willingness to go to prison for life on behalf of a nonracial democracy that might be the only hope for

Afrikaner survival, is an attitude that would improve the world greatly if it were more widely shared.[70]

The lesson for the United States might be that a larger number of Euro-Americans need to earn the respect of their African-American fellow citizens by showing through their actions that they are prepared to make major sacrifices for racial justice, not out of pure idealism but out of a recognition that America will be a decent and livable place, for people like themselves as well as people of color, only if race ceases to determine life chances and status—only, in other words, if black liberation is fully achieved.

Epilogue

The victory of Nelson Mandela and the African National Congress in the 1994 elections was the culmination of a remarkable series of events, beginning with Mandela's release from prison in 1990, that has brought an end to legalized white domination in South Africa. In the light of this enormous breakthrough, and taking into account the foregoing history of black liberation struggles in the United States and South Africa, what useful comparisons can be made between the current situation and future prospects of blacks in the two societies?

One possible assessment would celebrate the victory over apartheid in the 1990s as roughly equivalent to the triumph of the American civil rights movement over legalized segregation and de facto disfranchisement thirty years earlier, the operative assumption being that the American precedent was similarly successful. The result in both cases, according to this optimistic evaluation, was an end to official racism and the removal of the principal barriers to the achievement of a color-blind democratic society. From this vantage point, the essential struggles are over, and white racism is, if not quite dead, at least deprived of most of its power.

It would be difficult, however, to sell this triumphalist analogy to some of the most acute observers of black-white relations in the United States in the 1990s. A pessimistic view of black progress since the '50s has taken hold, not only among black intellectuals, but also among some of the most respected white students of the American race relations. The eminent sociologist Andrew Hacker concluded his bleak portrayal of the condition of African-Americans in 1992 by noting that "legal slavery may be in the past, but segregation and subordination have been allowed to persist." He concludes his horrendous account of black deprivation and disillusionment without offering even a glimmer of hope that the situation will improve: "A huge racial chasm remains, and there are few signs that the coming century will see it closed."[1]

In the light of this growing pessimism about the prospects for racial equality in the United States, a quite different comparative analysis suggests itself. South African blacks, it could be argued, have achieved something that has eluded African-Americans and will probably continue to elude them. Despite the problems that remain, black South Africans have thrown off the shackles of white domination and have achieved genuine self-determination, while African-Americans remain at the mercy of a white majority that remains racist—not in the old-fashioned sense of openly advocating the legal subordination of blacks, but in the new sense of denying the palpable fact that blacks as a group suffer from real disadvantages in American society and will continue to do so unless radical action is taken. When Nelson Mandela celebrated his electoral victory, he consciously echoed Martin Luther King, Jr., by exclaiming "Free at Last!" But King never used this cry, as Mandela did, to celebrate a victory already won. On the contrary, it was what he hoped blacks would be able to shout when, at some time in the near or distant future, they actually realized their dream of freedom and equality. If in fact this dream has permanently faded, a contrast between black South African winners and African-American losers can be made that would justify the disillusionment of many blacks with their prospects for equality in American society and encourage racial separatism and polarization.

This reversal of the comparative perspective of a few years ago—when it was possible to argue that African-American progress might be a model for black South Africans, but one that would be very difficult for them to emulate—may turn out to be valid. Yet there is a third way of making the comparison that, like the first, stresses similitude more than stark contrast but nevertheless acknowledges that racism is not dead or toothless. It is somewhat more hopeful in regard to African-American prospects than the winners-and-losers comparison but a bit less sanguine about the future of South Africa. It might well be the case, advocates of this view could maintain with some cogency, that the two liberation struggles are at a similar stage—significant progress has been made but major challenges still remain. Consequently they can learn important lessons from each other about how to proceed in the future.

There are a number of similarities between post-Jim Crow black America and post-apartheid black South Africa. Legalized segregation has been abolished for all time, just as racial slavery was in the previous century. The right of blacks to vote and hold office has been assured. But in both cases whites retain sufficient power to prevent either society from moving decisively and quickly beyond legal and political rights for all to the achievement of social and economic equality. This lack of substantive equality is most obvious in the case of the United States, where whites dominate the electorate, as well as the economy, and "the politics of race" is a fact of life pushing government social policy in a conservative direction. But it is also true in South Africa, despite the black-majority electorate. The negotiated settlement that led to the end of apartheid and the white monopoly on political power has left Europeans in control of the economy and has deprived the government of the constitutional authority to redistribute wealth in a radical or thoroughgoing way. "Growing the economy" through free-market mechanisms and the attraction of international investment may improve the average living standards of blacks to some extent, but it is unlikely to lead to a significant closing of the gap between an affluent white minority and an relatively impoverished African majority.

With a socialistic redistribution of wealth ruled out for the foreseeable future in both societies, the most obvious way to narrow the gap between privileged whites and underprivileged blacks is through some form of racial preference or affirmative action, but whites can be expected to invoke meritocratic criteria to prevent rapid black advancement, claiming—with some credibility given the way racism has limited black opportunities to acquire skills crucial to a technologically advanced society—that they are better qualified in terms of education and experience for the kinds of occupations and responsible positions in which blacks are currently underrepresented.

South Africa can learn from the experience of the United States since the 1960s that formal citizenship rights are not enough to overcome the effects of three hundred years of white supremacy. Some kind of crash program to compensate for inherited disadvantages is clearly required. Despite the built-in constraints previously described, a black-dominated government at least has the capacity to initiate compensatory educational programs and new policies to encourage self-help and entrepreneurship among blacks that might narrow the economic gap. Rather than being a manifestation of state socialism, the substantial land redistribution that would be necessary to give blacks a real stake in the economy would be simply a restoration of the original African property rights that have been flagrantly denied under the Natives' Land Act of 1913 and subsequent discriminatory legislation. If the African National Congress fails to adopt policies that improve the economic situation of blacks relative to whites within a few years, it will probably be replaced in some future election by another black party with a bolder program.

The lessons that the African-American freedom struggle might learn from recent developments in South Africa are less tangible and clear-cut but may be equally compelling. First of all, there is the message of hope. However unlikely giant strides in the direction of black liberation might seem to be, the South African experience of the past twenty years should tell us that they are not impossible. The "miracle" of apartheid's overthrow could serve as an antidote to the hopelessness and "nihilism" that observers have attributed to some segments of the African-American population.[2] If black South Africans, with all the oppression to which they have been subjected, could keep alive the hope of liberation and finally see it fulfilled to an extent that few detached observers would have thought possible a few years ago, then perhaps a "Third Reconstruction" (to use historian Manning Marable's phrase) is possible within the next few years, however unlikely it may seem today.[3] Nelson Mandela's unconquerable spirit during his twenty-seven years of imprisonment shows the value, in the face of seemingly insurmountable obstacles, of faith in ultimate freedom. White anti-racists looking for a role to play might also take inspiration from the career of those Europeans in the ANC who suffered imprisonment, exile, and the loss of family members or even their own extremities from letter bombs sent by government assassins, but never gave up the struggle and are now part of the government. What would have to be accomplished by a Third Reconstruction would not be entirely dissimilar from the challenge faced by the new government in South Africa. Some creative combination of affirmative action, government anti-poverty programs, and the encouragement of black self-help would be needed if a significant narrowing of the economic gap between whites and blacks is to be achieved.[4]

The black struggle in the United States might also learn something from the ideological and tactical flexibility that the ANC has demonstrated. "All-in" movements that incorporate as many shades of black opinion as possible can involve more people and exert more pressure than divisive, sectarian movements.[5] But, as the experience of the ANC also demonstrates, there is a difficult choice that sometimes has to be made between accommodating the full range of black opinion and cooperating with anti-racist whites. Some principled anti-racists have recently accused the NAACP of putting black solidarity ahead of the struggle against all forms of bigotry and discrimination in the United States. Inviting Louis Farrakhan of the black racialist and anti-Semitic Nation of Islam to a summit meeting of black leaders in 1994 was not the same as endorsing his views, but it did give him a measure of legitimacy and was deeply troubling to many of the association's white and integration-minded supporters. It might conceivably be contrasted with the decision made during the 1980s by the ANC and its domestic surrogate, the United Democratic Front, to give interracialism or "nonracialism" a higher priority than black unity. Besides welcoming the participation of anti-apartheid whites, the African leadership reached out to the Indian and Colored communities but made no concessions to Pan-Africanist and Black Consciousness hardliners who categorically rejected cooperation and reconciliation with whites. The mainstream liberation movement proclaimed that South Africa was a multi-racial and multi-ethnic society that could be fused into a single nation on the basis of shared democratic values.

Martin Luther King, Jr., articulated a similar vision of *American* nationality as the fulfillment of the democratic ideals expressed in principle by Jefferson and Lincoln but, more often than not, flagrantly denied when African-Americans laid claim to equal rights. King's dream of a future United States as "a beloved community" from which racism and racial exclusiveness had been banished is in danger of being lost in an era of identity politics when a "go-it-alone" mentality characterizes the thinking of status groups based on race, ethnicity, gender, and sexual orientation—to say nothing of the class-based interest groups that are self-seeking almost by definition. If the black freedom movement could regain or reemphasize a broadly inclusive and humane vision of a society that is multi-cultural but nonetheless unified in its basic commitment to democracy and human rights—the "nonracist, nonsexist" South African Bill of Rights with its prohibition of discrimination based on gender and sexual orientation as well as race could serve as a guide—it might serve as the catalyst for a new political majority composed of Americans who have been historically disadvantaged along with those from advantaged backgrounds who can be persuaded to sacrifice their own privileged status in order to live in a just and harmonious society. Such a majority could turn the American dream of "liberty and equality for all" into a reality. Then we would all be "free at last."

Notes

Introduction

1. George M. Fredrickson, *White Supremacy: A Comparative Study in American and South African History* (New York, 1981), xx.
2. For a theoretical conception of ideology consistent with the one employed in this study, see Martin Seliger, *Ideology and Politics* (London, 1979). According to Seliger, "An ideology is a group of beliefs and disbeliefs (rejections) expressed in value sentences, appeal sentences and explanatory sentences" that is "designed to serve on a relatively permanent basis a group of people to justify in reliance on moral norms and modicum of factual evidence and self-consciously rational coherence the legitimacy of the implements and technical prescriptions which are to ensure concerted action for the preservation, reform, destruction, or reconstruction of a given order" (120).
3. See Antonio Gramsci, *Selections from the Prison Notebooks,* Quentin Hoare and Geoffrey Nowell Smith, eds. and trans. (New York, 1971), 3–23. As will become evident, however, my "organic intellectuals" may speak for a Weberian "ethnic status group" as well as for a social class in the Marxian or Gramscian sense.
4. Significant works on connections between black America and black South Africa include J. Mutero Chirenje, *Ethiopianism and Afro-Americans in Southern Africa, 1883–1916* (Baton Rouge, 1987); James Campbell, "Our Fathers, Our Children: A History of the African Methodist Episcopal Church in the United States and South Africa" (Ph.D. diss., Stanford University, 1989); William Manning Marable, "African Nationalist: The Life of John Langalibele Dube" (Ph.D. diss., University of Maryland, 1976); Robert B. Edgar, ed., *An African American in South Africa: The Travel Notes on Ralph J. Bunche, 28 September, 1937—1 January, 1938* (Athens, Ohio, 1992); Donald B. Coplan, *In Township Tonight! South Africa's Black City Music and Theatre* (Johannesburg, 1985); Tim Couzens, "Moralizing Leisure Time: The Transatlantic Connection

and Black Johannesburg, 1918–1936," in *Industrialisation and Social Change in South Africa, 1870–1930,* Shula Marks and Richard Rathbone, eds. (London, 1982), 314–37; Robert A. Hill and Gregory A. Pirio, " 'Africa for the Africans': The Garvey Movement in South Africa," in *The Politics of Race, Class, and Nationalism in Twentieth-Century South Africa,* Shula Marks and Stanley Trapido, eds. (London, 1987), 209–53; Richard D. Ralston, "American Episodes in the Making of an African Leader: A Case Study of Alfred B. Xuma (1893–1962)," *International Journal of African Historical Studies* 6 (1973): 72–93; David H. Anthony III, "Max Yergan in South Africa: From Evangelical Pan-Africanist to Revolutionary Socialist," *African Studies Review* 34 (1991): 27–55; and R. Hunt Davis, Jr., "The Black American Education Component in African Responses to Colonialism in South Africa, *Journal of Southern African Studies* 3 (1978): 65–83. Two seminal articles of a more general nature that pioneered the study of the African-American impact on black South Africa are George Shepperson, "Notes on Negro American influences on the Emergence of African Nationalism," *Journal of African History* 1 (1960): 299–312, and Peter Walshe, "Black American Thought and African Political Attitudes in South Africa," *Review of Politics* 32 (1970): 51–77.

5. George M. Fredrickson, *The Arrogance of Race: Historical Perspectives on Slavery, Racism, and Social Inequality* (Middletown, Conn., 1988), 158–59, 216–24. For the relevant writings of Max Weber, see his *Economy and Society: An Outline of Interpretative Sociology,* Guenther Roth and Claus Witich, eds., 2 vols. (Berkeley, 1978), I: 385–98, II: 926–39, and passim.

Chapter 1

1. See George M. Fredrickson, *White Supremacy: A Comparative Study in American and South African History* (New York, 1981), 179–84.
2. On early American racial or ethnic conceptions of citizenship, see Lawrence J. Friedman, *Inventors of the Promised Land* (New York, 1975). Douglas A. Lorimer, *Colour, Class, and the Victorians: English Attitudes to the Negro in the Nineteenth Century* (Leicester, 1978), describes British race relations. On how African-American abolitionists could misread the British viewpoint, see Waldo E. Martin, *The Mind of Frederick Douglass* (Chapel Hill, 1984), 115–16. On Mill's ethnic qualification to liberal theory, see John Stuart Mill, "Considerations on Representative Government," in *Three Essays* (Oxford, 1975), 382, 408, and passim; and Dennis F. Thompson, *John Stuart Mill and Representative Government* (Princeton, 1976), 125–26, 146.
3. For a suggestive treatment of how black protest enlarged the conception of equality, see Celeste Michelle Condit and John Louis Lucaites, *Crafting Equality: America's Anglo-African Word* (Chicago, 1993). For differing views on the complex relationship between the development of capitalism and the rise of antislavery ideology, see David Brion Davis's *The Problem of Slavery in the Age of Revolution* (Ithaca, 1975) and Thomas Bender, ed., *The Antislavery Debate: Capitalism and Abolitionism as a Problem of Historical Interpretation* (Berkeley, 1988).
4. For a perceptive discussion of the problematic nature of the right to vote in nineteenth-century liberal thought, see W. R. Brock, *An American Crisis: Congress and Reconstruction, 1865–1867* (London, 1963), 293–98. Brock

shows that even the Radical Republicans had difficulty defending black suffrage as a matter of natural or inalienable right.

5. On the history of the suffrage and popular politics in the United States, see Chilton Williamson, *American Suffrage: From Property to Democracy* (Princeton, 1960), and Michael McGerr, *The Decline of Popular Politics: The American North, 1865–1928* (New York, 1988).
6. See Brock, *American Crisis,* 288, 297–98.
7. Philip S. Foner, ed., *The Life and Writing of Frederick Douglass,* vol. 4 (New York, 1955), 149.
8. Among the more valuable studies that shed light on black loss of voting rights in the South are William Gillette, *Retreat from Reconstruction, 1869–1879* (Baton Rouge, 1979), and J. Morgan Kousser, *The Shaping of Southern Politics: Suffrage Restriction and the Establishment of the One-Party South, 1880–1910* (New Haven, 1974).
9. J. L. McCracken, *The Cape Parliament, 1854–1910* (Oxford, 1967). The quotation from Porter is on p. 65. See also Stanley Trapido, "White Conflict and Non-white Participation in the Politics of the Cape of Good Hope" (Ph.D. diss., University of London, 1969).
10. See note 9 above; Stanley Trapido, " 'The Friends of the Natives': Merchants, Peasants, and the Political and Ideological Structure of Liberalism in the Cape, 1854–1910," in Shula Marks and Anthony Atmore, eds., *Economy and Society in Pre-industrial South Africa* (London, 1980), 247–74.
11. On the origins of the slogan, see Fredrickson, *White Supremacy,* 185.
12. On the situation of northern free blacks, see Leon F. Litwack, *North of Slavery: The Negro in the Free States, 1790–1860* (Chicago, 1961), and Leonard P. Curry, *The Free Black in Urban America* (Chicago, 1981). On southern "free Negroes," see Ira Berlin, *Slaves Without Masters: The Free Negro in the Antebellum South* (New York, 1972).
13. Insight into the political ideas and actions of the free black elite can be derived from Benjamin Quarles, *The Black Abolitionists* (New York, 1969); Jane H. Pease and William H. Pease, *They Who Would Be Free: Blacks' Search for Freedom, 1830–1861* (New York, 1974); Sterling Stuckey, ed., *The Ideological Origins of Black Nationalism* (Boston, 1972); and David E. Swift, *Black Prophets of Justice: Activist Clergy Before the Civil War* (Baton Rouge, 1989). On Douglass's prewar career and ideas, see especially William S. McFeely, *Frederick Douglass* (New York, 1991), 91–182, and Martin, *The Mind of Frederick Douglass, passim.*
14. Joel Schor, *Henry Highland Garnet: A Voice of Black Radicalism in the Nineteenth Century* (Westport, Conn., 1977), 142–44, 161. Garnet's American nationalism was most strongly affirmed in his *The Past and Present Condition and the Destiny of the Negro Race* (1848); see Howard Brotz, ed., *Negro Social and Political Thought, 1850–1920: Representative Texts* (New York, 1966), 199–200.
15. Pease and Pease, *They Who Would Be Free,* 288–93. For a general discussion of predominance of the "vanguard" style of black leadership, see John Brown Childs, *Leadership, Conflict, and Cooperation in Afro-American Social Thought* (Philadelphia, 1989).
16. For a juxtaposition of the viewpoints of Delany and Whipper, see Stuckey, ed., *Ideological Origins,* 195–236, and 252–60 (quote from Delany on 196–

97). See also Martin R. Delany, *The Condition, Elevation, Emigration, and Destiny of the Colored People of the United States* (Philadelphia, 1852).

17. Leonard I. Sweet, *Black Images of America, 1784–1870* (New York, 1976), 172 and passim. On the duality of black and American nationalism in Frederick Douglass, see David W. Blight, *Frederick Douglass' Civil War: Keeping Faith in Jubilee* (Baton Rouge, 1989), 101–21. Douglass wavered briefly in his opposition to black expatriation when he endorsed some emigration to Haiti on the eve of the Civil War.
18. Sweet, *Black Images,* 101. On the self-help ideology among antebellum free blacks, see especially Frederick Cooper, "Elevating the Race: The Social Thought of Black Leaders, 1827–1850," *American Quarterly* 24 (1972): 604–25. Martin Delany was one of the few who dissented from the view that condition and not color was the problem. Unlike most other black abolitionists, he tended to see racism as deeply rooted and probably ineradicable.
19. See George M. Fredrickson, *The Black Image in the White Mind: The Debate on Afro-American Character and Destiny, 1817–1914* (rev. ed., Middletown, Conn., 1987), 33–42, and *The Arrogance of Race: Historical Perspectives on Slavery, Racism, and Social Inequality* (Middletown, Conn., 1988), esp. 216–35.
20. Blight, *Frederick Douglass' Civil War,* 116–20; Cyril E. Griffith, *The African Dream: Martin R. Delany and the Emergence of Pan-African Thought* (University Park, Pa., 1975), 83–88; Benjamin Quarles, *The Negro in the Civil War* (Boston, 1953).
21. See Thomas Holt, *Black Over White: Negro Political Leadership in South Carolina During Reconstruction* (Urbana, 1977), 73–77 and passim, and John Hope Franklin, *Reconstruction: After the Civil War* (Chicago, 1961), 86–92.
22. Foner, ed., *Life and Writings,* 4: 31, 164 (quote); Martin, *The Mind of Douglass,* 71–72.
23. Some of the works that have influenced my view of southern Reconstruction, in addition to those already cited, are Eric Foner, *Reconstruction: America's Unfinished Revolution* (New York, 1988); Michael Perman, *The Road to Redemption: Southern Politics, 1869–1879* (Chapel Hill, 1984); Mark W. Summers, *Railroads, Reconstruction, and the Gospel of Prosperity: Aid Under the Radical Republicans* (Princeton, 1984); and Eric Anderson and Alfred A. Moss, eds., *The Facts of Reconstruction: Essays in Honor of John Hope Franklin* (Baton Rouge, 1991). For a perceptive analysis of the problematic circumstances of black politicians, see W. McKee Evans, *Ballots and Fence Rails: Reconstruction on the Lower Cape Fear* (New York, 1974), 155–57.
24. On the Republican party and blacks after Reconstruction, see Vincent P. De Santis, *Republicans Face the Southern Question—The New Departure Years, 1877–1897* (Baltimore, 1959), and Stanley Hirshon, *Farewell to the Bloody Shirt: Northern Republicans and the Southern Negro, 1877–1893* (Bloomington, 1962). For an example of how black loyalty to the Republican party encouraged equal accommodations laws in the North, see David A. Gerber, *Black Ohio and the Color Line, 1860–1915* (Urbana, 1976), 46.
25. Quoted in Emma Lou Thornbrough, *T. Thomas Fortune: Militant Journalist* (Chicago, 1972), 59–60. See also Bess Beatty, *A Revolution Gone Backwards: The Black Response to National Politics, 1876–1896* (Westport, Conn., 1987), 48–49. Fortune repeated this anti-Republican rhetoric in his book *Black and White, Land, Labor, and Capital in the South* (1884; New York, 1968), 126.

26. On the politics of the suffrage question in 1890, see Hirshon, *Bloody Shirt,* 190–235. The standard work on the disfranchisement process in the South is Kousser, *Shaping of Southern Politics.*
27. Fortune, *Black and White,* 147–58, 170–74 (quote on 174; emphasis in the original); Thornbrough, *Fortune,* 89.
28. Fortune, *Black and White,* 180–85 (quotes on 180–81).
29. Ibid., 193–94. Other anticipations of Washington can be found on pp. 73, 80–81 (where he attacks classical education and advocates industrial schools), 87–90, and 128.
30. Ibid., 56, 106.
31. Quoted in Beatty, *Revolution Gone Backward,* 58. For the context, see Peter Gilbert, ed., *Selected Writings of John Edward Bruce* (New York, 1971), 23.
32. George Washington Williams, *A History of the Negro Race from 1619 to 1880,* 2 vols. (New York, 1882), II: 527–28. For information on Williams, see John Hope Franklin, *George Washington Williams: A Biography* (Chicago, 1985).
33. Booker T. Washington, *The Future of the American Negro* (New York, 1900), 132–33, 141, 151. On Washington's career up to the turn of the century, see Louis R. Harlan, *Booker T. Washington: The Making of a Black Leader, 1856–1901* (New York, 1972). For an analysis of Washington's ideas in the context of the black thought of the time, see August Meier, *Negro Thought in America, 1880–1915* (Ann Arbor, 1963).
34. This interpretation of Washington draws on William Toll, *The Resurgence of Race: Black Social Theory from Reconstruction to the Pan-African Conferences* (Philadelphia, 1979). On self-help and the Jewish model, see Meier, *Negro Thought,* 105–6. For more on Washington's influence on Africans, see Chapter 3 below.
35. W. E. B. Du Bois, *Souls of Black Folk* (Chicago, 1903), 11. Meier, *Negro Thought,* and Toll, *Resurgence of Race,* provide insights into the Washington-Du Bois controversy. For a fuller discussion of the controversy—with an emphasis on the segregation issue—see Chapter 3 below.
36. Quoted in Trapido, "White Conflict and Black Participation," 139–40.
37. See Fredrickson, *White Supremacy,* 162–98, and idem, *Arrogance of Race,* 236–53.
38. Fredrickson, *White Supremacy,* 131–34; J. S. Marais, *The Cape Coloured People, 1652–1937* (1939; Johannesburg, 1968), passim; Gavin Lewis, *Between the Wire and the Wall: A History of South African "Coloured" Politics* (Cape Town, 1987), 6–10.
39. Fredrickson, *White Supremacy,* 184–85; T. R. H. Davenport, *The Afrikaner Bond: The History of a South African Political Party* (Cape Town, 1966), 118–23.
40. My understanding of these mission-trained political leaders draws heavily on two works by André Odendaal, *Vukani Bantu! The Beginnings of Black Protest Politics in South Africa to 1912* (Cape Town, 1984), and "African Political Mobilization in the Eastern Cape, 1880–1910" (D.Phil. diss., University of Cambridge, 1983). These works overlap to a degree, but each contains significant material that is not in the other.
41. Colin Bundy, *The Rise and Fall of the South African Peasantry* (2nd ed., Cape Town, 1988), 32–43.

42. Still valuable as a source of information on missionary activities and attitudes is J. DuPlessis, *A History of Christian Missions in South Africa* (London, 1911). Norman Etherington's *Preachers, Peasants, and Politics in Southeast Africa, 1835–1880* (London, 1978) focuses on Natal rather than the eastern Cape, but it sheds important new light on the kind of impact missions were having throughout southern Africa.
43. Trapido, " 'The Friends of the Natives.' " See also Bundy, *Rise and Fall,* 65–108.
44. Bundy describes the process of mobility from peasantry to educated elite in *Rise and Fall,* 140.
45. Donovan Williams, *Umfundisi: A Biography of Tiyo Soga, 1829–1871* (Lovedale, 1978), 95–97, and passim. See also J. Mutero Chirenje, *Ethiopianism and Afro-Americans in Southern Africa, 1883–1916* (Baton Rouge, 1987), 15–16. For more on A. K. Soga, see below, this chapter.
46. Increasing discrimination against African clergy was a main cause of the Ethiopian movement, which is treated in Chapter 2 below. On the rise of the Bond, see Davenport, *Afrikaner Bond,* and Leonard Thompson, "Great Britain and the Africaner Republics, 1870–1899," in Monica Wilson and Leonard Thompson, eds., *The Oxford History of South Africa,* vol. II (Oxford, 1971), 302–7.
47. Odendaal, *Vukani Bantu!,* 8–9.
48. Ibid., 11–16; Sheridan Johns III, *Protest and Hope,* vol. 1 of Thomas Karis and Gwendolen Carter, eds., *From Protest to Challenge: A Documentary History of African Politics in South Africa,* (Stanford, Calif., 1972), 12–17 (quote on 14–15).
49. Odendaal, *Vukani Bantu!,* 16; Johns, *Protest and Hope,* 9; McCracken, *Cape Parliament,* 97; Fredrickson, *White Supremacy,* 85.
50. Peter Norwich, *Black People and the South African War, 1899–1902* (Johannesburg, 1983), 110–14, 174–78; Odendaal, *Vukani Bantu!,* 16; Fredrickson, *White Supremacy,* 195 (quote).
51. Johns, *Protest and Hope,* 18, 22–23, 28.
52. A work that focuses on the long-term costs of British generosity to the defeated Afrikaners and corresponding neglect of African interests is Nicholas Mansergh, *South Africa, 1906–1961: The Price of Magnanimity* (London, 1962).
53. Odendaal, *Vukani Bantu!,* xi (quote) and passim; and idem, "African Political Mobilization," 43–44, 93, and passim.
54. Lewis, *Between the Wire and the Wall,* 9–13; Fredrickson, *White Supremacy,* 132–34.
55. My view of early Colored politics derives primarily from Lewis, *Between the Wire and the Wall,* but see also Stanley Trapido, "The Origins and Development of the African Political Organization," *Collected Seminar Papers on the Societies of Southern Africa in the Nineteenth and Twentieth Centuries,* I (London: Institute of Commonwealth Studies, 1970), and H. R. van der Ross, "A Political and Social History of the Cape Coloured People," 4 vols. (typescript at the Centre for Intergroup Studies, University of Cape Town, 1976).
56. Lewis, *Between the Wire and the Wall,* 16–18; Christopher C. Saunders, "F. Z. S. Peregrino and the *South African Spectator,*" *Quarterly Bulletin of the South African Library* 32 (1977–78): 82–87. The Indianapolis *Freeman* of May 14, 1898, reprinted a Peregrino article from the Buffalo *Evening Times.* Copies of the *Times* containing articles by Peregrino have not been located.

57. Johns, *Protest and Hope,* 36.
58. For a good summary of how the African-American example inspired and influenced educated Africans at the beginning of the twentieth century, see Peter Walshe, *The Rise of African Nationalism in South Africa: The African National Congress, 1912–1952* (Berkeley, 1971), 12–14. See also Chirenje, *Ethiopianism and Afro-Americans,* passim.
59. Chirenje, *Ethiopianism and Afro-Americans,* 32–34 (quote on 33).
60. The five-part series "Call the Black Man to Conference" appeared in the *Colored American Magazine* for December 1903, and January, February, March, and April 1904. The biographical information on Soga is from Sarah A. Allen, "Mr. Alan Kirkland Soga," *Colored American Magazine* (Feb. 1904): 114–16. The information in this biographical sketch was probably provided by Soga himself. Soga's authorship of the SANC petition of 1903, which was quoted extensively earlier in this chapter, seems evident from the fact that he reprinted it under his own name in another series of articles for the *Colored American Magazine* that appeared in May/June, July, and August 1903. I am indebted to Professor Robert Edgar of Howard University for supplying me with photocopies of the articles by or about Soga that appeared in the magazine.
61. Johns, *Protest and Hope,* 1, 20, 21, 28.
62. *Colored American Magazine* (March 1904): 197–98.
63. Thornbrough, *Fortune,* 187, 243–44, and passim.
64. *Colored American Magazine* (Dec. 1903): 868–73.
65. Ibid. (March 1904): 198. The Fortune quotation is from the *New York Age* for July 2, 1903.
66. *Colored American Magazine* (Feb. 1904), 93; (March 1904): 200–201; (April 1904): 252.
67. On Miller and Walters, see Meier, *Negro Thought,* 172, 213–18. The radical, neo-abolitionist black intellectuals will be discussed in Chapter 3, below. For a modern defense of such mutualistic approaches, see Childs, *Leadership, Conflict, and Cooperation.*
68. *Colored American Magazine* (Feb. 1904): 115–16 (emphasis added).
69. On Jabavu's maneuverings, see Johns, *Protest and Hope,* part one, passim. On the refusal of Douglass and other prominent black Republicans to join the Afro-American League, see Thornbrough, *Fortune,* 120.
70. For an analysis of some aspects of this political-constitutional contrast, see Fredrickson, *Arrogance of Race,* 254–69, and Chapter 6, below.
71. On the "civilizationism" that predominated in black thought of the period, see Wilson Jeremiah Moses, *The Golden Age of Black Nationalism* (Hamden, Conn., 1978).
72. Thornbrough, *Fortune,* 356–58.
73. Quoted in Odendaal, "African Political Mobilization," 314.

Chapter 2

1. The comparable traditions of Christian humanitarianism that challenged slavery and legal inequality in the United States and South Africa in the early to mid-nineteenth century are discussed in George M. Fredrickson, *White Supremacy: A Comparative Study in American and South African History* (New York, 1981), 165–66. On black "civilizationism" and its relation to Christian-

ity, see Wilson Jeremiah Moses, *The Golden Age of Black Nationalism, 1850–1925* (Hamden, Conn., 1978), 20–31.

2. Good general treatments of independent black churches in the two societies are C. Eric Lincoln and Lawrence H. Mamiya, *The Black Church in the American Experience* (Durham, N.C., 1990), and B. G. M. Sundkler, *Bantu Prophets in South Africa* (2nd ed., London, 1961). In 1960, about 20 percent of Africans in South Africa belonged to independent churches as compared with 46 percent who were members of the "historic" churches originally founded by whites and brought to South Africa by European missionaries and settlers. See the table in Monica Wilson and Leonard Thompson, eds., *The Oxford History of South Africa,* vol. II (New York, 1971), 475.
3. See especially St. Clair Drake, *The Redemption of Africa and Black Religion* (Chicago, 1970), and J. Mutero Chirenje, *Ethiopianism and Afro-Americans in South Africa, 1883–1916* (Baton Rouge, 1987).
4. Lincoln and Mamiya, *Black Church,* 12–13. See below for discussions of Garnet, Turner, and King. For a sense of the bewildering variety of religio-political responses in Africa, see Terrence Ranger, "Religious Movements and Politics in Sub-Saharan Africa," *African Studies Review* 29 (March 1986): 1–69.
5. Debates on the relation of black religion to political activism in the United States are surveyed in Lincoln and Mamiya, *Black Church,* 199–212, and Hans A. Baer and Merrill Singer, *African-American Religion in the Twentieth Century* (Knoxville, 1992), 21–27. On the similar debates about faith and politics in black Africa, see Ranger, "Religious Movements and Politics."
6. On the often troubled relations between black and white abolitionists within the interracial antislavery movement, see Benjamin Quarles, *Black Abolitionists* (New York, 1969), and Jane H. and William H. Pease, *They Who Would Be Free: Blacks' Search for Freedom, 1830–1861* (New York, 1974).
7. Drake, *The Redemption of Africa.* Another general account of Ethiopianist ideology is Albert J. Raboteau, *"Ethiopia Shall Stretch Forth Her Hands": Black Destiny in Nineteenth Century America* (Tempe, Ariz.: Department of Religious Studies, Arizona State University, 1983). On the religious culture of the slaves, see especially Albert J. Raboteau, *Slave Religion: The Invisible Institution in the Antebellum South* (New York, 1979), and Eugene Genovese, *Roll, Jordan, Roll: The World the Slaves Made* (New York, 1976).
8. Henri Baudet, *Paradise on Earth: Some Thoughts on European Images of Non-European Man* (New Haven, 1965), 13–17.
9. George M. Fredrickson, *The Black Image in the White Mind: The Debate on Afro-American Character and Destiny, 1817–1914* (1971; Middletown, Conn., 1987), 97–129. See also Phillip Curtin, *The Image of Africa: British Ideas and Action, 1780–1850* (Madison, Wisc., 1964), 26–27.
10. In addition to Drake, *Redemption of Africa* and Raboteau, "*Ethiopia,*" see Gayraud S. Wilmore, *Black Religion and Black Radicalism* (Garden City, N.J., 1972), 136–86; Wilson Jeremiah Moses, *Black Messiahs and Uncle Toms: Social and Literary Manipulation of a Religious Myth* (University Park, Pa., 1993), 6–7, 49–66, passim; and idem, *Golden Age,* 23–25, 162–69, and passim.
11. Reprinted in Sterling Stuckey, *The Ideological Origins of Black Nationalism* (Boston, 1972), 30–38 (quotes on 35 and 37).
12. Reprinted in ibid., 39–117 (quote on 40).
13. Ibid., 63, 55–57.

14. See Floyd J. Miller, *The Search for a Black Nationality: Colonization and Emigration, 1787–1863* (Urbana, 1975).
15. Henry Highland Garnet, *The Past and Present Condition of the Colored Race: A Discourse* . . . (1848; Miami, 1969), 11, 26–29.
16. Joel Schor, *Henry Highland Garnet: A Voice of Black Radicalism in the Nineteenth Century* (Westport, Conn., 1977), 150–74 (quote on 161).
17. Victor Ullman, *Martin R. Delany: The Beginnings of Black Nationalism* (Boston, 1971), 225 and passim. Martin R. Delany, *The Condition, Elevation, Emigration, and Destiny of the Colored People of the United States* (1852; New York, 1968).
18. Quote on Ethiopia is taken from Gayraud S. Wilmore, *Black Religion and Black Radicalism,* 155; Delany's statement on emigration appears in *Condition of the Colored People,* 183.
19. Delany, *Condition of the Colored People,* 37–38; 42–43; Miller, *Search for Black Nationality,* 122–23, 171, 178; Ullman, *Delany,* 221 and passim.
20. For biographical information on Blyden, see Hollis Lynch, *Edward Wilmot Blyden: Pan-Negro Patriot, 1832–1912* (New York, 1970).
21. For information on Crummell, see Wilson Jeremiah Moses, *Alexander Crummell: A Study of Civilization and Discontent* (New York, 1989).
22. Edward W. Blyden, "The Call of Providence to the Descendents of Africa in America" in Blyden, *Liberia's Offering* (New York, 1862), as reprinted in Howard Brotz, ed., *Negro Social and Political Thought* (New York, 1966), 112–26 (quote on 117); Alexander Crummell, *Africa and America: Addresses and Discourses* (New York, 1969), 405–53. See also Hollis Lynch, ed., *Black Spokesman: Selected Published Writings of Edward Wilmot Blyden* (New York, 1971), passim; and Moses, *Crummell,* 79–80, 133, 220, 278.
23. Moses, *Crummell,* 207–9, 251, 295. For a good example of Crummell's thinking in his last years, see his "Civilization, the Primal Need of the Race" (1897) in John H. Bracey, August Meier, and Elliott Rudwick, eds., *Black Nationalism in America* (Indianapolis, 1970), 139–53. On Blyden's emerging Africanist perspective see Lynch, *Blyden,* 121 and passim.
24. Moses, *The Golden Age of Black Nationalism,* 20–21, 59–61, and passim; Fredrickson, *Black Image in the White Mind,* 101–9 and passim.
25. Lynch, *Blyden,* 58–83, 138–39; Moses, *Crummell,* 255–56, 269. Moses reveals that in the 1890s Crummell was privately anxious about the extent of mulatto influence in African-American affairs but was nevertheless willing to accept Du Bois and other "distinguished men of mixed race" into the Negro Academy.
26. Lynch, *Blyden,* 61–62.
27. Ibid., 67–77. For a general account of the development of Pan-Africanist ideas in the twentieth century, see Immanuel Geiss, *The Pan-African Movement: A History of Pan-Africanism in America, Europe, and Africa* (New York, 1974).
28. Crummell, *Africa and America,* 46–51.
29. Walker quote is from Stuckey, *Ideological Origins,* 108.
30. David Levering Lewis, *W. E. B. Du Bois: The Biography of a Race, 1868–1919* (New York, 1993), 101. Post-millennialism was the belief that the Second Coming of Christ would take place *after* a thousand years of peace. Since human betterment to a point of virtual perfection would precede the millenium, this viewpoint was compatible with optimistic reformism—in contrast to premillennialism, which envisioned Christ returning to judge a world that had be-

come increasingly sinful. On Du Bois's attraction to Hegelian dialectics, with its view of human history as the struggle of contrary ideas toward a higher synthesis, see especially Joel Williamson, *The Crucible of Race: Black-White Relations in the American South Since Emancipation* (New York, 1984), 399–413.

31. Philip S. Foner, ed., *W. E. B. Du Bois Speaks: Speeches and Addresses, 1890–1919* (New York, 1970), 76–81.
32. W. E. B. Du Bois, *Against Racism: Unpublished Essays, Papers, and Addresses,* Herbert Aptheker, ed. (Amherst, 1985), 47. For a provocative view of Du Bois as a racial Hegelian, see Williamson, *The Crucible of Race,* 399–413.
33. Raboteau, *"Ethiopia,"* 5 and passim; St. Clair Drake, *The Redemption of Africa,* 47–50; Wilmore, *Black Religion,* 166–68.
34. Walter L. Williams, *Black Americans and the Evangelization of Africa, 1877–1900* (Madison, 1982), 44, 7, 161, and passim (*Church Review* quoted on 161).
35. See Waldo Martin, *The Mind of Frederick Douglass* (Chapel Hill, 1984), 100–102.
36. James Campbell, "Our Fathers, Our Children: The African Methodist Episcopal Church in the United States and South Africa," (Ph.D. diss., Stanford University, 1989), 34–80. See also Clarence Walker, *A Rock in a Weary Land: The African Methodist Episcopal Church During the Civil War and Reconstruction* (Baton Rouge, 1982).
37. On Turner, see Stephen Ward Angell, *Bishop Henry McNeal Turner and African-American Religion in the South* (Knoxville, 1992); Edward S. Redkey, *Black Exodus: Black Nationalist and Back-to-Africa Movements, 1890–1910* (New Haven, 1969), 24–46, 170–194, and passim; and idem, ed., *Respect Black: The Writings and Speeches of Henry McNeal Turner* (New York, 1971).
38. Redkey, *Black Exodus,* 24–27; Angell, *Turner,* 7–140.
39. Redkey, ed., *Respect Black,* 156.
40. See George Tindall, *South Carolina Negroes, 1877–1890* (Columbia, S.C., 1952), 153–58; Nell Irwin Painter, *The Exodusters: Black Migration to Kansas after Reconstruction* (New York, 1977); and Angell, *Turner,* 134–41, 155–56.
41. Redkey, ed., *Respect Black,* 76, 143. Turner even conceded at one point that there might be some truth in the contention that lynching was caused by an increasing tendency of black males to rape white women. He vigorously denied, however, that such behavior could reveal anything about the innate character of the black race. African societies, he pointed out, had always protected female honor, and there was no record of white teachers and missionaries being assaulted in Africa (*Respect Black,* 150–51).
42. Ibid., 124, 147–55. For a provocative analysis of the paradoxes and ambiguities of Turner's thought see James Campbell, "Our Fathers, Our Children," 54–60.
43. The quotations are from Campbell, "Our Fathers, Our Children," 56–60. See also Angell, *Turner,* 214–34.
44. See Chirenje, *Ethiopianism,* 13–19, and Campbell, "Our Fathers, Our Children," 82–97. On Tiyo Soga, see Chapter 1 above.
45. Chirenje, *Ethiopianism,* 23, 58, 69–71; Campbell, "Our Fathers, Our Children," 97–99. On Mzimba's earlier political conservatism, see Chapter 1 above.
46. See Christopher C. Saunders, "Tile and the Thembu Church: Politics and Independency on the Cape Frontier in the Late Nineteenth Century," *Journal of African History* 11 (1970): 553–70.

47. Chirenje, *Ethiopianism,* 42–45; Campbell, "Our Fathers, Our Children," 103–7. Chirenje's assertion that the Ethiopians planned to send missionaries throughout Africa is not well documented, but there is nothing improbable about it.
48. The fullest accounts are Veit Erlmann, "A Feeling of Prejudice: Orpheus McAdoo and the Virginian Jubilee Singers in South Africa," *Journal of Southern African Studies* 14 (1988): 331–50; and Campbell, "Our Fathers, Our Children," 119–30.
49. Campbell, "Our Fathers, Our Children," 129–30; Chirenje, *Ethiopianism,* 50–53.
50. From the *Voice of Missions,* March 1896, as quoted in Chirenje, *Ethiopianism,* 53–54.
51. Chirenje, *Ethiopianism,* 62–64; quote from Williams, *Black Americans and Evangelization,* 170. Williams's source for this quote (which sounds very much like Turner) is Daniel Thwaite, *The Seething Pot: A Study of Black Nationalism, 1882–1935* (London, 1936), 37–38. Thwaite's somewhat sensationalistic work is not annotated.
52. On Booth's activities, see George Shepperson and Thomas Price, *Independent African: John Chilembwe and the Nyasaland Rising of 1915* (1958; Edinburgh, 1987), 70–81, 109–12, and passim (quote on 75); and Shula Marks, *Reluctant Rebellion: The 1906–8 Disturbances in Natal* (Oxford, 1970), 60.
53. See Campbell, "Our Fathers, our Children," 135–88, and Chirenje, *Ethiopianism,* 85–111. The testimony of the AME ministers appears in Sheridan Johns III, *Protest and Hope, 1882–1934,* vol. 1 of Thomas Karis and Gwendolen M. Carter, eds., *From Protest to Challenge: A Documentary History of African Politics in South Africa* (Stanford, Calif., 1972), 34–36.
54. Campbell, "Our Fathers, Our Children," 160–88; Beulah M. Fournoy, "The Relationship of the African Methodist Church to Its South African Members," *Journal of African Studies* 2 (Winter 1975–76): 529–45.
55. On the violence of 1921 see Chapter 4 below. On the role of Ethiopianists in the Natal uprising see Marks, *Reluctant Rebellion,* 334. For a valuable general discussion of the relation between Ethiopianism and nationalist politics, see Barbara Lahouel, "Ethiopianism and African Nationalism in South Africa Before 1937," *Cahiers d'Etudes africaines* 104, XXVI-4 (1986): 681–88.
56. William Beinart and Colin Bundy, *Hidden Struggles in Rural South Africa: Politics and Popular Movements in the Transkei and Eastern Cape, 1890–1930* (London, 1987), 114–18 and passim.
57. André Odendaal, *Vukani Bantu! The Beginnings of Black Protest Politics in South Africa to 1912* (Cape Town, 1984), 82–84.
58. Ibid., 84–85.
59. On Mtimkulu, see the biographical sketch in Gail M. Gerhart and Thomas Karis, *Political Profiles, 1882–1964,* vol. 4 of Thomas Karis and Gwendolen Carter, eds., *From Protest to Challenge: A Documentary History of African Politics in South Africa* (Stanford, Calif., 1977), 106. For an example of Pan-Africanist rhetoric from a minister of a mainline church see the quotations from the Rev. Z. R. Mahabane in Chapter 4, below.
60. A survey of the capsule biographies of African political leaders between 1882 and 1964 in Gerhart and Karis, *Political Profiles,* generated 44 whose religious affiliations were mentioned, could be inferred, or were known to the author. Among these, 28 were affiliated with mission churches and 16 with African

independent churches. None belonged to churches of the Zionist type. It was noticeable also that most of the leaders who were identified with the independent churches were active in the period before World War II. On the overall decline of Ethiopianism and the rise of Zionism, see Jean Comaroff, *Body of Power, Spirit of Resistance: The Culture and History of a South African People* (Chicago, 1985), 191–92 and passim. See below for an interpretation of these developments that differs from Comaroff's.

61. Quoted in Odendaal, *Vukani Bantu!,* 110–11.
62. Chapters 4 and 7 below deal with later manifestations of Africanism.
63. Beinart and Bundy, *Hidden Struggles,* 246–47. See Chapter 4 below for an extensive treatment of black populism in the 1920s.
64. B. G. M. Sundkler, *Bantu Prophets in South Africa* (2nd ed., London, 1961), 56–59, 297, and passim.
65. Comaroff, *Body of Power, Spirit of Resistance,* 192.
66. The debate on these issues in the context of the history of African resistance movements is surveyed in Ranger, "Religious Movements and Politics." Unlike Ranger, however, I am willing to take Zionists at their word when they claim to be apolitical. Of the historians Ranger cites, I find myself agreeing most with the views of H. W. Turner (p. 4). For Max Weber's basic views on power and domination, see *Economy and Society: An Outline of Interpretive Sociology,* Guenther Roth and Claus Wittich, eds. (Berkeley, 1978), I: 53–56, II: 941–48, and passim.
67. On the political acquiescence of the Zionist churches in the apartheid era see Louise Kretzchmar, *The Voice of Black Theology in South Africa* (Johannesburg, 1986), 52–54. For a brief account of the 1985 endorsement of the South African government by the three-million-strong Zion Christian Church, see Ranger, "Religious Movements and Politics," 20. Pierre van den Berghe forcefully made the point about the irrelevance of cultural-hegemonism to the understanding of white domination in South Africa in *South Africa: A Study in Conflict* (Berkeley, 1967).
68. Drake, *Redemption of Africa,* 73–74.
69. Randall K. Burkett, *Garveyism as a Religious Movement: The Institutionalization of a Black Civil Religion* (Methuen, N.J., 1978), 34–35, 85–86, 122, 125, 135–36. In Chapter 4, below, I will deal with secular influences on the movement and also show how Garveyism departed from traditionalist Ethiopianist thinking about race.
70. Drake, *The Redemption of Africa,* 63. On the differences between mainstream and Holiness Pentecostal churches in extent of political involvement, see Lincoln and Mamiya, *Black Church,* 221–27. For an earlier and less nuanced view of the relation between different modes of black religiosity and political activism see Gary T. Marx, "Religion: Opiate or Inspiration of Civil Rights Militancy?," in Hart M. Nelson et al., eds., *The Black Church in America* (New York, 1971), 150–60.
71. For more on King's reformulation of black religious values, see Chapter 6 below.

Chapter 3

1. See John W. Cell, *The Highest Stage of White Supremacy: The Origins of Segregation in South Africa and the American South* (Cambridge, Eng., 1982),

2–3; and Saul Dubow, *Racial Segregation and the Origins of Apartheid in South Africa, 1919–1936* (New York, 1989), 22–25.

2. George Wilson Pierson, *Tocqueville and Beaumont in America* (New York, 1938), 512–15; Richard C. Wade, *Slavery in the Cities: The South, 1820–1860* (New York, 1964), 266–71; Leon Litwack, *North of Slavery: The Negro in the Free States, 1790–1860* (Chicago, 1961), passim. On the rise of a pattern of customary segregation during Reconstruction, see Joel Williamson, *After Slavery: The Negro in South Carolina, 1861–1877* (Chapel Hill, 1965), 274–99, and Howard N. Rabinowitz, *Race Relations in the Urban South, 1865–1890* (New York, 1978), 127–254. It is Rabinowitz who calls attention to the relatively benign segregation of public facilities during Reconstruction, which he describes as a movement from "exclusion to segregation."
3. See George M. Fredrickson, *White Supremacy: A Comparative Study in American and South African History* (New York, 1981), 177, 185–86, 257–68.
4. For a good summing-up of the urbanist interpretation of segregation, see Cell, *Highest Stage,* 131–35.
5. See Pierre van den Berghe, *Race and Racism: A Comparative Perspective* (New York, 1967), 25–34, and Edna Bonacich, "A Theory of Ethnic Antagonism: The Split Labor Market," *American Sociological Review* 37 (1972): 549, 553–58.
6. For a persuasive development of this thesis, see Herbert Blumer, "Industrialization and Race Relations," in Guy Hunter, ed., *Industrialization and Race Relation: A Symposium* (London, 1965), 220–53.
7. C. Vann Woodward, *The Strange Career of Jim Crow* (3rd rev. ed., New York, 1974), 31–109. See also Edward L. Ayers, *The Promise of the New South: Life After Reconstruction* (New York, 1992), 132–59.
8. See Vincent P. De Santis, *Republicans Face the Southern Question—The New Departure Years, 1877–1897* (Baltimore, 1959), and Stanley Hirshon, *Farewell to the Bloody Shirt: Northern Republicans and the Southern Negro, 1877–1893* (Bloomington, 1962). On the original Republican party, see especially Eric Foner, *Free Soil, Free Labor, Free Men: The Ideology of the Republican Party Before the Civil War* (New York, 1970).
9. See Woodward, *Strange Career,* 67–109; Ayers, *Promise of the New South,* 249–309; and Gavin Wright, *Old South, New South: Revolutions in the Southern Economy Since the Civil War* (New York, 1986). Lillian Smith, in *Killers of the Dream* (New York, 1963), provides an suggestive interpretation of Jim Crow as based on a bargain between "Mr. Rich White" and "Mr. Poor White" (pp. 154–68).
10. For evidence of black resistance to customary or privately enforced segregation and data on black landowning, see Ayers, *Promise of the New South,* 140–46, 208, 513–14n. On the Tennessee law of 1881, see Joseph H. Cartwright, *The Triumph of Jim Crow: Tennessee Race Relations in the 1880s* (Knoxville, 1976), 106 and passim.
11. Rabinowitz, *Race Relations,* 333–38; August Meier and Elliott Rudwick, "The Boycott Movement Against Jim Crow Streetcars in the South, 1900–1906," in Meier and Rudwick, *Along the Color Line: Explorations in the Black Experience* (Urbana, 1976), 267–89.
12. On the thinking behind South African segregation, see Cell, *Highest Stage,* 192–229, and Dubow, *Racial Segregation,* 21–29.

13. Cell, *Highest Stage,* 196–210; *South African Native Affairs Commission, 1903–1905,* 5 vols. (Cape Town, 1905).
14. The fullest account of the development of segregation in the 1920s and '30s is Dubow, *Racial Segregation.*
15. See Leonard M. Thompson, *The Unification of South Africa, 1902–1910* (Oxford, 1960).
16. On the significance of the Land Act, see Colin Bundy, *The Rise and Fall of the South African Peasantry* (2nd ed., Cape Town, 1988), 213–14.
17. Ibid., 230–32 and passim.
18. See Fredrickson, *White Supremacy,* 230–34, 241–42.
19. In *White Supremacy,* I stressed the basic differences between Jim Crow and "Native Segregation" (pp. 239–54). John Cell in *The Highest Stage of White Supremacy* emphasized the parallels. I now believe that I was right to contrast the objective or structural aspects of these two systems of racial domination, but that, culturally and ideologically speaking, there was more similarity in the segregationist impulse—and in the black response to it—than I had allowed for. Cell's emphasis on the way segregation—especially if it was allegedly "separate but equal"—could mask or obscure the brutality of the system and actually convince moderate to liberal whites that it was a humane and enlightened race policy adds an important dimension to our understanding of the segregationist mentality in the early twentieth century.
20. Shula Marks, *The Ambiguities of Dependence in South Africa: Class, Nationalism, and the State in Twentieth-Century Natal* (Baltimore, 1986), 5.
21. Louis R. Harlan, *Booker T. Washington: The Making of a Black Leader, 1856–1901* (New York, 1972), 218, 236–37, 230–31.
22. W. E. B. Du Bois, *Souls of Black Folk* (New York: New American Library, 1982), 80, 88.
23. These observations about Washington's influence are based on Louis R. Harlan's magisterial two-part biography, *Booker T. Washington: The Making of a Black Leader* (New York, 1972) and *Booker T. Washington: The Wizard of Tuskegee* (New York, 1983). On the receptivity to Washington's doctrines of an emerging business class, see August Meier, *Negro Thought in America, 1880–1915* (Ann Arbor, 1963), 139–57.
24. David Levering Lewis, *W. E. B. Du Bois: Biography of a Race* (New York, 1992), 281–82; Du Bois, *Souls of Black Folk,* 45–46. See Chapter 2, above, for a discussion of Du Bois's racialism of the late 1890s.
25. Otto Olsen, *The Carpetbagger's Crusade: A Life of Albion Winegar Tourgée* (Baltimore, 1965), 353; W. E. B. Du Bois, *Against Racism: Unpublished Essays, Papers, and Addresses,* Herbert Aptheker, ed. (Amherst, 1985), 81–83; Philip S. Foner, ed., *W. E. B. Du Bois Speaks: Speeches and Addresses, 1890–1919* (New York, 1970), 173.
26. See Stephen R. Fox, *The Guardian of Boston: William Monroe Trotter* (New York, 1971), 97–99, 140–141.
27. Elliott W. Rudwick, *W. E. B. Du Bois: Propagandist of the Negro Protest* (New York, 1968), 100–102; Fox, *Guardian,* 124–26.
28. W. E. B. Du Bois, "The Talented Tenth," in *The Negro Problem: A Series of Articles by Representative American Negroes of Today* (New York, 1903), 50–54.
29. Du Bois, *Against Racism,* 60; Arnold Rampersad, *The Art and Imagination of W. E. B. Du Bois* (Cambridge, Mass., 1976), 87.

30. Meier, *Negro Thought,* 178–80; Rudwick, *Du Bois,* 118; Fox, *Guardian,* 113–14.
31. "The Niagara Movement Platform," in August Meier, Elliott Rudwick, and Francis L. Broderick, eds., *Black Protest Thought in the Twentieth Century* (2nd ed., Indianapolis, 1971), 61.
32. Fox, *Guardian,* 44–45; Meier, *Negro Thought,* 214–17 and passim. For an eloquent recent statement of the compatibility of self-help and protest, see Roy L. Brooks, *Rethinking the American Race Problem* (Berkeley, 1990).
33. Rudwick, *Du Bois,* 103–19; Fox, *Guardian,* 101–14; Lewis, *Du Bois,* 376–77.
34. Charles Flint Kellogg, *NAACP: A History of the National Association for the Advancement of Colored People, 1909–1920* (Baltimore, 1920), 10–15; Rudwick, *Du Bois,* 120–29; Meier, *Negro Thought,* 180–82.
35. Kellogg, *NAACP,* 15–90 and passim; Nancy J. Weiss, *The National Urban League, 1910–1940* (New York, 1974), 47–70 and passim. For a critique of the "noneconomic liberalism" of the NAACP, see B. Joyce Ross, *J. E. Spingarn and the Rise of the NAACP, 1911–1939* (New York, 1972).
36. Fox, *Guardian,* 127–29; Kellogg, *NAACP,* 21–22; Ida B. Wells, *Crusade for Justice: The Autobiography of Ida B. Wells* (Chicago, 1970), 324–28; Rudwick, *Du Bois,* 124.
37. Kellogg, *NAACP,* 103; Rudwick, *Du Bois,* 151–83; Lewis, *Du Bois,* 472, 494–95.
38. On how Africans attempted to unify before the establishment of the South African Union, see André Odendaal, *Vukani Bantu! The Beginnings of Black Protest Politics in South Africa to 1912* (Cape Town, 1984), 64–196.
39. Ibid., 167–80 (quote on 168).
40. Ibid., 175 (quote), 178.
41. Ibid., 193, 197–227.
42. Ibid., 284; André Odendaal, "African Political Mobilization in the Eastern Cape" (D.Phil. diss., Cambridge University, 1983), 262–64.
43. Odendaal, *Vukani Bantu!,* 256–79; Peter Walshe, *The Rise of African Nationalism in South Africa: The African National Congress, 1912–1952* (Berkeley, 1971), 30–31.
44. Walshe, *African Nationalism,* 31–33.
45. Sheridan Johns III, *Protest and Hope,* vol. 1 of Thomas Karis and Gwendolen Carter, eds., *From Protest to Challenge: A Documentary History of African Politics in South Africa, 1882–1964* (Stanford, Calif., 1972), 69–71. For biographical information on Seme, see Gail M. Gerhart and Thomas Karis, *Political Profiles, 1882–1964,* vol. 4 of Karis and Carter, eds., *From Protest to Challenge* (Stanford, Calif., 1977), 137–39.
46. From *Imvo Zabatsundu,* Oct. 24, 1911, reprinted in Johns, *Protest and Hope,* 71–73.
47. Ibid., 92.
48. Seme quoted in Walshe, *African Nationalism,* 34; Louis R. Harlan, "Booker T. Washington and the White Man's Burden," *American Historical Review* 71 (1966): 464.
49. See R. Hunt Davis, Jr., "John L. Dube: A South African Exponent of Booker T. Washington," *Journal of African Studies* 2 (Winter 1975–76): 497–528.
50. Quoted in ibid., 497–98. Dube's links to Washington are further explored in

Manning Marable, "African Nationalist: The Life of John Langalibele Dube" (Ph.D. diss., University of Maryland, 1976), 93–120 and passim.

51. Brian Willan, *Sol Plaatje: South African Nationalist, 1876–1932* (Berkeley, 1984), 110–11, 271–72, 278–79, 303.
52. Harlan, "Washington and the White Man's Burden," 461; Marable, "African Nationalist," 121.
53. "Resolution Against the Natives Land Act of 1913 and the Report of the Natives Land Commission," Oct. 2, 1916, reprinted in Johns, *Protest and Hope,* 86–88.
54. "The South African Races Congress," inaugural address by J. Tengu Jabavu, April 2, 1912, in Johns, *Protest and Hope,* 73–75.
55. These issues will be dealt with more fully in subsequent chapters, especially Chapters 4 and 7.
56. See Walshe, *African Nationalism,* 33–40. Extracts from the SANNC constitution are reprinted in Johns, *Protest and Hope,* 76–82.
57. On the professional composition of the leaderships, see Walshe, *African Nationalism,* 36, and Weiss, *Urban League,* 58–60.
58. See Chapter 1, above.
59. Du Bois, who supported the socialist party in 1910, was a partial exception to the prevalence of this liberal capitalist outlook among the black founders of the NAACP and SANNC.
60. S. M. Molema, *The Bantu Past and Present* (Edinburgh, 1920), 260, 263–64, 275.
61. Ibid., 317–18.
62. See Chapter 4 below for a discussion of these upheavals.
63. See Chapter 6 below on the role of nonviolence in the American and South African struggles.
64. Johns, *Protest and Hope,* 221. For a more extensive discussion of these differences, see George M. Fredrickson, *The Arrogance of Race: Historical Perspectives on Slavery, Racism, and Social Inequality* (Middletown, Conn., 1988), 254–69.
65. The figures are from Lawrence H. Fuchs, *The American Kaleidoscope: Race, Ethnicity, and American Culture* (Hanover, N.H., 1991), 99.
66. Fox, *Guardian,* 168–87; Rudwick, *Du Bois,* 158–64; Nancy J. Weiss, "The Negro and the New Freedom: Fighting Wilsonian Segregation," *Political Science Quarterly* 84 (1969): 61–79.
67. See Jack Temple Kirby, *Darkness at the Dawning: Race and Reform in the Progressive South* (Philadelphia, 1972), 119–30; and Jeffrey J. Crow, "An Apartheid for the South: Clarence Poe's Crusade for Rural Segregation," in Jeffrey J. Crow, Paul D. Escott, and Charles L. Flynn, Jr., eds., *Race, Class, and Politics in Southern History* (Baton Rouge, 1989), 216–59.
68. Kirby, *Darkness at Dawning,* 24–25; Roger L. Rice, "Residential Segregation by Law, 1910–1917," *Journal of Southern History* 24 (1968): 179–99.
69. William Pickens, "The Ultimate Effects of Segregation and Discrimination," pamphlet of 1915, reprinted in full in Herbert Aptheker, ed., *A Documentary History of the Negro People in the United States, 1910–1932* (Secaucus, N.J., 1973), 78–87 (quote on p. 82).
70. From *New Republic,* Dec. 4, 1915; reprinted in ibid., 117–20.
71. A good account of the Louisville case appears in the biography of the

lawyer who argued it: William B. Hixon, *Moorfield Story and the Abolitionist Tradition* (New York, 1972), 139–42. See also Rice, "Residential Segregation."

72. The congress's first secretary, Sol Plaatje, wrote a vivid account of the effects of the Land Act in the Orange Free State. See his *Native Life in South Africa* (London, 1916).
73. John Dube to Members of the American Committee of the Zulu Christian Industrial School, Jan. 27, 1912, copy in the Booker T. Washington Papers, Library of Congress; Petition to the Prime Minister from the Rev. John L. Dube, Feb. 14, 1914, reprinted in Johns, *Protest and Hope,* 84–86 (quote on 85). See also Walshe, *African Nationalism,* 47–49.
74. Walshe, *African Nationalism,* 50–51; "Natives' Land Act—No. 27 of 1913. Native Protest," undated document signed by John L. Dube and Walter Rubusana in Anti-slavery Society Papers, Brit. Emp. S22 G203, Rhodes Library, Oxford University. (This is apparently the petition that the delegation brought to England in 1914.)
75. Willan, *Plaatje,* 174–204.
76. Selope Thema to Travers Buxton, undated (received March 28, 1917), Buxton to Thema, March 29, 1917, and Thema to Buxton, Aug. 15, 1917, Antislavery Society Papers, Brit. Emp. S22 G203, Rhodes Library, Oxford University; Walshe, *African Nationalism,* 52–61.
77. For a strong statement of Plaatje's advocacy of public equality and opposition to the social mingling of the races, see Willan, *Plaatje,* 111. The statement was made in 1902, but as Willan points out, it sustained Plaatje "for many years afterward."
78. On the training camp controversy, see Ross, *Spingarn,* 84–97; Rudwick, *Du Bois,* 198–201; Fox, *Guardian,* 217–19; Jane Lang Scheiber and Harry N. Scheiber, "The Wilson Administration and the Wartime Mobilization of Black Americans," *Labor History* 10 (1969): 442–45.
79. *The Crisis* 16 (July 1918): 111.
80. Albert Grundlingh, *Fighting Their Own War: South African Blacks and the First World War* (Johannesburg, 1987), 12–13, 105–6, and passim; Johns, *Protest and Hope,* 86–88.
81. Weiss, "The Negro and the New Freedom," 67, 69 (quote from Miller on 69). Weiss's article stresses the failure of the effort to stop segregation in the civil service, but her own data supports the conclusion that worse was possible and averted.
82. Quoted in Gunnar Myrdal, *An American Dilemma* (New York, 1944), 832.
83. See Scheiber and Scheiber, "The Wilson Administration and Black Americans."
84. Grundlingh, *Fighting Their Own War,* 132–41, 155–56.

Chapter 4

1. For a discussion of Black Marxism, see Chapter 5 below.
2. My conception of populism derives principally from Margaret Carnavon, *Populism* (New York, 1981), and Lawrence Goodwyn, *Democratic Promise: The Populist Movement in America* (New York, 1976). Goodwyn's celebration of the American Populist movement of the 1890s leads to a definition that is

somewhat too narrow and ideologically charged for my purposes, but his conception of the kind of economic program that a populist movement is likely to endorse in the face of capitalist and socialist alternatives has proven quite useful. Carnavon was especially helpful for her broader canvas and sensitivity to the ethnic or nationalist component of many populisms.

3. On populism as a "third way," see Goodwyn's *Democratic Promise* and his shorter version of the same work, *The Populist Moment* (New York, 1978). On the South African usage of the term in the 1980s, see William Cobbett and Robin Cohen, eds., *Popular Struggles in South Africa* (London, 1988), 14–16, 217–19, 224.
4. Carnavon, *Populism,* 120–21.
5. On aspects of the American side of this story, see James R. Grossman, *Land of Hope: Chicago, Black Southerners, and the Great Migration* (Chicago, 1989). On parallel developments in South Africa, see Albert Grundlingh, *Fighting Their Own War: South African Blacks and the First World War* (Johannesburg, 1987), 146–53. For a discussion of the immediate postwar strike activity in the United States, see Judith Stein, *The World of Marcus Garvey: Race and Class in Modern Society* (Baton Rouge, 1986), 54–56.
6. Johnson quoted in Stein, *World of Garvey,* 58; Mvabaza quoted in Philip Bonner, "The Transvaal Native Congress, 1917–1920: The Radicalisation of the Black Petty Bourgeoisie on the Rand," in Shula Marks and Richard Rathbone, eds. *Industrialisation and Social Change in South Africa: African Class Formation, Culture, and Consciousness* (London, 1982), 293. On Mvabaza, see Gail M. Gerhart and Thomas Karis, *Political Profiles, 1882–1964,* vol. 4 of Thomas Karis and Gwendolen M. Carter, eds., *From Protest to Challenge: A Documentary History of African Politics in South Africa* (Stanford, Calif., 1977), 106–7.
7. Cheryl Walker, *Women and Resistance in South Africa* (New York, 1991), 27–32; Bonner, "Transvaal Native Congress," 300–303.
8. P. Wickens, *The Industrial and Commercial Workers Union of South Africa* (Capetown, 1978), 43–44, 61–63; Sheridan Johns III, *Protest and Hope,* vol. 1 of Karis and Carter, eds., *Protest to Challenge* (Stanford, Calif., 1972), 154–55, 317–20; Bonner, "Transvaal Congress," 304–6; Stein, *World of Garvey,* 131–32.
9. See David Montgomery, *The Fall of the House of Labor: The Workplace, the State, and American Labor Activism, 1865–1925* (Cambridge, Eng., 1987), and William H. Harris, *Keeping the Faith: A. Philip Randolph, Milton P. Webster, and the Brotherhood of Sleeping Car Porters* (Urbana, 1971).
10. Richard C. Cortner, *A Mob Intent on Death: The NAACP and the Arkansas Riot Cases* (Middletown, Conn., 1985), 1–30, 196; Herbert Shapiro, *White Violence and Black Response: From Reconstruction to Montgomery* (Amherst, 1988), 197–99.
11. Bonner, "Transvaal Congress," 303–6 and passim.
12. See B. Joyce Ross, *J. E. Spingarn and the Rise of the NAACP, 1911–1939* (New York, 1972).
13. William Toll, *The Resurgence of Race: Black Social Theory from Reconstruction to the Pan-African Conferences* (Philadelphia, 1979), 148–53; Mvabaza quoted in Bonner, "Transvaal Congress," 293.
14. See Sterling Stuckey, *Slave Culture: Nationalist Theory and the Foundations of Black America* (New York, 1987); Joel Williamson, *New People: Miscegena-*

tion and Mulattoes in the United States (New York, 1980); and George M. Fredrickson, *White Supremacy: A Comparative Study in American and South African History* (New York, 1981), 129–31.

15. Magema M. Fuze, *The Black People, and Whence They Came: A Zulu View,* H.C. Lugge, trans., A. T. Cope, ed. (1922; Pietermaritzburg and Durban, 1979), 8. (First published in Zulu in 1922 but apparently written shortly after the turn of the century.)
16. There was also a Marxist version of Pan-Africanism that emerged late in the 1920s and will be treated in Chapter 5 below.
17. Louis T. Harlan, "Booker T. Washington and the White Man's Burden," *American Historical Review* 71 (1966): 275 and passim. Booker T. Washington to Moussa-Mangounsled, Nov. 12, 1906, Washington Papers, Library of Congress.
18. Kenneth J. King, *Pan-Africanism and Education: A Study of Race Philanthropy and Education in the Southern States of America and East Africa* (Oxford, 1971), 17–18.
19. Ibid., 98, 128, and passim; Edwin W. Smith, *Aggrey of Africa: A Study in Black and White* (London, 1929), 119–22 and passim.
20. Smith, *Aggrey,* 121–24. On AMEZ activities in West Africa and the significance of Aggrey's recruitment, see Walter L. Williams, *Black Americans and the Evangelization of Africa, 1877–1900* (Madison, Wisc., 1982), 150.
21. Quoted in Smith, *Aggrey,* 188.
22. Ibid., 171–77. On Christian nonviolence see Chapter 6 below.
23. On the CIC, see Morton Sosna, *In Search of the Silent South: Southern Liberals and the Race Issue* (New York, 1977), 20–41, and Wilma Dykeman and James Stokley, *Seeds of Southern Change: The Life of Will Alexander* (Chicago, 1962). On the Joint Councils, see Paul Rich, *White Power and the Liberal Conscience: Racial Segregation and South African Liberalism* (Johannesburg, 1984), 4–5, 11–32.
24. Robert R. Moton, "Installation Address at Tuskegee," in Herbert Aptheker, ed., *A Documentary History of the Negro People of the United States, 1910–1932* (Secaucus, N.J., 1973), 126; D. D. T. Jabavu, *The Black Problem* (Lovedale, 1920), 58.
25. Jabavu, *Black Problem,* 63–64, 153, and passim (quote on 153).
26. D. D. T. Jabavu, *The Segregation Fallacy and Other Papers* (Lovedale, 1928), 17, 19–20, 38–39, 75, 85, 124–25.
27. Owen Charles Mathurin, *Henry Sylvester Williams and the Origins of the Pan-African Movement, 1869–1911* (Westport, Conn., 1976), 71–76 and passim; J. Ayodelde Langley, *Pan-Africanism and Nationalism in West Africa* (Oxford, 1973), 27–29.
28. W. E. B. Du Bois, *The Negro* (New York, 1915), 241–42; Toll, *Resurgence of Race,* 168–69.
29. Du Bois, *Negro,* 242.
30. Aptheker, ed., *Documentary History, 1910–1932,* 248–52. On "civilizationism," see Wilson Jeremiah Moses, *The Golden Age of Black Nationalism, 1850–1925* (Hamden, Conn., 1978).
31. W. E. B. Du Bois, *Darkwater* (1920; New York, 1975), 56–74. For an account of Du Bois's difficulties with Franco-African assimilationists, see Elliott M. Rudwick, *W. E. B. Du Bois: Propagandist of the Negro Protest* (New York, 1968), 208–35.

32. Du Bois, *Darkwater,* 70; Herbert Aptheker, ed., *Newspaper Columns by W. E. B. Du Bois* (New York, 1986), I: 66.
33. Rudwick, *Du Bois,* 223, 231.
34. Ibid., 211–12, 235; Johns, *Protest and Hope,* 68; Brian Willan, *Sol Plaatje: South African Nationalist* (Berkeley, 1984), 229–30, 271–73.
35. See Lawrence W. Levine, "Marcus Garvey and the Politics of Revitalization," in John Hope Franklin and August Meier, eds., *Black Leaders of the Twentieth Century* (Urbana, 1982), 105–38 (for Garvey's membership claims, see p. 121). The plausible estimate can be found in Emory Tolbert, *The UNIA and Black Los Angeles: Ideology and Community in the American Garvey Movement* (Los Angeles, 1980), 3.
36. Marcus Garvey, "The Negro's Greatest Enemy," in Robert A. Hill, ed., *The Marcus Garvey and Universal Negro Improvement Association Papers* (Berkeley, 1983), I: 5–6.
37. Ibid., 135; Levine, "Garvey," 116–17; Stein, *World of Garvey,* 41.
38. On the rise of the UNIA, see Stein, *World of Garvey,* 38–88, and David Cronon, *Black Moses: The Story of Marcus Garvey and the Negro Improvement Association* (Madison, Wisc., 1969), 39–72.
39. See Robert A. Hill, "General Introduction" to *Garvey Papers,* I: lxvii–lxxviii.
40. Hill, ed., *Garvey Papers,* I: lxvi–lxvii, 212–20, 377.
41. From the *Negro World,* April 5, 1919, in Hill, ed., *Garvey Papers,* I: 397.
42. Amy Jacques Garvey, ed., *Philosophy and Opinions of Marcus Garvey,* 2 vols. (1925; New York, 1969), I: 53. On Garvey's abandonment of the black American equal rights struggle, his relations with the Klan and his increasing emphasis on race purity, see Cronon, *Black Moses,* 187–95.
43. Garvey, *Philosophy and Opinions,* II: 35.
44. George Rudé, *Ideology and Popular Protest* (London, 1980). The best study of local Garveyism is Tolbert, *UNIA and Black Los Angeles.* See also Kenneth Kusmer, *A Ghetto Takes Shape: Black Cleveland, 1870–1930* (Urbana, 1976), 229–32.
45. Garvey, *Philosophy and Opinions,* I: 63; William H. Ferris, "The World in Which We Live," *Negro World,* Nov. 6, 1920, reprinted in Randall K. Burkett, ed., *Black Redemption: Churchmen Speak for the Garvey Movement* (Philadelphia, 1978), 73.
46. Hill, ed., *Garvey Papers,* I: 198; Garvey, *Philosophy and Opinions,* I: 90, 12, 14, 63–64, 72; II: 46.
47. Garvey, *Philosophy and Opinions,* II: 21, 42–43, 56–60.
48. The key documents of the Du Bois-Garvey feud can be found in Theodore G. Vincent, ed., *Voices of a Black Nation: Political Journalism in the Harlem Renaissance* (1973; Trenton, N.J., n.d.), 93–101.
49. For a discussion of the Garvey movement's appropriation of some of the texts and rhetoric of the Ethiopianist religious tradition, see Chapter 2 above. Wilson Jeremiah Moses provides a perceptive discussion of Garvey's relationship to the nineteenth-century civilizationist and Ethiopianist traditions in his *The Golden Age of Black Nationalism, 1850–1925* (Hamden, Conn., 1978), 262–68. Randall Burkett in *Garveyism as a Religious Movement: The Institutionalization of a Black Civil Religion* (Methuen, N.J., 1978) demonstrates that the movement adopted the trappings of Ethiopianism but fails to perceive its ethical and cultural deviations from earlier manifestations of black religious nationalism. It might be going too far to say that Garvey broke deci-

sively with the Christian ethical tradition, black or white, to indulge in an amoral worship of power and strength; but it *is* true that he later professed admiration for Mussolini and fascism, even claiming that he and his followers "were the first Fascists" (see Cronon, *Black Moses,* 197–98).

50. See Stein, *World of Garvey,* passim, for the view that Garveyism was a bourgeois, pro-capitalist movement. Stein's argument is not new, but in fact recapitulates the standard Marxist or Communist condemnation of the UNIA in the 1920s and '30s. My view, as developed below, is that the movement represented a populist alternative to classic forms of capitalism and socialism.
51. The argument that Garvey, rather than being simply an advocate of black capitalism, "leaned toward reforms of a social democratic nature" and "tried to walk a precarious balance between capitalism and communism" is made by Tony Martin in *Race First: The Ideological and Organizational Struggles of Marcus Garvey and the Universal Negro Improvement Association* (Westport, Conn., 1976), 53–55. Martin draws attention to Garvey's way of "organizing his businesses along cooperative lines and . . . placing limits on the number of shares any one person could own." On Garvey's attitude toward unions, see Cronon, *Garvey,* 195–96; Stein, *World of Garvey,* 261–62; Martin, *Race First,* 233–35.
52. Cronon, *Black Moses,* 73–137; Stein, *World of Garvey,* 186–208.
53. Stein, *World of Garvey,* 223–37 and passim; Tolbert, *UNIA and Black Los Angeles,* 90–96 and passim; Kusmer, *A Ghetto Takes Shape,* 230–32.
54. On Garvey's failure to protest racist violence after 1921, see Herbert Shapiro, *White Violence and the Black Response* (Amherst, 1988), 166–69.
55. On the Garvey movement in South Africa, see Robert A. Hill and Gregory A. Pirio, " 'Africa for the Africans': The Garvey Movement in South Africa, 1920–1940," in Shula Marks and Stanley Trapido, eds., *The Politics of Race, Class, and Nationalism in Twentieth Century South Africa* (London, 1987), 209–53; Tony Martin, *The Pan-African Connection: From Slavery to Garvey and Beyond* (Dover, Mass., 1983), 133–47; and Arnold Hughes, "Africa and the Garvey Movement in the Interwar Years," in Rupert and Maureen Warner Lewis, eds., *Garvey, Africa, Europe, and the Americas* (Jamaica, 1986), 111–35.
56. Martin, *Pan-African Connection,* 134–35; Hill and Pirio, " 'Africa for the Africans,' " 216–17.
57. Hill and Pirio, " 'Africa for the Africans,' " 230–38; Gerhart and Karis, *Political Profiles,* 154–55.
58. Johns, *Protest and Hope,* 290–94 (long quotation on 294); Hill and Pirio, " 'Africa for the Africans,' " 233.
59. Johns, *Protest and Hope,* 296; Walshe, *African Nationalism,* 99, highlights the generally moderate orientation of Mahabane, as does the biographical sketch in Gerhart and Karis, *African Profiles,* 65–66.
60. J. T. Campbell, "T. D. Mweli Skota and the Making and Unmaking of a Black Elite" (unpublished paper, University of the Witwatersrand History Workshop, 1987), 19 and passim; Mary Benson, *South Africa: The Search for a Birthright* (New York, 1969), 46; Hill and Pirio, " 'Africa for the Africans,' " 236; Martin, *Pan-African Connection,* 136–37, 141; Willan, *Sol Plaatje,* 264–68. Plaatje's account of his trip to America has not been published but can be found in the Silas T. Molema and Solomon T. Plaatje Papers, University of the Witwatersrand Library, Johannesburg.

61. See Walshe, *African Nationalism,* 166–69. On Gumede's pro-Soviet radicalism, see Chapter 5 below.
62. On the conditions of rural South Africans by the 1920s, see especially Colin Bundy, *The Rise and Fall of the South African Peasantry* (2nd ed., Cape Town, 1988), 146–54, 221–36; William Beinart and Colin Bundy, *Hidden Struggles in Rural South Africa: Politics and Popular Movements in the Transkei and Eastern Cape, 1890–1930* (London, 1987), 1–46, 222–29; and Helen Bradford, *A Taste for Freedom: The ICU in Rural South Africa, 1924–1930* (New Haven, 1987), 21–62.
63. Bradford, *Taste for Freedom,* 48–49. On the AME Church's early work in rural areas and its involvement in local politics, see James Campbell, "Our Fathers, Our Children: The African Methodist Episcopal Church in the United States and South Africa" (Ph.D. diss. Stanford University, 1989), 219–59.
64. Richard Newman, "Archbishop Daniel William Alexander and the African Orthodox Church," *International Journal of African Historical Studies* 16 (1983): 615–30. For an account of the Israelite massacre, see Robert Edgar, *They Chose the Plan of God* (Johannesburg, 1988).
65. Bradford, *Taste of Freedom,* 214–18; Smith, *Aggrey,* 180–81.
66. This account is based primarily on Robert Edgar's "Garveyism in Africa: Dr. Wellington and the American Movement in the Transkei," *Ufahumu* 6 no. 1 (1976): 31–57. A shorter version appears in *Collected Seminar Papers,* South African Studies Seminar, Vol. 6 (London: Institute of Commonwealth Studies, 1976), 100–110. See also Beinart and Bundy, *Hidden Struggles,* 250–55.
67. Edgar, "Garveyism in Africa," 47.
68. Ibid., 33; Bradford, *Taste of Freedom,* 217–18.
69. Hill and Pirio, " 'Africa for the Africans,' " 214–42 (quote on 215); Bradford, *Taste of Freedom,* 4–5, 126. A good general account of the ICU is Sheridan Johns III, "Trade Union, Political Pressure Group or Mass Movement? The Industrial and Commercial Workers' Union of South Africa," in Robert I. Rotberg and Ali A. Mazrui, eds., *Protest and Power in Black Africa* (New York, 1970), 695–754.
70. Hill and Pirio, " 'Africa for the Africans,' " 217; *Worker's Herald,* April 2, 1925; Clements Kadalie, *My Life and the ICU: The Autobiography of a Black Trade Unionist in South Africa,* Stanley Trapido, ed. (London, 1970), 220–21.
71. From *Messenger,* July 1924, reprinted in Vincent, ed., *Voices of a Black Nation,* 288–94.
72. *Workers' Herald,* May 15, 1925, quoted in Johns, "The ICU of South Africa," 695.
73. For more on the relations between the Communist party and the ICU, see Chapter 5 below.
74. This is the picture of the ICU at the grass roots that emerges from Bradford's *Taste of Freedom.* Membership information is drawn from the survey of growth on pp. 1–20.
75. See Johns, "The ICU of South Africa," 729–45; and Wickens, *ICU,* 167–86.
76. Beinart and Bundy, *Hidden Struggles,* 282–83, 291–93, and 270–320, passim.
77. Tolbert, *UNIA and Black Los Angeles,* 94.
78. Bradford, *Taste of Freedom,* 63–74.
79. For perspectives on the frustration of black South Africans with white liberal-

ism as a force for gradual reform, see Walshe, *African Nationalism,* 61–114, passim; and Saul Dubow, *Racial Segregation and the Origins of Apartheid in South Africa, 1919–1936* (New York, 1988).

80. Cronon, *Garvey,* 138–69; Johns, "The ICU of South Africa," 731–32, 748, 753.
81. For evidence of Garvey's autocratic and sectarian tendencies, see especially Stein, *World of Garvey,* 140–42, 153–70. For more on the radicals who were purged by the UNIA and the ICU, see Chapter 5 below.
82. Kadalie's anti-white rhetoric is quoted Johns, *Protest and Hope,* 301.
83. These were essentially the conclusions that Du Bois, who had criticized Garvey for his separation in the 1920s, came to during in the 1930s. See his *Dust of Dawn: An Essay Toward an Autobiography of a Race Concept* (1940; New York, 1968). For an incisive recent exposition of white working-class racism, see David R. Roediger, *The Wages of Whiteness: Race and the Making of the American Working Class* (London, 1991).

Chapter 5

1. Famous black intellectuals and artists were a problem only for the American party, since black South Africa had not yet produced a cultural elite with some prestige in the white community. Two African-American creative geniuses whom the party treated with kid gloves have been the subjects of excellent recent biographies. See Martin Bauml Duberman, *Paul Robeson* (New York, 1988), and Arnold Rampersad, *The Life of Langston Hughes,* vol. 1, *I Too Sing America* (New York, 1986).
2. See Bert Cochran, *Labor and Communism: The Conflict That Shaped American Unions* (Princeton, N.J., 1977), 94–102.
3. Evidence of the generally favorable image of Communists in the black community in the early depression years can be found in the responses to a survey of black newpaper editors conducted by Du Bois in 1932. See *The Crisis* 39 (April and May 1932): 117–19, 154–56, 170. See below for more discussion and analysis of African-American attitudes toward Communism.
4. For a good account of the shifting Comintern positions on various issues, including the liberation of colonized and racially oppressed peoples, see Fernando Claudin, *The Communist Movement from Comintern to Cominform,* Brian Pearce, trans., 2 vols. (New York, 1975).
5. Among the significant revisionist studies that deal partly or mainly with CP relations with the black community are Maurice Isserman, *Which Side Were You On? The American Communist Party During the Second World War* (Middletown, Conn., 1982); Mark Naison, *Communists in Harlem During the Depression* (Urbana, Ill., 1983); and Robin D. G. Kelley, *Hammer and Hoe: Alabama Communists During the Great Depression* (Chapel Hill, 1990). The older interpretation stressing Soviet manipulation of the American party received its class formulation in Theodore Draper, *American Communism and Soviet Russia* (New York, 1960), and was reaffirmed in Harvey Klehr, *The Heyday of American Communism: The Depression Decade* (New York, 1984). A recent history of the post-World War II freedom struggle that stresses the positive contributions of Communists is Manning Marable, *Race, Reform, and Revolution: The Second Reconstruction in Black America* (Jackson, Miss., 1991).
6. Stephen P. Possany, ed., *The Lenin Reader* (Chicago, 1966), 246; Karl Marx,

"British Rule in India," in David Fernbach, ed., *Surveys from Exile* (New York, 1974), II: 307. For an incisive brief critique of Marx's views on ethnic nationalism, see Isaiah Berlin, *Karl Marx* (Oxford, 1978), 147–48.

7. W. E. B. Du Bois, *Souls of Black Folk* (1903; New York: Signet Classics, 1982), xi.
8. Philip S. Foner, *American Socialism and Black Americans: From the Age of Jackson to World War II* (Westport, Conn., 1977), 92–127 (Debs quoted on 104).
9. Quoted in H. J. and R. E. Simons, *Class and Colour in South Africa, 1850–1950* (Harmondsworth, Eng., 1969), 155.
10. Ibid., 162–65, 271–99, and passim. The American Socialist party also produced some outright segregationists in its early days, at least in Louisiana. See Foner, *American Socialism,* ch. 6.
11. See V. I. Lenin, *National Liberation, Socialism, and Imperialism* (New York, 1968), 110–21, 165–71, and passim.
12. Martin Legassick, *Class and Nationalism in South African Protest: The South African Communist Party and the "Native Republic"* (Syracuse, 1973), 39–41.
13. Joseph Stalin, *Marxism and the National and Colonial Question* (New York, 1934), 214–18.
14. Ibid., 8, 168.
15. *Lenin on the United States* (Moscow: Progress Publishers, 1967), 305–6.
16. The South African Communist party and the Industrial and Commercial Workers' Union of Africa (ICU) actually supported the Afrikaner Nationalist/ South African Labor party coalition in 1924 on the assumption that it was an anti-capitalist front. In the early '30s, the South African Communists briefly embraced an ethnic pluralist version of self-determination (see below, this chapter).
17. Philip S. Foner and James S. Allen, eds., *American Communism and Black Americans: A Documentary History, 1919–1929* (Philadelphia, 1987), 28–30. (The proposed conference was never actually held.) Throughout the 1920s Communists emphasized the vanguard role of African-Americans. See Wilson Record, *The Negro and the Communist Party* (1951; New York 1971), 15, 24, 28–29, 58.
18. Foner and Allen, *Communism and Black Americans,* 53–65, 88 (quote), 109–29; Record, *The Negro and the Party,* 31–51. Estimates of black membership in the CP are from Klehr, *Heyday of Communism,* 5. According to Record, ". . . by 1928 there were probably less than two hundred Negro Communists in the United States" (52).
19. Edward Roux, *Time Longer Than Rope: A History of the Black Man's Struggle for Freedom in South Africa* (Madison, Wisc., 1964), 161–69, 198–217; Simons and Simons, *Class and Colour,* 324–27, 352–56, 386–88, 392–93.
20. Foner and Allen, *Communism and Black Americans,* 16–23 (quote on 19). Theodore Draper appears to be responsible for discovering that Briggs anticipated the self-determination doctrine (*American Communism and the Soviet Union,* 322–26).
21. On the South African side the main debaters have been white veterans of the CPSA. Roux in *Time Longer Than Rope* sees the policy as an imposition and part of a pattern of disastrous meddling by the Comintern (p. 256). Simons and Simons in *Class and Colour* attempt to show that the South African Colored Communist James La Guma played a major role in

persuading the Comintern to adopt this policy and that, in doing so, he spoke for the sentiments of radical blacks in South Africa (pp. 386–415). The most authoritative presentation of the case that the self-determination policy demonstrated the American party's slavish and self-destructive subservience to the Comintern is Harvey Klehr and William Thompson, "Self-Determination in the Black Belt: Origins of a Communist Policy," *Labor History* 30 (1989): 354–66. There is to my knowledge no sustained effort by an historian to establish the indigenous origins of the policy, but Philip S. Foner and Herbert Shapiro have disagreed with Klehr and Thompson's contention that the policy hindered the work of the party among southern blacks. They view it rather as an efficacious revival of the Reconstruction-era demand for "land to the tiller" (*American Communism and Black Americans: A Documentary History, 1930–1934* (Philadelphia, 1991), xi, xxv). The basis for arguing that at least one black American made an important contribution to the development of the policy can be found in Harry Haywood, *Black Bolshevik: Autobiography of an Afro-American Communist* (Chicago, 1978).

22. Haywood, *Black Bolshevik.*
23. Ibid., 132–33, 158 (quote), 218–30; Draper, *American Communism and Soviet Russia,* 334 (Hall on Stalin); Cedric Robinson, *Black Marxism: The Making of the Black Radical Tradition* (London, 1983), 306–7.
24. Haywood was out of favor during the Popular Front era and World War II, returning to prominence in the party when it revived the self-determination policy in the late '40s. Haywood's *Negro Liberation* (New York, 1948) was the definitive restatement of the case for the right to establish a black nation in the American South.
25. Haywood, *Black Bolshevik,* 235–37. The biographical information on La Guma is from Gail M. Gerhart and Thomas Karis, *Political Profiles, 1882–1964,* vol. 4 of Thomas Karis and Gwendolen M. Carter, eds., *From Protest to Challenge: A Documentary History of African Politics in South Africa, 1882–1964* (Stanford, Calif., 1977), 53–54.
26. Ibid. (both sources); Simons and Simons, *Class and Colour,* 390–412, passim; Legassick, *Class and Nationalism,* 11–12.
27. Theses and Resolutions Adopted by the Sixth World Congress, summary as reported in *International Press Correspondence,* Dec. 12, 1928, in Foner and Allen, *Communism and Black Americans,* 197–98.
28. Ibid., 198.
29. For arguments of black American Communists against self-determination, see Foner and Allen, *Communism and Black Americans,* 164–72, 180–88. The arguments of the South African delegates (all white) to Sixth World Congress can be found in the Bunting Papers, Cullen Library (A949), University of the Witwatersrand. Accounts of the reactions of American and South African Communists to the policy include Record, *The Negro and the Party,* 54–71; Klehr and Thompson, "Self-Determination in the Black Belt"; Simons and Simons, *Class and Colour,* 386–97; Sheridan Johns III, "Marxism, Leninism in a Multi-Racial Environment: The Origins and Early History of the Communist Party of South Africa, 1914–1932" (Ph.D. diss., Harvard University, 1965), 431–33, 444–54, and passim; Edward Roux, *S. P. Bunting: A Political Biography* (Cape Town, 1944), 86–100.
30. Foner and Allen, *Communism and Black Americans,* 197–98.

31. Klehr and Thompson, "Self-Determination in the Black Belt," 360–66.
32. See Johns, "Marxism, Leninism," 444–54; Draper, *Russia and American Communism,* 348–49; J. Grossman, "Class Relations and the Policies of the Communist Party of South Africa" (D.Phil. diss., University of Warwick, 1985), 146–61.
33. For insight into these twists and turns, see Claudin, *Communist Movement,* 88–89, 242–71.
34. Legassick, *Class and Nationalism,* 17–20 (quote on 20); On the league, see also Simons and Simons, *Class and Colour,* 417–25; Roux, *Time Longer Than Rope,* 226–27.
35. On the LSNR, see Wilson Record, *Race and Radicalism: The NAACP and the Communist Party in Conflict* (Ithaca, N.Y., 1964), 59–61 and Mark Naison, *Communists in Harlem,* 42.
36. Legassick, *Class and Nationalism,* 50–52; Simons and Simons, *Class and Colour,* 439, 459, 473; Roux, *Bunting,* 139.
37. Roux, *Time Longer Than Rope,* 255–86; Simons and Simons, *Class and Colour,* 438–61; T. R. H. Davenport, *South Africa: A Modern History* (Toronto, 1977), 212–17.
38. See Naison, *Communists in Harlem,* 3–126 passim, and St. Clair Drake and Horace R. Cayton, *Black Metropolis: A Study of Negro Life in a Northern City* (1946; New York, 1961), 85–88, 734–36.
39. Kelley, *Hammer and Hoe,* esp. 92–116; Foner and Shapiro, *Communism and Black Americans, 1930–34,* xxv, xxvii; W. E. B. Du Bois, "Postscript," *The Crisis* 38 (Sept. 1931): 314.
40. The standard work on the Scottsboro case is Dan T. Carter, *Scottsboro: a Tragedy of the American South* (Baton Rouge, La., 1969). But see also James Goodman, *Stories of Scottsboro* (New York, 1994), an innovative study that looks at the case from a variety of perspectives. For evidence of the Communist effort to make the Scottsboro case symbolize the need for self-determination, see Foner and Shapiro, *Communism and Black Americans, 1930–34,* 119–20. Mark Naison shows that Communist involvement in the case made Harlem blacks view the party much more favorably than they had previously (*Communists in Harlem,* 57–94).
41. See Claudin, *Communist Movement,* 171–99.
42. Ibid., 293–94; Record, *The Negro and the Party,* 128–34; Simons and Simons, *Class and Colour,* 471–85, passim. The most notable defection from Communism because of its lack of anti-imperialist militancy was that of George Padmore, the Afro-West Indian executive secretary of the Comintern's International Trade Union Committee of Negro Workers, who became a staunchly anti-Communist leader of the Pan-African movement (see Klehr, *Heyday of Communism,* 340).
43. Carter, *Scottsboro,* 330–35; Naison, *Communists in Harlem,* 169–254, passim; Record, *Race and Radicalism,* 84–109.
44. My account of the NNC draws on Raymond Wolters in *The Negro and the Great Depression: The Problem of Economic Recovery* (Westport, Conn., 1970), 353–82; and John B. Kirby, *Black Americans in the Roosevelt Era: Liberalism and Race* (Knoxville, Tenn., 1980), 155–75.
45. Sumner M. Rosen, "The CIO Era, 1935–1955," in Julius Jacobson, ed., *The Negro and the American Labor Movement* (New York, 1967), 188–208. See

also Philip S. Foner, *Organized Labor and the Black Worker, 1619–1981* (New York, 1981), 215–37.

46. See Kelley, *Hammer and Hoe,* 159–92.
47. My principal source on the crisis of 1935–37 is Thomas Karis, *Hope and Challenge, 1935–1952,* vol. 2 of Karis and Carter, eds., *From Protest to Challenge,* (Stanford, Calif., 1973), 3–66.
48. Ibid., 6–10, 31–46.
49. For somewhat differing views of what happened, see ibid., 8, and Mary Benson, *South Africa: The Struggle for a Birthright* (Minerva Press, n.p., 1969), 66–69.
50. Karis, *Hope and Challenge,* 8–10; Benson, *South Africa,* 70.
51. Simons and Simons, *Class and Colour,* 480–85, 500–506; Brian Bunting, *Moses Kotane: South African Revolutionary* (London, 1974), 78–83; Naboth Mokgatle, *The Autobiography of an Unknown South African* (Berkeley, 1971), 188–201, and passim. In 1938, the Communists were also involved in forming the Non-European United Front, an organization bringing together African, Indian, and Colored radicals.
52. My treatment of Kotane is based on Bunting, *Kotane,* a work that has its obvious biases but is convincing in its estimate of Kotane's significance (Comintern Special Resolution on South Africa (1928) is quoted on p. 33).
53. Ibid., 4, 46, 52–53, 64–65 (quote). See also Gerhart and Karis, *Political Profiles,* 50–52.
54. Bunting, *Kotane,* 89 and passim.
55. Roux, *Time Longer Than Rope,* 232–37.
56. Eslanda Robeson (wife of Paul Robeson) reported in the diary of her trip to South Africa in 1936 that discussions with prominent African academics at Fort Hare Native College had led to general agreement on the proposition that the Soviet Union had solved the race problem. For the first time, "men and women and children of all races, colors, and creeds walk the streets and work out their lives in dignity, safety, and comradeship." *African Journey* (New York, 1945), 49.
57. Kelley, *Hammer and Hoe,* 107–8; Nell Irvin Painter, *The Narrative of Hosea Hudson: His Life as a Communist in the South* (Cambridge, Mass., 1979), 75–109.
58. See Angelo Herndon, *Let Me Live* (New York, 1937), for a graphic account of how much persecution and pain a black Communist in the South might have to endure.
59. Du Bois's thinking of the 1930s was summed up in *Dusk of Dawn: An Essay Toward an Autobiography of a Race Concept* (New York, 1940). Historical accounts of the controversies involving Du Bois and the NAACP during the early '30s include Wolters, *Negroes and the Depression,* 231–63; Harvard Sitkoff, *A New Deal for Blacks: The Emergence of Civil Rights as a National Issue: The Depression Decade* (New York, 1978), 244–56; and B. Joyce Ross, *J. E. Spingarn and the Rise of the NAACP, 1911–1939* (New York, 1972), 186–98.
60. A recent work that provides new insights into the political attitudes of the African elite during this period is Alan G. Cobley, *Class and Consciousness: The Black Petty Bourgeoisie in South Africa, 1924 to 1950* (New York, 1990).
61. On Haywood's fall from grace, see Naison, *Communists in Harlem,* 128–29,

and Klehr, *Heyday of Communism,* 342. For more on black nationalist charges of excessive white influence in the CPSA, see Chapters 6 and 7 below.

62. See Paul Rich, *White Power and the Liberal Conscience: Racial Segregation and South African Liberalism* (Johannesburg, 1984), 1–76.
63. This overview draws mainly on Sitkoff, *New Deal for Blacks* and Kirby, *Black Americans in the Roosevelt Era.*
64. For reactions to Stalin-Hitler Pact, see Record, *The Negro and the Party,* 184–208; Naison, *Communists in Harlem,* 287–320; Tom Lodge, "Class Conflict, Communal Struggle and Patriotic Unity: The Communist Party of South Africa During the Second World War" (unpublished paper, African Studies Institute, University of the Witwatersrand, 1985), 3; Mokgatle, *Autobiography,* 212–13.
65. Naison, *Communists in Harlem,* 273; Drake and Cayton, *Black Metropolis,* 737–38.
66. For somewhat differing accounts of these developments, see Kirby, *Black Americans in the Roosevelt Era,* 165–70; and Wolters, *Negroes and the Depression,* 368–76. My version follows Kirby more closely than Wolters. On Randolph's personal involvement with the NNC, see Paula F. Pfeffer, *A. Philip Randolph, Pioneer of the Civil Rights Movement* (Baton Rouge, 1990), 33–43. On the divided response of the left to nationalistic consumer boycotts, see Naison, *Communists in Harlem,* 50–51, 100–102.
67. Naison, *Communists in Harlem,* 288–89; Kelley, *Hammer and Hoe,* 195–219.
68. Pfeffer, *Randolph,* 45–58. See also Herbert Garfinkel, *When Negroes March: The March on Washington Movement in the Organizational Politics for FEPC* (Glencoe, Ill., 1959), passim.
69. Garfinkel, *When Negroes March,* 47–53; Pfeffer, *Randolph,* 46–50.
70. Grossman, "Class Relations," 333–38; Lodge, "Class Conflict," 3. The CPSA's position during the pact period is presented in *Must We Fight? Yes! For Our Rights,* Communist party pamphlet (Cape Town, n.d.), copy in the SAIRR Collection University of Witwatersrand Library.
71. Essop Pahad, "The Development of Indian Political Movements in South Africa" (D.Phil. thesis, University of Sussex, 1972), 151–72.
72. On the ANC's response to the war, see Peter Walshe, *The Rise of African Nationalism in South Africa: The African National Congress, 1912–1952* (Berkeley, 1971), 263–64. The ANC-AAC Joint Resolution on the War of July 7, 1940, is reprinted in Karis, *Hope and Challenge,* 339–40.
73. Isserman, *Which Side Were You On?,* 117–19 (Ford quoted on 119); Record, *The Negro and the Party,* 209–26.
74. Isserman, *Which Side Were You On?,* 141–43, 166–69; Randolph quoted in Record, *The Negro and the Party,* 225.
75. Lodge, "Class Conflict," 5–12; Walshe, *African Nationalism,* 271–79.
76. Lodge, "Class Conflict," passim (quote on 12). For a somewhat different view, see Baruch Hirson, *Yours for the Union: Class and Community Struggles* (London, 1989), 158, 161, and passim. Hirson, a Trotskyist, believes the Communists lost the initiative or missed their chances when it came to township community organizing.
77. On New York and Chicago, see Isserman, *Which Side Were You On?,* 209–10, and Cayton and Drake, *Black Metropolis,* 111.
78. Tom Lodge, *Black Politics in South Africa Since 1945* (London, 1983), 19–22;

Hirson, *Yours for the Union,* 183–90. The attitudes and actions of the Youth League will be examined more closely in Chapters 6 and 7 below.

79. For the new position on the race issue, see Haywood, *Negro Liberation.* There is relatively little scholarship on black-Communist relations in the immediate postwar period beyond two works by Gerald Horne: *Black and Red: W. E. B. Du Bois and the Afro-American Response to the Cold War* (Albany, 1986) and *Communist Front? The Civil Rights Congress, 1946–1956* (Rutherford, N.J., 1988). These works are valuable for the wealth of factual information they present, but in my opinion they are marred by a tendency to treat the Communist party in an uncritical and celebratory fashion.
80. On Robeson, see Martin Bauml Duberman, *Paul Robeson* (New York, 1988), and Sterling Stuckey, *Slave Culture: Nationalist Theory and the Foundations of Black America* (New York, 1987), 303–58.
81. See Chapter 6 below for more on the early successes of the civil rights movement.
82. See Rich, *Liberal Conscience,* 77–122. On Hofmyer, see Alan Paton, *Hofmyer: A South African Tragedy* (London, 1964).
83. See Chapter 7 below for a discussion of postwar black radical movements.

Chapter 6

1. The literature on nonviolence is vast, but there is surprisingly little social-scientific or comparative historical work on the subject. Among the books that have informed my understanding of nonviolent protest are M. K. Gandhi, *Non-violent Resistance* (New York, 1961); Richard B. Gregg, *The Power of Non-violence* (Philadelphia, 1934); Reinhold Niebuhr, *Moral Man and Immoral Society* (New York, 1932), ch. 9; Judith M. Brown, *Gandhi: Prisoner of Hope* (New Haven, 1989); Erik Erikson, *Gandhi's Truth: On the Origins of Militant Nonviolence* (New York, 1969); and Joan V. Bondurant, *Conquest of Violence: The Gandhian Philosophy of Conflict* (Berkeley, 1965).
2. Niebuhr, *Moral Man and Immoral Society,* 242.
3. Examples of such third parties might be the influential segment of British domestic opinion that sympathized with Gandhi's campaigns in India and the liberal elements in the northern United States who supported the southern civil rights movement in the 1960s and were instrumental in getting Congress to pass the Civil Rights Acts.
4. See M. K. Gandhi, *Satyagraha in South Africa* (Ahmedabad, 1961); Maureen Swan, *Gandhi: The South African Experience* (Johannesburg, 1985); and Robert A. Huttenback, *Gandhi in South Africa: British Imperialism and the Indian Question* (Ithaca, 1971).
5. Benedict Anderson, *Imagined Communities: Reflections on the Origins and Spread of Nationalism* (London, 1983); Gandhi, *Non-violent Resistance,* 386.
6. See Brown, *Gandhi,* 77–88, and passim.
7. Ibid., 386–88 and passim. A spirited defense of Gandhian nonviolence on utilitarian or pragmatic grounds is Gregg, *Power of Non-violence.*
8. Gandhi, *Non-violent Resistance,* 70–71, 173. For a full list of the methods that can be employed in a nonviolent campaign, see Bondurant, *Conquest of Violence,* 40–41.
9. Gandhi, *Non-violent Resistance,* 174, 222.

10. Ibid., 110, 171, 100.
11. Sheridan Johns III, *Protest and Hope, 1882–1934,* vol. 1 of Thomas Karis and Gwendolen Carter, eds., *From Protest to Challenge: A Documentary History of South African Politics, 1882–1964* (Stanford, Calif., 1972), 62, 66, 108.
12. Msimang is quoted in Manning Marable, "African Nationalist: The Life of John Langaliblele Dube" (Ph.D. diss., University of Maryland, 1976), 243. The discussions of passive resistance at the Non-European Conference and Abdurahman's comments on it can be found in Johns, *Protest and Hope,* 268–69.
13. E. S. Reddy, ed., "Indian Documents on South Africa, 1927–1948" (Parts IV, V, and VI), (undated typescript, Indian Documentation Centre, University of Durban-Westville (1254/493–613)), 418–22.
14. See August Meier and Elliott Rudwick, *Along the Color Line: Explorations in the Black Experience* (Urbana, Ill., 1976), 309–44. For an example of Communist criticism of Gandhi in the early '30s, see Philip S. Foner and Herbert Shapiro, eds., *American Communism and Black Americans, 1930–1934* (Philadelphia, 1991), 70–78.
15. Johnson as quoted in Meier and Rudwick, *Along the Color Line,* 345; Sudarshan Kapur, *Raising Up a Prophet: The African-American Encounter with Gandhi* (Boston, 1992), 67 and 23–100, passim.
16. My understanding of Randolph's role derives principally from Paula F. Pfeffer, *A. Philip Randolph: Pioneer of the Civil Rights Movement* (Baton Rouge, 1990).
17. For biographical information on Randolph, see also William H. Harris, *Keeper of the Faith: A. Philip Randolph, Milton P. Webster, and the Brotherhood of Sleeping Car Porters, 1925–1937* (Urbana, Ill., 1977), and Jervis Anderson, *A. Philip Randolph: A Biographical Portrait* (1973; Berkeley, Calif., 1986). On Randolph's atheism and attraction to labor militancy as a model for the civil rights movement, see Pfeffer, *Randolph,* 58, 63–64.
18. A. Philip Randolph, "Keynote Address to the Policy Conference of the March on Washington Movement" (1942), in August Meier, Elliott Rudwick, and Francis L. Broderick, eds., *Black Protest Thought in the Twentieth Century* (Indianapolis, 1971), 227.
19. A. Philip Randolph, "March on Washington Movement Presents a Program for the Negro," in Rayford W. Logan, ed., *What the Negro Wants* (Chapel Hill, 1944), 154–55. See also Pfeffer, *Randolph,* 55–58.
20. Meier, Rudwick, and Broderick, *Black Protest Thought,* 230–33.
21. Logan, ed., *What the Negro Wants,* 133–62; Pfeffer, *Randolph,* 87.
22. August Meier and Elliott Rudwick, *CORE: A Study in the Civil Rights Movement, 1942–1968* (New York, 1968), 35–38. On the prison strikes, see James Tracy, "Forging Dissent in an Age of Consensus: Radical Pacifism in America, 1940 to 1970" (Ph.D. diss. Stanford University, 1993), 69–110.
23. "Testimony of A. Philip Randolph, . . . Prepared for Delivery Before the Senate Armed Services Committee, Wednesday, March 31, 1948," in Meier, Rudwick, and Broderick, *Black Protest Thought,* 276–77; Pfeffer, *Randolph,* 133–68; Anderson, *Randolph,* 274–82.
24. Meier and Rudwick, *CORE,* 40–71.
25. Meier and Rudwick, *Along the Color Line,* 363; idem, *CORE,* 63–71; Pfeffer, *Randolph,* 108.
26. Manning Marable is one of the first historians to examine the effect of the

Cold War on black protest in the 1940s and '50s. See *Race, Reform, and Reconstruction: The Second Reconstruction in Black America* (2nd ed., Jackson, Miss., 1991), 13–60. In my opinion, however, Marable's bias against anti-Stalinist social democrats like Randolph leads him to understate their achievement and overemphasize the dampening effect of the Cold War on the civil rights movement. McCarthyite hysteria began to decline by the mid-'50s, and thereafter the cold war was clearly more of an advantage than a hindrance to the movement. (Unless of course one defines the objectives of the civil rights movement differently from the way contemporaries did.)

27. For a provocative discussion of the significance of black economic gains in the period during and immediately after World War II, see William Julius Wilson, *Power, Racism, and Privilege: Race Relations in Theoretical and Sociohistorical Perspectives* (New York, 1973), 22–27.
28. Frene N. Ginwala, "Class, Consciousness, and Control: Indian South Africans, 1860–1946" (D.Phil. thesis, University of Oxford, 1974), 366–412; Essop Pahad, "The Development of Indian Political Movements in South Africa, 1924–1946" (D.Phil. thesis, University of Sussex, 1972), 40–156. Unfortunately, there is no adequate published history of the South African Indian community or of its political activities.
29. *Dr. Yusuf Mohamed Dadoo: His Speeches, Articles, and Correspondence with Mahatma Gandhi [1939–1983]*, E. S. Reddy, comp. (Durban, 1991), 55–67 (quote on 64), 366–85, and passim; Ginwala, "Class, Consciousness, and Control," 412–13; Pahad, "Indian Political Movements," 157–58.
30. *Monty Speaks: Speeches of Dr. GM (Monty) Naicker, 1945–1963,* E. S. Reddy, comp. (Durban, 1991), 54 and passim; Dowlat Bagwandeen, *A People on Trial—For Breaching Racism: The Struggle for Land and Housing of the Indian People of Natal, 1940–1946* (Durban, 1991), 167–68 and passim.
31. Bagwandeen, *People on Trial,* 165–92; *Dadoo: Speeches,* 14–122; *Monty Speaks,* 25–56.
32. For the text of the Doctor's Pact of 1947 and reactions to the riots of 1949, see Thomas Karis, *Hope and Challenge, 1935–1953,* vol. 2 of Karis and Carter, eds., *From Protest to Challenge* (Stanford, Calif., 1973), 272–73, 285–88. For a detailed discussion of the movement to joint action, see Dilshad Nomenti Cachalia, "The Radicalization of the Transvaal Indian Congress and the Moves to Joint Action, 1946–1952" (B.A. honors diss., University of the Witwatersrand, 1981).
33. See Peter Walshe, *Rise of African Nationalism: The African National Congress, 1912–1952* (Berkeley, Calif., 1971), 262–81; Karis, *Hope and Challenge,* 69–98, 168–300, passim.
34. *Amplification of Notes: A.N.C. Youth League* (interview with the president-general, Dr. A. B. Xuma) (21/2/44), ANC collection, University of the Witwatersrand Library, AD 11 89, Box 9, LA III-6. For the text of ANCYL manifesto of 1944 and other documents concerning the early history of the organization, see Karis, *Hope and Challenge,* 300–339. See also Walshe, *Rise of African Nationalism,* 281–85, 312–14, 349–61, and Gail M. Gerhart, *Black Power in South Africa: The Evolution of an Ideology* (Berkeley, 1978), 45–84.
35. In addition to the sources cited in the previous note, see Tom Lodge, *Black Politics in South Africa Since 1945* (London, 1983), 20–27.
36. Karis, *Hope and Challenge,* 337–38.
37. See Gerhart, *Black Power in South Africa,* 75–76 and passim.

38. Nelson Mandela to Francis (?), Oct. 6, 1949, carbon of typescript, University of the Witwatersrand Library, AD 1189, ANC Associated Organizations, LA IV 2; ANCYL correspondence, 1949. On the religious affiliations of the Youth Leaguers, see Gerhart, *Black Power in South Africa,* 52, 125 (on Lembede and Mda); Gail M. Gerhart and Thomas Karis, *Political Profiles, 1882–1964,* vol. 4 of Karis and Carter, eds., *From Protest to Challenge* (Stanford, Calif., 1977), 143–45, 151–53 (on Sisulu and Tambo); and Fatima Meer, *Higher Than Hope: The Authorized Biography of Nelson Mandela* (New York, 1988), 62. For a broader discussion of the role of independent black Christianity in protest politics, see Chapter 2 above.
39. Karis, *Hope and Challenge,* 475.
40. Ibid., 403–40; Lodge, *Black Politics,* 33–66. Still valuable for its detailed account of the rise of apartheid and the African response to it is Gwendolen M. Carter, *The Politics of Inequality: South Africa Since 1948* (2nd ed., New York, 1959).
41. Karis, *Hope and Challenge,* 410–16, 458–82.
42. Ibid., 458–66 (quotes on 461–62); Lodge, *Black Politics,* 36–66.
43. Lodge, *Black Politics,* 43–45; See also Karis, *Hope and Challenge,* 419–23.
44. Chief A. J. Lutuli, "The Road to Freedom Is Via the Cross," statement of Nov. 12, 1952, in Karis, *Hope and Challenge,* 487–88.
45. Naboth Mokgatle, *The Autobiography of an Unknown South African* (Berkeley, 1971), 307.
46. Letter from Bokwe "Joe" Matthews to Z. K. Matthews, 16 Sept. 1952, Z. K. Matthews Papers (B2.82), University of Cape Town (microfilm; original at the University of Botswana). For the ANC membership figures, see Lodge, *Black Politics,* 61, and Karis, *Hope and Challenge,* 427. On the Non-European Unity Movement, see ibid., 112–20, 494–508. With its base mainly among Colored intellectuals in the Western Cape, the NEUM was a source of insights into the system of racial oppression in South Africa that were rigorous and often perceptive. But its uncompromising commitment to noncooperation with segregation went to the point of refusing to work for anything less than its immediate and total overthrow. Such an orientation precluded most forms of political action and mobilization.
47. On the repression of the '50s, see Thomas Karis and Gail M. Gerhart, *Challenge and Violence, 1953–1964,* vol. 3 of Karis and Carter, eds., *From Protest to Challenge* (Stanford, Calif., 1977), 6, 36, 80–82, and passim. See Brown, *Gandhi,* 386–88, for a discussion of the limitations of Indian nonviolence.
48. See Lodge, *Black Politics,* 91–187; Karis and Gerhart, *Challenge and Violence,* 19–35 and passim; and Edward Feit, *African Opposition in South Africa: The Failure of Passive Resistance* (Stanford, 1967).
49. Cheryl Walker, *Women and Resistance in South Africa* (New York, 1991), 165–226, passim. Lodge, *Black Politics,* 142–46.
50. "War Against the People," *Liberation* 34 (Dec. 1958): 4–6. For discussions of the "M Plan," see Karis and Carter, *Challenge and Violence,* 35–40, and Lodge, *Black Politics,* 75–77. The Treason Trail, beginning in late 1956, kept black leaders in the dock for more than four years before they were found not guilty.
51. *Luthuli: Speeches of Chief Albert John Luthuli* (Durban, 1991), 7, 44, 71, 146–47, and passim. See also Albert Luthuli, *Let My People Go: An Autobiog-*

raphy (London, 1962). Despite the spelling in both of these titles, "Lutuli" is now generally accepted as correct.

52. Some of these issues will be dealt with more extensively in Chapter 7 below. My understanding of the ANC's post-Sharpeville attitude toward violence has been informed by Steven Mufson, *The Fighting Years: Black Resistance and the Struggle for a New South Africa* (Boston 1990), 96 and passim, and Stephen M. Davis, *Apartheid's Rebels: Inside South Africa's Hidden War* (New Haven, 1987). For Frantz Fanon's advocacy of violence, see especially *The Wretched of the Earth* (New York, 1966).
53. Quoted in *Asian Times* (London), June 26, 1987. (Clipping found in Reddy collection, Indian Documentation Centre, University of Durban-Westville.)
54. On the CAA petition, see Gerald Horne, *Black and Red: W. E. B. Du Bois and the Afro-American Response to the Cold War* (Albany, 1986), 185. For an account of the ASAR, see George Houser, *No One Can Stop the Rain: Glimpses of Africa's Liberation Struggle* (New York, 1989), 12–20. A review of *The Crisis* for 1952 and 53 (vols. 59, 60) turned up several references to the injustices of apartheid, but the only way readers would have known about the Defiance Campaign was from the publication in January 1953 of the text of the petition to the United Nations. See *The Crisis* 60 (1953): 38. There may have been some coverage in black weekly newspapers, but if so it does not appear to have influenced political thought and leadership in a significant way.
55. The Wofford speech, which was shown to King by E. D. Nixon, is printed for the first time in David Garrow, ed., *We Shall Overcome: The Civil Rights Movement in the 1950's and 60's* (Brooklyn, 1989), III: 1151–162 (reference to South Africa on 1160).
56. Among the works that have provided a localistic perspective on the movement are William H. Chafe, *Civilities and Civil Rights: Greensboro, North Carolina and the Black Struggle for Freedom* (New York, 1980); David R. Colburn, *Racial Change and Community Crisis: St. Augustine, Florida, 1877–1980* (New York, 1985); and Robert J. Norrell, *Reaping the Whirlwind: The Civil Rights Movement in Tuskegee* (New York, 1985). A work that effectively demonstrates the connections between local "movement centers" and the regional or national movement is Aldon D. Morris, *The Origin of the Civil Rights Movement: Black Communities Organizing for Change* (New York, 1984).
57. On the Baton Rouge boycott, see Morris, *Origins of the Civil Rights Movement,* 17–26.
58. Pfeffer, *Randolph,* 172–73; David Garrow, ed., *Walking City: The Montgomery Bus Boycott, 1955–1956* (Brooklyn, 1989), 202–8, 345–46, 545–51, and passim.
59. See J. Mills Thornton, "Challenge and Response in the Montgomery Bus Boycott and 1955–56," in Garrow, ed., *Walking City,* 367, for a persuasive description of how King grew during the boycott. On the context of "massive resistance," see especially Nunan V. Bartley, *The Rise of Massive Resistance: Race and Politics in the South During the 1950s* (Baton Rouge, 1969).
60. Martin Luther King, Jr., *Stride Toward Freedom* (New York, 1958), 76–86; David Garrow, *Bearing the Cross: Martin Luther King, Jr., and the Southern Christian Leadership Conference* (New York, 1986), 66–68; Morris, *Origins of the Civil Rights Movement,* 157–62; Adam Fairclough, *To Redeem the Soul of*

America: The Southern Christian Leadership Conference and Martin Luther King, Jr. (Athens, Ga., 1987), 23–26; Bayard Rustin, *Down the Line* (Chicago, 1971), passim.

61. Quoted in Garrow, *Bearing the Cross,* 68.
62. Taylor Branch, *Parting the Waters: America in the King Years, 1954–63* (New York, 1988), 82–87.
63. Ibid., 87; Niebuhr, *Moral Man and Immoral Society,* 247, 251–53, and passim.
64. Documents reprinted in Meier, Rudwick, and Broderick, *Black Protest Thought,* 300, 303, 306.
65. Garrow, *Bearing the Cross,* 273–74; Martin Luther King, Jr., *Why We Can't Wait* (New York, 1964), 80; Fairclough, *To Redeem the Soul of America,* 52 (1957 quote).
66. On King's "Hegelian" tendency, see John J. Ansbro, *Martin Luther King, Jr.: The Making of a Mind* (Maryknoll, N.Y., 1982), 122–23 and passim. On the philosophical and theological aspects of King's nonviolence, see James P. Hanigan, *Martin Luther King, Jr., and the Foundations of Nonviolence* (Lanham, Md., 1984).
67. William H. Chafe, "The End of One Struggle, the Beginning of Another," in Charles W. Eagles, ed., *The Civil Rights Movement in America* (Jackson, Miss., 1986), 129–30; Nathan Huggins, "Commentary," in Peter J. Albert and Ronald Hoffman, eds., *We Shall Overcome: Martin Luther King, Jr., and the Black Liberation Struggle* (New York, 1990), 91.
68. Inge Powell Bell, *CORE and the Strategy of Non-violence* (New York, 1968), 36.
69. On the contributions of SNCC and CORE to the nonviolent movement of the early '60s, see especially Clayborne Carson, *In Struggle: SNCC and the Black Awakening of the 1960s* (Cambridge, Mass., 1979), 9–30; and Meier and Rudwick, *CORE,* 135–58.
70. See Morris, *Origins of the Civil Rights Movement,* 40–76.
71. See Fairclough, *To Redeem the Soul,* 13–19, and Morris, *Origins of the Civil Rights Movement,* 30–39. Fairclough seems to me to have captured the relation between lay and clerical leaders quite effectively. For a critique of Morris's apparent tendency to exaggerate a self-generated clerical activism, see Clayborne Carson's review of Morris's book in *Constitutional Commentary* 3 (Summer 1986): 616–21. Carson takes Morris to task for failing to note "the large number—perhaps a majority—of southern black clergymen who did not become active in the civil rights movement or allow their churches to be used for civil rights meetings" (620–21).
72. For a broader historical discussion of the relationship of black religion to the struggle for freedom and equality, see Chapter 2 above.
73. See Morris, *Origins of the Civil Rights Movement,* 158–61, and Fairclough, *To Redeem the Soul,* 26, on the limited appeal of absolute nonviolence in the southern black community.
74. Fairclough, *To Redeem the Soul,* is especially good on the tactic of nonviolently risking violence. See pp. 227–78 and passim. For an eloquent and persuasive presentation of the case that the nonviolent civil rights movement was genuinely radical, see Anne Braden, "The Southern Freedom Movement in Perspective," *Monthly Review* 17 (Aug.–Sept. 1965): 74–84.
75. On King's recognition of the limitations of civil rights approaches and the

need for empowering the poor, see especially Garrow, *Bearing the Cross,* 431–624, and Thomas F. Jackson, "Recasting the Dream: Martin Luther King, Jr., African-American Political Thought and the Third Reconstruction, 1955–1968" (Ph.D. diss., Stanford University, 1993).

76. It is difficult to prove a negative, but for what it's worth, I can affirm that these observations are based on an extensive reading in the South African protest literature of the 1950s and '60s.
77. For an incisive account of the radicalization of SNCC, see Carson, *In Struggle,* 96–190. On the relationship of King and Stanley Levison, see Garrow, *Bearing the Cross,* passim.
78. Bokwe "Joe" Matthews to Z. K. Matthews, 20 Nov. 1952, Matthews Papers (B2, 89). For a good example of ANC anti-Americanism, see Nelson Mandela, "A New Menace in Africa," *Liberation* (March 1958): 22–26. The full extent of American support for the apartheid regime in its early years has recently been revealed in Thomas Borstelmann, *Apartheid's Reluctant Uncle: The United States and Southern Africa in the Early Cold War* (New York, 1993).
79. For King's early identification with liberation struggles, see Garrow, *Bearing the Cross,* 63 (quote), 71, 91.
80. Lewis quoted in Carson, *In Struggle,* 94; Lutuli is quoted from the statement he made after being removed from his chieftainship in 1952, as reprinted in Karis, *Hope and Challenge,* 487; King quoted in David Levering Lewis, *King: A Biography* (Urbana, Ill., 1978), 93.
81. See Leo Kuper, *An African Bourgeoisie: Race, Class, and Politics in South Africa* (New Haven, 1965), 101–3, and passim, and Alan Cobley, *Class and Consciousness: The Black Petty Bourgeoisie in South Africa, 1924 to 1950* (New York, 1990).
82. Steven M. Millner, "The Montgomery Bus Boycott: A Case Study in the Emergence and Career of a Social Movement," in Garrow, ed., *The Walking City,* 512–13. For an analysis of the middle-class aspirations of the student protesters of the early '60s, see Carson, *In Struggle,* 12–15.
83. On the role of boycotts in the South African struggle, see Lodge, *Black Politics,* 181–82 and passim. The significance of consumer boycotts for the success of desegregation campaigns in southern cities is clear from all accounts of the movement, but few historians have generalized about them or reflected on their importance. For the rudiments of such a discussion, see Fred Powledge, *Free at Last? The Civil Rights Movement and the People Who Made It* (Boston, 1991), 83, 369.
84. The localized basis for the southern movement is set forth effectively in Morris, *Origins of the Civil Rights Movement.* On the peculiarities of the Eastern Cape, see Lodge, *Black Politics,* 45–60 and passim. For a valuable analysis of the organizational assets that southern African-American urban communities brought to the civil rights struggle, see Doug McAdam, *Political Process and the Development of Black Insurgency, 1930–1970* (Chicago, 1982), 94–106.
85. See Lodge, *Black Politics,* 170–71, on the ANC's failure in Alexandria. My understanding of how SCLC operated is based primarily on Morris, *Origins of the Civil Rights Movement,* and Fairclough, *To Redeem the Soul.* On the role of community organizations in South Africa during the 1980s, see Mufson, *Fighting Years,* 323–24.

86. On the PAC's overtures to the independent black churches, see Lodge, *Black Politics,* 81.
87. King is quoted in Garrow, *Bearing the Cross,* 92. For an earlier formulation of these contrasts of political context, see George M. Fredrickson, "The South and South Africa," in idem, *The Arrogance of Race: Historical Perspectives on Slavery, Racism, and Social Inequality* (Middletown, Conn., 1988), 254–69.
88. See Jackson "Recasting the Dream," 108–10, and Renée Romano, "Burden on Our Backs: Domestic Racism and American Foreign Policy" (unpublished seminar paper, Stanford University, 1992).
89. Sheridan Johns and R. Hunt Davis, Jr., eds., *Mandela, Tambo, and the African National Congress: The Struggle Against Apartheid, 1948–1990* (New York, 1991), 173–74 (excerpted from the *Washington Times,* Aug. 22, 1985).
90. Tom Lodge, *Black Politics,* 196–99. On the revival of sabotage in the late '70s, see Anthony Marx, *Lessons of Struggle: South African Internal Opposition* (New York, 1992), 93–94.
91. George M. Houser, "Freedom's Struggle Crosses Oceans and Mountains: Martin Luther King, Jr., and the Liberation Struggles in Africa and America," in Albert and Hoffman, eds., *We Shall Overcome,* 189–91; Lewis, *King,* 259; Branch, *Parting the Waters,* 599.
92. Martin Luther King, Jr., "Address on South African Independence," London, England, Dec. 7, 1964. Martin Luther King, Jr., Papers, Library, and Archives, Martin Luther King, Jr., Center for Nonviolent Change, Atlanta, Georgia.
93. Ibid., 2; Address of Dr. Martin Luther King on Dec. 10, 1965, for the benefit of the American Committee on Africa, Hunter College, New York City—Martin Luther King, Jr., Papers, Library, and Archive, Atlanta, Georgia.
94. Good accounts of developments in South Africa in the 1980s are Marx, *Lessons of Struggle,* Mufson, *The Fighting Years,* and Richard Price, *The Apartheid State in Crisis, 1975–1990* (New York, 1991). Oliver Tambo, speaking on behalf of the ANC, explicitly repudiated "passive resistance" in 1966 (Johns and Davis, *Mandela, Tambo,* 134). I learned first-hand about the extent to which the UDF had revived the Gandhian tradition of noncooperation on two recent visits to South Africa. In April 1989 I saw the Mass Democratic Movement in action and interviewed Allan Boesak about it. In July 1992 I interviewed Mewa Ramgobin, a leader of the Indian community of Natal and a former UDF treasurer. For him, the domestic protest of the 1980s, which featured a boycott of elections and leaders going to jail without paying fines or making bail, was a vindication of Gandhism.

Chapter 7

1. See Jean Paul Sartre, *Black Orpheus,* S. W. Allen, trans. (Paris, n.d.), 59.
2. Immanuel Geiss, *The Pan African Movement: A History of Pan Africanism in America, Europe, and Africa* (New York, 1974), 399–408 (quote on 403); Peter Walshe, *The Rise of African Nationalism in South Africa: The African National Congress, 1912–1952* (Berkeley, Calif., 1971), 337–38.
3. Walshe, *Rise of African Nationalism,* 337; Thomas Karis and Gail M. Gerhart, *Challenge and Violence, 1953–1964,* vol. 3 of Thomas Karis and Gwendolen Carter, eds., *From Protest to Challenge: A Documentary History*

of African Politics in South Africa, 1882–1964 (Stanford, Calif., 1977), 320–23; Harold R. Isaacs, *The New World of Negro Americans* (New York, 1964), 95, 276, 290, 292, 306; Robert G. Weisbord, *Ebony Kinship: Africa, Africans, and the Afro-American* (Westport, Conn., 1973), 184.

4. This account is based mainly on Hollis Lynch, *Black American Radicals and the Liberation of Africa: The Council on African Affairs, 1937–1955* (Cornell University African Studies and Research Center, 1978). On Max Yergan, see David H. Anthony, "Max Yergan in South Africa: From Evangelical Pan-Africanist to Revolutionary Socialist," *African Studies Review* 34 (1991): 27–55.
5. See note 4. On the American Committee on Africa, see Chapter 6 above.
6. Richard Wright, *Black Power: A Record of Reactions in a Land of Pathos* (New York, 1954), 348 and passim. Wright expressed similar opinions in *White Man, Listen!* (New York, 1957).
7. Gail M. Gerhart, *Black Power in South Africa: The Evolution of an Ideology* (Berkeley, 1978), 45–76; Thomas Karis, *Hope and Challenge,* vol. 2 of Karis and Carter, eds., *From Protest to Challenge* (Stanford, Calif., 1973), 300–322 (quotes on 308 and 317).
8. Gerhart, *Black Power in South Africa,* 60. For a provocative recent treatment of the cultural and psychological basis of nationalism, see Benedict Anderson, *Imagined Communities: Reflections on the Origin and Spread of Nationalism* (London, 1983). On Lembede's Africanist precursors, see Chapters 2 and 4 above.
9. Quotes from Karis, *Hope and Challenge,* 326–30. See also Gerhart, *Black Power in South Africa,* 64–75.
10. Karis, *Hope and Challenge,* 330; Nelson Mandela to Francis (?), Oct. 6, 1949, carbon of typescript, University of Witwatersrand Library, ANC associated organizations, LA 1V 2. For more on how and why non-Communists in the ANC came to ally themselves with Communists and to side with the Soviet Union against the United States, see Chapters 5 and 6 above.
11. Karis and Gerhart, *Challenge and Violence,* 11–14, 63–64, 105, 205 (quote from Matthews on 12).
12. Ibid., 205–8; Gail Gerhart, *Black Power in South Africa,* 124–172, passim; George Padmore, *Pan-Africanism or Communism?* (1956; New York, 1971), 339–40. For other presentations of the Communist conspiracy theory of the Freedom Charter, see also Jordan Ngubane, *An African Explains Apartheid* (New York, 1963), 164; and Richard Gibson, *African Liberation Movements: Contemporary Struggles Against White Minority Rule* (London, 1972), 50–55.
13. See Gerhart, *Black Power in South Africa,* 151–64.
14. For Sobukwe's views as head of the PAC, see the 1959 documents reprinted in Karis and Gerhart, *Challenge and Violence,* 506–17. For biographical information on Sobukwe, see Benjamin Pogrund, *Sobukwe and Apartheid* (New Brunswick, N.J., 1991).
15. Sobukwe quote is from Karis and Gerhart, *Challenge and Hope,* 514.
16. Ibid., 516.
17. See Gerhart, *Black Power in South Africa,* 221–24, 317–19; Leo Kuper, *An African Bourgeoisie: Race, Class, and Politics in South Africa* (New Haven, 1965), 364–87.
18. Karis and Gerhart, *Challenge and Violence,* 669–71; Tom Lodge, *Black Politics in South Africa Since 1945* (London, 1983), 240–55.

19. W. E. B. Du Bois, *The Crisis* 54 (Dec. 1947): 362–63. See Chapters 2 and 4 above for discussions of earlier manifestations of black nationalism or separatism.
20. E. U. Essien-Udom, *Black Nationalism: A Search for an Identity in America* (Chicago, 1962), 46–48, 55–60, 83–87. Essien-Udom's work remains the best account of the Black Muslims before the 1960s. But see also C. Eric Lincoln, *The Black Muslims in America* (Boston, 1961).
21. Ibid., 312 and passim.
22. Malcolm X, *Autobiography* (New York, 1965). Bruce Perry's deeply researched *Malcolm: The Life of a Man Who Changed Black America* (Barrytown, New York, 1992) adds many details and corrects some errors or misrepresentations.
23. Perry, *Malcolm,* 211–12 and passim.
24. On Malcolm's thought in 1964–65, see Perry, *Malcolm,* 288–367, passim, and Eugene Victor Wolfenstein, *The Victim of Democracy: Malcolm X and the Black Revolution* (Berkeley, 1981), 300–346. Perry stresses the conflicts and confusions in Malcolm's thinking, while Wolfenstein finds the consistent advocacy of "an anti-imperialist struggle for national liberation" (p. 300). These viewpoints can perhaps be reconciled if one acknowledges that Malcolm knew what he wanted—the liberation of blacks from white supremacy—but remained uncertain as to how to attain it—or indeed whether it was even possible—under the conditions that existed in the United States during the 1960s.
25. Julius Lester, *Look Out, Whitey! Black Power's Gon' Get You Mama* (New York, 1968), 91.
26. The fullest published account of King's campaigns for economic justice can be found in David J. Garrow, *Bearing the Cross: Martin Luther King, Jr., and the Southern Christian Leadership Conference* (New York, 1986), 527–625. For an incisive analysis of the thought behind this effort, see Thomas F. Jackson, "Recasting the Dream: Martin Luther King, Jr., African-American Political Thought and the Third Reconstruction" (Ph.D. diss., Stanford University, 1993).
27. The shifting attitudes in SNCC are well described and analyzed in Clayborne Carson, *In Struggle: SNCC and the Black Awakening of the 1960s* (Cambridge, Mass., 1981), 111–211, passim. On CORE's similar evolution toward separatism and away from nonviolence, see August Meier and Elliott Rudwick, *CORE: A Study in the Civil Rights Movement, 1942–1968* (New York, 1973), 374–408.
28. A good account of the Meredith march can be found in Carson, *In Struggle,* 206–11. Carmichael did not actually invent the term Black Power, even in the context of the mid-'60s. Adam Clayton Powell, for one, had used it earlier. Carmichael was not even the first to use it on the Meredith march; but his usage was the first to be widely publicized. The quote on the rejection of nonviolence is from James Boggs in Floyd Barbour, ed., *The Black Seventies* (Boston, 1970), 35.
29. Lester, *Look Out, Whitey!,* 100.
30. Gayraud S. Wilmore and James H, Cone, eds., *Black Theology: A Documentary History* (Maryknoll, N.Y., 1979), 27; Nathan S. Wright, Jr., *Black Power and Urban Unrest: Creative Possibilities* (New York, 1967), 61.
31. Wright, *Black Power and Urban Unrest,* 7; Wright, "The Crisis Which Bred

Black Power," in Floyd Barbour, ed., *The Black Power Revolt* (Boston, 1968), 116–17.

32. Stokely Carmichael and Charles V. Hamilton, *Black Power: The Politics of Liberation in America* (New York, 1967), 44–45.
33. Stokely Carmichael, *Stokely Speaks: From Black Power to Pan-Africanism* (New York, 1971), 35, 97.
34. For Carmichael's revolutionism of 1968, see ibid., 134–36.
35. This discussion is based mainly on Manning Marable, *Race, Reform, and Revolution: The Second Reconstruction in Black America* (Jackson, Miss., 1991), 86–148 (quote on 110); John T. McCartney, *Black Power Ideologies: An Essay in African-American Political Thought* (Philadelphia, 1992); and William L. Van Deburg, *New Day in Babylon: The Black Power Movement in American Culture, 1965–1975* (Chicago, 1992), 112–91. Conspicuous separatists (or in William Van Deburg's terminology, "territorial nationalists"), in addition to those named above, included the poet Imamu Baraka (Leroi Jones) and Imari Obadele I (Richard Henry), founder of a sect called the Republic of New Africa. Prominent among those that political scientist John McCartney labels "counter-communalists"—but whom I prefer to call, in accordance with the terminology of the late '60s and Van Deburg's classifications, "revolutionary nationalists"—were (in addition to Newton and other Black Panther leaders such as Eldridge Cleaver) James Foreman, the former SNCC leader, and Robert L. Allen, author of the book that made the strongest case for a black-led revolution against American capitalism: *Black Awakening in Capitalist America* (New York, 1969).
36. On Chisholm's significance among "the Black Power pluralists," see McCartney, *Black Power Ideologies,* 151–65. For a more general discussion of the pluralist tendency, see Van Deburg, *New Day in Babylon,* 113–29.
37. See Van Deburg, *New Day in Babylon,* 193–291, for an extensive treatment of the impact of Black Power on African-American and American culture.
38. On the "hiatus of the 1960s," see Gerhart, *Black Power in South Africa,* 251–59.
39. On the Liberal party, see Douglas Irvine, "The Liberal Party, 1953–1958," in Jeffrey Butler, Richard Elphick, and David Welsh, eds., *Democratic Liberalism in South Africa: Its History and Prospect* (Middletown, Conn., 1987), 116–33; Lodge, *Black Politics in South Africa,* 210–12, 240–41, 307–10 (on Duncan's career in the PAC), and passim; and Janet Robertson, *Liberalism in South Africa, 1948–1963* (Oxford, 1971).
40. On NUSAS in the '60s, see Gerhart, *Black Power in South Africa,* 257–59; Lodge, *Black Politics in South Africa,* 322–23; and Baruch Hirson, *Year of Fire, Year of Ash: The Soweto Revolt: Roots of a Revolution* (London, 1979), 65–68. Jordan Ngubane's special synthesis of liberalism and African nationalism is expounded in *An African Explains Apartheid.*
41. Gerhart, *Black Power in South Africa,* 259–70; Hirson, *Year of Fire,* 68–84; Robert Fatton, *Black Consciousness in South Africa: The Dialectics of Ideological Resistance to White Supremacy* (Albany, 1986), 63–80; N. Barney Pityana, Mamphela Ramphele, Malus Mpumlwana, and Lindy Wilson, eds., *The Bounds of Possibility: The Legacy of Steve Biko and Black Consciousness* (Cape Town, 1991), 154–78 and passim.
42. Gerhart, *Black Power in South Africa,* 288–90; Aelred Stubbs, ed., *Steve Biko—I Write What I Like* (1978; San Francisco, 1986), 80–86 (quote on 86).

43. Stubbs, ed., *Biko—I Write What I Like,* 67; Gerhart, *Black Power in South Africa,* 259. It is also worth noting, however, that BC, like the PAC, was willing to tolerate the permanent presence of whites in South Africa. According to the SASO policy manifesto of 1971, "South Africa is a country in which both black and white live and shall continue to live together" (B. A. Khoapa, ed., *Black Review,* 1972 (Johannesburg, 1972), 40).
44. Stubbs, ed., *Biko—I Write What I Like,* 49–53; Khoapa, ed., *Black Review, 1972,* 42–43.
45. See Gerhart, *Black Power in South Africa,* 277, for an analysis of this similarity. Gerhart, however, creates confusion when she writes that "the term black by the late 1960s in the United States had become a loose synonym for 'nonwhite'-a new catchall term encompassing all victims of racial discrimination." Clearly one had to have some specifically African ancestry to qualify as "black." Other nonwhites, such as Asians and American Indians, have never been so designated. Hence, curiously enough, the South African designation has become broader than the American. It parallels in its usage the newer American designation "people of color," which was popularized by the multicultural movement of the 1980s.
46. Lodge in *Black Politics in South Africa* provides the basis for this comparison, although he does not actually make it explicit (see pp. 83–86 and 323–24).
47. Sipho Buthelezi, "The Emergence of Black Consciousness: An Historical Appraisal," in Pityana et al., eds., *Bounds of Possibility,* 124–28.
48. On the religious character and associations of BC, see especially Fatton, *Black Consciousness in South Africa,* 107–19, and Hirson, *Year of Fire,* 78–81. On the role of religion in Biko's life and thought, see Lindy Wilson, "Bantu Steve Biko: A Life," in Pityana et al., eds., *Bounds of Possibility,* 20, 43–44. N. Barney Pityana, the second most important of the original student leaders, became a clergyman and eventually the director of the World Council of Churches' Program to Combat Racism.
49. The best source on the development of Black Theology is Wilmore and Cone, *Black Theology: A Documentary History.* Among its major expressions were Albert B. Cleage, *The Black Messiah* (New York, 1968); James H. Cone, *Black Theology and Black Power* (New York, 1969), *A Black Theology of Liberation* (Philadelphia, 1970), and *God of the Oppressed* (New York, 1972); and Gayraud S. Wilmore, *Black Religion and Black Radicalism* (New York, 1972). A work that shows the connections between American and South African versions is Dwight N. Hopkins, *Black Theology: USA and South Africa* (Maryknoll, N.Y., 1989).
50. James H. Cone, *A Black Theology of Liberation,* 7. Buthelezi quoted in Louise Kretzschmar, *The Voice of Black Theology in South Africa* (Johannesburg, 1986), 62.
51. See Kretzschmar, *Voice of Black Theology,* 43–70.
52. Cone, *Black Theology of Liberation,* 6; Buthelezi quoted in Hopkins, *Black Theology: USA and South Africa,* 99; Cone quoted in Basil Moore, ed., *The Challenge of Black Theology* (Atlanta, 1973), 48; Allan Boesak, *Farewell to Innocence: A Socio-Ethical Study of Black Theology and Black Power* (Johannesburg, 1976), 78. For a discussion of the differences, see Kretzschmar, *Voice of Black Theology,* 65–68.
53. Khoapa, ed., *Black Review, 1972,* 42.
54. Gerhart, *Black Power in South Africa,* 276. Emphasis added.

55. Stubbs, ed., *Biko—I Write What I Like,* 69; Steve Biko, *Black Consciouness in South Africa* (New York, 1978), 99.
56. Stubbs, ed., *Biko-I Write What I Like,* 132–36.
57. See Anthony Marx, *Lessons of Struggle: South African Internal Opposition, 1960–1990* (New York, 1992), 39–60, 194–95, and Geoff Budlender, "Black Consciousness and the Liberal Tradition," in Pityana et al., eds., *Bounds of Possibility,* 234–35. For a good example of white leftist criticism of BC, see Hirson, *Year of Fire,* passim.
58. See Marx, *Lessons of Struggle,* 64–72, and Lodge, *Black Politics in South Africa,* 328–39.
59. Keith Mokoape, Thenjiwe Mtintso, and Welile Nhlapo, "Towards the Armed Struggle," in Pityana et al., eds., *Bounds of Possibility,* 142–43; Marx, *Lessons of Struggle,* 91–105.
60. Lodge, *Black Politics in South Africa,* 344–46; Marx, *Lessons of Struggle,* 87–91.
61. Budlender, "Black Consciousness and Liberalism," in Pityana et al., eds, *Bounds of Possibility,* 235.
62. Good accounts of black politics in South Africa in the 1980s can be found in Marx, *Lessons of Struggle,* 106–234; Robert M. Price, *The Apartheid State in Crisis: Political Transformation in South Africa, 1975–1990* (New York, 1991), 152–219; and Steven Mufson, *The Fighting Years: Black Resistance and the Struggle for a New South Africa* (Boston, 1980).
63. Revealing statements of former Black Consciousness supporters who embraced nonracialism as a more advanced form of struggle can be found in Julie Frederickse, *The Unbreakable Thread: Non-racialism in South Africa* (Bloomington, Ind., 1990), 114–15, 134–35, 161–62.
64. Nelson Mandela, *Nelson Mandela Speaks: Forging a Democratic, Nonracial South Africa* (New York, 1993), 39.
65. See Lodge, *Black Politics in South Africa,* 223, and Frederikse, *Unbreakable Thread,* 100, 242, on the gradual implementation of nonracialism in the ANC.
66. See Price, *Apartheid State,* 166–67, 251, and passim.
67. On the debates within the Black Consciousness movement in the mid-'70s, see Marx, *Lessons of Struggle,* 75–85.
68. For the text of McKay's poem, as well as some commentary on it, see Nathan Irvin Huggins, *Harlem Renaissance* (New York, 1971), 71–72. On Newton's concept of "revolutionary suicide," see McCartney, *Black Power Ideologies,* 139–40, and Newton's book *Revolutionary Suicide* (New York, 1973). Newton distinguished revolutionary suicide from "reactionary suicide," the throwing away of one's life out of despair without engaging in direct resistance to the oppressor.
69. Quoted in Mufson, *Fighting Years,* 70. A recent book that recounts in detail the history of white South African involvement in the anti-apartheid movement is Joshua M. Lazerson, *Against the Tide: Whites in the Struggle Against Apartheid* (Boulder, Colo., 1994).
70. Price, *Apartheid State,* 166; Gibson, *African Liberation Movements,* 25, 30.

Epilogue

1. Andrew Hacker, *Two Nations: Black and White, Separate, Hostile, Unequal* (New York, 1992), 218–19.

2. See Cornell West, *Race Matters* (New York, 1994), 17–31.
3. See Manning Marable, *Race, Reform, and Rebellion* (Jackson, Miss., 1991), 227–30.
4. For an imaginative effort to reconceive the African-American struggle in a way that combines pressure for government action and self-help, see Roy L. Brooks, *Rethinking the American Race Problem* (Berkeley, 1990).
5. The case for such inclusiveness in African-American political thought is made in John Brown Childs, *Leadership, Conflict, and Cooperation in Afro-American Social Thought* (Philadelphia, 1989).

Index

AAC (All African Convention), 205–7
ABB (African Blood Brotherhood), 192, 193
Abdurhahman, Abdul, 47, 48, 115, 230
Abolitionism: and black nationalism, 26–27; and black-white relations, 23–24, 60–61; and class issues, 25; failures of, 64; and free blacks, 23–24; and integration, 26–27; missions/missionaries compared with, 39; and nonviolent resistance, 231; protest tradition of, 108, 109; and racism, 27; and Reconstruction, 29; and religion/Ethiopianism, 57, 60, 62–63, 64; and suffrage, 48. *See also specific person*
Accomodationism. *See specific person or organization*
Affirmative action, 321, 322
"Africa for the Africans," 138, 161–62, 164, 242, 280, 281
African Americans: and African independence, 154, 166, 167, 168, 198; and African redemption, 84, 89, 190; Africans as superior to, 78–79; as civilized, 48; in the communist vanguard, 190; and the definition of nation/nationality, 188; election to public office of, 313; failures of, 320–23; future of, 320–23; influence in South Africa of, 37, 48–49, 53–54, 278–79, 297–98, 306–8, 313–14; liberation of, 154; as losers, 319–20; and millennialism, 164, 165; as a minority community, 315; as models of black progress, 48; as an oppressed nation, 188, 193–94, 196; paternalism of, toward Africans, 85; redemption of, 288; as special, 151; superiority of, toward Africans, 85–86. *See also specific organization or topic*
African Blood Brotherhood (ABB), 192, 193
"The African Christian Union," 84
African Civilization Society, 65
African Communities League, 153
African independence, 49, 279, 287, 307; and African Americans, 154, 166, 167, 168, 198; and civilization, 151; and communism, 192, 195, 196, 198, 200, 208–9, 273; geopolitical context of, 272–73; leadership of movement for, 280; and Pan-Africanism, 151, 152, 174, 206–7, 277, 280, 281, 282; and religion, 44, 243–44; and the Wellington movement, 167. *See also specific person or organization*
African Jubilee Choir, 82
African Methodist Episcopal (AME) Church, 165, 168, 244, 262; and Ethiopianism, 75, 76, 80, 81–83, 84, 85–86, 87, 89, 93
African Methodist Episcopal Zion Churches, 146
African National Congress (ANC): and African independence, 200; American help for the, 253; anti-Americanism of the, 267; and apartheid, 245, 300, 311; banning of the, 251, 265, 271, 274, 275, 286, 298; and Black Consciousness, 298, 300, 309, 310, 316, 322; and black solidarity/pride, 322; and the Civil Rights

African National Congress (ANC) (*continued*)
Movement, 265; and class issues, 142–43; and Coloreds, 194–95, 312, 322; and communism, 183, 191, 195, 198–99, 200, 205, 207–8, 209, 218, 219, 241, 242, 265, 266, 281–82, 286, 299; credibility of the, 245, 248; and cultural orientation, 243–44; decline of the, 249; and Ethiopianism, 87, 88; and Europeans, 321; founding of the, 144; and the Freedom Charter, 283; funding for the, 253; and the future of South Africa, 321; as a government-in-waiting, 316; guerrilla army of the, 252, 273–74, 275, 285, 310, 311; and the ICU, 173; ideological flexibility of the, 312–13, 322; and Indians, 240–41, 245–46, 282, 312, 322; King's influence on the, 265; and labor issues, 142–43, 285, 311; and liberalism, 266, 284, 313; and the M Plan, 250–51; membership of the, 245, 248–49, 268, 285, 286, 312; motto of the, 163; multi-racialism/nonracialism of the, 278, 300, 309, 312, 317, 322; NAACP compared with the, 266; and nationalism, 312–13; and nonviolent resistance, 229–30, 231, 237, 240–41, 244–45, 246–51, 252, 266, 267; and the PAC, 284–86, 299, 311, 316, 318, 322; and Pan-Africanism, 149, 152, 278, 280–81, 282, 312–13; and populism, 143, 171; and the post-apartheid South African constitution, 312; and the Programme of Action, 242–43, 245, 250; as a racially exclusive organization, 285; radicalization of the, 163, 164; and religion, 313; and segregation, 207, 240, 242, 244–45; and self-determination, 242; shift in tactics of the, 252; strength/visibility of the, 311; and students, 310; as a success, 316; and suffrage, 205, 206, 240, 268; and the UDF, 311, 322; and the UNIA, 162, 163–64, 176; victory of the, 319; and whites, 298, 309, 312, 318, 322. *See also* Defiance Campaign; South African Native National Congress; *Umkonto We Sizwe* organization; Youth League
African National Congress (ANC)—leadership of the: and alliances with other groups, 245, 301, 317; background of the, 268–69, 271, 285; and communism, 195, 241, 242; comparison of NAACP and, 266; and ideology, 142–43; and militancy, 285–86; and nonviolent resistance, 230, 231, 242, 243, 251; and populism, 163–64; and religion/Ethiopianism, 88, 271; repression of the, 251, 274; and self-determination, 195
African Orthodox Churches, 165–66
African People's Organization, 87
African Political Organization (APO), 46, 47, 115
African redemption, 65–66, 67, 68–74, 75–80, 84, 89, 190, 287
African Resistance Movement, 298, 299
African Rights Movement: culture/ethos of the, 271–72; failure of the, 273–74; King's influence on the, 274–75, 276; and nonviolent resistance, 237–52; and the sanctions movement, 275; structure/organization in the, 268–69. *See also* Black Consciousness; *specific person or organization*
African socialism, 56
The African World (ICU newspaper), 162
African Yearly Register (Skota), 164
Africans: and African-American superiority, 85–86; as clergymen, 83, 86; as superior to African Americans, 78–79. *See also specific organization or topic*
Afrikaner Bond, 37, 38, 41, 42, 43
Afrikaner Nationalists, 200, 280, 298, 306; and apartheid, 223, 244–45; and Coloreds, 245; coming to power of the, 185; and communism, 205, 245; and the Defiance Campaign, 246; and the election of 1948, 223, 224, 237, 240, 244; and Indians, 240, 245; and suffrage, 224; and the United party, 205; and World War II, 214, 217
Afrikaners, 20, 37–47, 95, 116, 189. *See also* Afrikaner Bond; Afrikaner Nationalists; South African War
Afro-American Council, 51, 53, 54, 121
Afro-American League, 51, 54
Aggett, Neil, 317
Aggrey, J.E.K., 145, 146–47, 149, 150, 166, 173
Alexander, Daniel William, 165–66
Ali, Noble Drew, 287
All African Convention (AAC), 205–7
AME. *See* African Methodist Episcopal (AME) Church
American Colonization Society, 65, 67, 70, 78
American Committee on Africa, 253, 275, 279
American Federation of Labor (AFL), 204
American Freedom Struggle, 276

American Labor party, 220
American Negro Academy, 68
American Negro Labor Congress, 190–91
American redemption, 74, 93
Americans for South African Resistance, 253
Anderson, Benedict, 227
Anglican Church, 81, 87
ANLC (American Negro Labor Congress), 190–91
Anti-racist racism, 278
Anti-Segregation Council, 239
Anti-Slavery and Aborigines' Protection Society, 131
Antislavery movement, 23–24, 273
Apartheid, 223, 244–45, 277, 300, 309, 310, 320–21. *See also specific person or organization*
APO (African Political Organization), 46, 47, 115
Appeal . . . to the Colored Citizens of the World (Walker), 63–64
Asiatic Land Tenure Act, 240
Assimilation: cultural, 45–46. *See also* Integration; *specific person or organization*
"Atlanta Compromise" speech (Washington), 105
Azanian People's Organization (AZAPO), 310, 311

Bach, Lazar, 200, 208–9
"Back to Africa" movement, 154, 169, 190
Ballinger, William, 171–72
Bantu Authorities Act, 246
Bantu Education Act (1953), 249
Bantu Methodist Churches, 87, 244
The Bantu Past and Present (Molema), 123–24
Bantu Presbyterian Churches, 244
Bantu Prophets (Sundkler), 90
Baptist Churches, 58, 75, 80, 86, 93, 165, 262
Baraka, Imamu, 297
Baton Rouge, Louisiana, 253–54, 255
BCP (Black Community Programmes), 300
Beaumont Commission, 131
Beinart, William, 89–90
Bell, Inge Powell, 260
"Beloved community" (King), 257, 258, 262, 264, 312, 317, 322–23
Bible, 57, 61, 63, 75, 89, 92, 93, 306
Biko, Steve, 299–301, 303, 304, 307–9, 315
Birmingham, Alabama, 258, 263, 270, 271
Black: definition of, 301–2
Black Capitalism, 314
Black Community Programmes (BCP), 300
Black Consciousness: African-American influence on, 278, 297–98; and African independence, 307; and apartheid, 300; and Black Power, 298, 304, 306–8, 313–18; and black solidarity/pride, 302, 313; characteristics of, 47; and civil disobedience, 308; and class issues, 309, 312, 314; and Coloreds, 47, 310–11; and communism, 309, 311–12; and community organization, 313–14; and cultural nationalism, 308–9; and the definition of black, 301–2; and economic issues, 303; emergence of, 277–86, 300; and Europeanization, 47; as a failure, 315–16; fate of, 310–13; and Indians, 310–11; and inferiority of blacks, 313; and labor issues, 309; leadership of, 303, 307, 314; and liberalism, 301, 302, 303; as a "majority philosophy," 308; and missions/missionaries, 305–6; and nationalism, 302, 304; and nonracialism, 312; origins of the movement for, 41, 241, 301, 307; and Pan-Africanism, 301, 302; and political issues, 303, 309–10; and power, 315; and radicals, 301, 302, 303, 309; and religion, 303, 305–6, 314–15; and revolution, 308; and self-help, 300, 308–9, 313–14; suppression of, 303, 314–15; and violence/intimidation, 298; and whites, 301, 302, 303, 309, 310–11, 312, 317–18. *See also specific person or organization*
Black Eagle Flying Corps, 166
"Black is beautiful," 297, 302, 313
The Black Man (ICU newspaper), 169
"Black man, you are on your own," 300
Black messianism, 93
Black migration, 103, 112, 142, 196
Black Muslims. *See* Malcolm X; Nation of Islam
Black Panthers, 224, 296, 308–9, 314
Black People's Convention (1972), 300, 303, 308
Black populism. *See* Industrial and Commercial Workers' Union of Africa (ICU); Millennialism; Populism; Universal Negro Improvement Association (UNIA)
Black Power: and African independence, 279; and Black Consciousness, 298, 304, 306–8, 313–18; and black solidarity/pride, 297, 302–3, 313; and capitalism, 296–97, 314; and the Civil Rights Movement, 278; and class issues, 296;

Black Power (*continued*)
and colonialism, 295–96; and community organization, 313–14; and cultural nationalism, 296, 297, 314; emergence of, 292–97; and entrepreneurs, 296; and ethnicism, 297, 308, 314; first use of term, 279; and imperialism, 295, 314; and inferiority of blacks, 313; and liberalism, 293, 294, 296, 297; meanings of, 293–94, 296, 297; and the media, 315; and nationalism, 278, 296, 297, 314; and nonviolent resistance, 295; and political parties, 296–97; and racial exclusivity, 293; and religion, 294, 304, 306; rise of, 286–97; and romanticism, 279; and self-determination, 294; and self-help, 295, 296, 313–14; and separatism, 296, 314; as a success, 315; and violence/intimidation, 293, 314; and whites, 292–97, 301. *See also specific person or organization*
Black Power (Carmichael and Hamilton), 295–96, 306
The Black Problem (Jabavu), 148
Black redemption: and accommodationism, 262; and emigration, 61, 66; and millennialism, 165, 168; and missions/missionaries, 313; and Pan-Africanism, 146–47; and religion/Ethiopianism, 61, 73, 74, 76, 90; and self-help, 66; and the UNIA, 155, 168; and the Wellington movement, 167. *See also* African redemption
Black solidarity/pride: and black vs. black, 52, 54; and Coloreds, 47; and cultural assimilation, 45–46; and the Du Bois-Washington controversy, 110; and Ethiopianism, 68, 84, 89; importance of, 26, 55–56; and integration, 25–27; and Islam, 68; and leadership, 45–46; and nationalism, 144, 145; and nonviolent resistance, 233–34, 259; and Pan-Africanism, 47, 144, 145–46, 174; and racism, 278; and students, 309–10; and suffrage, 55; and the Wellington movement, 167. *See also* Black Consciousness; Black Power; *specific person or organization*
Black Star Steamship Line, 154, 159, 160, 166
Black supremacy, 304
Black Theology, 304–6
Black Theology (Cone), 304
Black and White: Land, Labor, and Capital in the South (Fortune), 32, 33
Blackness: as an ontological symbol, 305
Blacks: character of, 68, 70, 150; and Jews, 35, 304; as role models for whites, 93; specialness of, 93, 262; stereotypes of, 260, 278; as superior to whites, 64, 65, 260; Westernization/Europeanization of, 41, 56. *See also* African Americans; Africans; Inferiority of blacks
Blaming the victims, 27
Blyden, Edward, 67–71, 73, 75, 157, 161, 301
Boer War. *See* South African War
Boesak, Allan, 276, 306, 317
Boksburg Native Vigilance Association, 89
Booth, Joseph, 84–85, 86, 87
Boston riot (1903), 52, 54, 108
Botha, Louis, 116, 130
Boycotts: and Black Consciousness, 311; bus, 237, 252, 253–56, 257, 260, 261, 262, 265, 267, 269, 270; consumer, 270
Bradford, Helen, 165, 168
Branch, Taylor, 256–57
Briggs, Cyril, 192
British Labor party, 171–72
Brotherhood of Sleeping Car Porters, 141, 177, 203, 233, 254
Brower, Earl, 221
Brown, John, 107, 317
Brown, Rap, 304
Brown, William Wells, 24
Brown vs. Board of Education, 237, 255
Bruce, John Edward, 33, 173
Buddhists, 60
Bukharin, Nikolai, 195
Bunche, Ralph, 203–4, 212, 215
Bunting, Bryan, 209
Bunting, S. P., 196–97
Burkett, Randall, 92
Bus boycotts, 237, 252, 253–56, 257, 260, 261, 262, 265, 267, 269, 270
Butelezi, Elias Wellington (aka Dr. Wellington), 166–67
Buthelezi, Chief Gatsha, 300
Buthelezi, Bishop Manas, 305, 306
Buxton, Travers, 131–32

Cachalia, Yusuf, 246
Calvin, John, 60
Campbell, James, 79
Cape African Voters' Convention, 199, 206
Cape Colony/Province: ANC in, 194; British administration of, 42; Coloreds in, 194; Ethiopianism/religion in, 85, 87; land issues in, 39, 40; Mfengu people in, 39; mining in, 40; nonviolent resistance in, 238, 246–47, 248, 270; political

parties in, 43; protest movements in, 87; and segregation, 95, 96, 104, 105, 207; suffrage in, 19–23, 36–56, 101, 105, 116, 205–6, 223, 245; traditional authority in, 104; and the UNIA, 162. *See also* Cape liberalism
Cape liberalism, 40, 47, 54, 205, 206; and Pan-Africanism, 148, 149, 150, 298; and segregation, 96, 121, 122–23, 131; and suffrage, 20, 21, 41, 43, 45, 52–53
Cape Province African Congress, 162
Capitalism, 16, 56, 67, 112, 139, 159, 170, 172; and Black Power, 296–97, 314; and communism/Marxism, 32–33, 186, 188, 191–92, 194, 195, 198, 202, 211, 222, 223, 224; and Marxism, 32–33; and nonviolent resistance, 266, 267; and Pan-Africanism, 150, 152; racial, 310; and segregation, 96–97
Carmichael, Stokely, 93, 293, 295–96, 297, 298, 304, 306–8
Carnavon, Margaret, 138, 139
Cell, John, 103
Chafe, William, 259
Charterists. *See* Congress Alliance; Freedom Charter
Chilembwe, John, 86
Chisholm, Shirley, 297
Christian Institute, 303
Christian perfectionism, 24
Christian redemption, 62, 64, 71–72
Christianity, 57, 60, 68
Church of England, 166
Church of God and Saints of Christ, 166
Churches. *See* Religion; Separatist churches; *specific denomination*
CIO (Congress of Industrial Organizations), 180, 204, 215, 220
Ciskei region, 20, 38
Civil disobedience, 125, 219, 308; Gandhian, 226, 228, 229; individual vs. mass, 229; King's views of, 272; in South Africa, 226, 237, 240, 241, 242, 246, 247, 249, 250, 251, 272, 286; in the U.S., 234, 235, 254; and World War II, 235. *See also* Strikes
Civil Rights Act (1866), 29
Civil Rights Act (1875), 31, 33, 77, 97
Civil Rights Acts (1964, 1965), 14, 236, 263, 273
Civil Rights Congress, 221
Civil Rights Movement: African influence on the, 252–53; and Black Power, 278; and black racism, 289; and class issues, 269–70; culture/ethos of the, 271–72; evaluation of the, 264–65; governmental resistance to the, 272; and the inevitability of winning, 273–74; influence on the ANC of the, 265; King's importance to the, 253, 260, 271; leadership of the, 268, 269–70, 273; and liberalism, 273; and "local movement centers," 261; and the media, 272; and nonviolent resistance, 231, 252–65; origins of the, 217, 253; and Pan-Africanism, 277; and religion, 261–62, 269, 271–72; structure/organization of the, 270; and suffrage, 268, 269; and violence/intimidation, 295; and whites, 273, 289, 292–93, 295. *See also specific person, organization, or event*
Civil service, 49–50, 126–27, 134–35
Civil War, 24, 28
Civilization: and African independence, 151; and capitalism, 67; characteristics of, 39–40; and Coloreds, 44; and education, 39–41, 49; and equal rights, 44, 56, 57; and liberalism, 43, 56; and Marxism, 184; and missions/missionaries, 40–41, 49; and Pan-Africanism, 149, 150–51; and political issues, 48–49; and property issues, 48; and racism, 15; and religion/Ethiopianism, 48, 56, 57, 67, 68; and self help, 49; and suffrage, 20, 39–41, 44–45, 48–49; universal standards for, 48; Western, as a model for blacks, 56, 151
Civilizationism, 69–74, 78, 79–80
Class issues, 265, 296, 310; and abolitionism, 25; and Black Consciousness, 312, 314; among blacks, 25; and the Civil Rights Movement, 269–70; and communism/Marxism, 184–85, 189–90, 203, 215, 284; and economic issues, 25, 32–33; and education, 25; and Ethiopianism, 91, 92; and ethnicity, 184; and Pan-Africanism, 144, 145, 150; and political issues, 32–33; and property issues, 25; and race issues, 27, 150, 169–70, 176–77, 184–85; and segregation, 96–97, 98, 123, 124, 125; and suffrage, 15, 19–20. *See also* Black middle class; Labor issues; Populism; *specific person or organization*
Clausewitz, Karl von, 226
Cleage, Albert, 304
Cleaver, Eldridge, 298, 304
Clergy, 87, 88, 303; and the black middle class, 75; and Black Power, 294, 304; and the Civil Rights Movement, 261–62, 269,

Clergy (*continued*)
271; discrimination against black, 80–81; and Ethiopianism, 74–80, 87, 92–93; and inferiority of blacks, 80–81; and nonviolent resistance, 261–62, 269, 270, 271; ordination of black/African, 41, 83, 86; in the Republican Party, 77; in the Southern United States, 77; and the UNIA, 92–93. *See also specific person*
Cleveland, Grover, 31
Colonialism, 151, 267, 305; and Black Power, 295–96; and civilizationism, 69; and communism/socialism, 184, 186–87, 188–89, 195–96, 203, 208, 217; and Ethiopianism, 69, 86–87
Colonization movement, 27, 62, 64–67. *See also* Emigration; Liberia; *specific person or organization*
Colored American Magazine, 50, 51, 53
Colored Press Association, 31
Coloreds: and African consciousness, 47; and the ANC, 194–95, 312, 322; and Black Consciousness, 47, 310–11; and black solidarity/pride, 47; and Cape liberalism, 47; and civilization, 44; and communism, 217; consciousness movement among, 46–47; and the definition of black, 162, 301–2; and Europeanization, 47; Great Britain's relationship with, 46–47; and the ICU, 194–95; leadership of, 47; and Pan-Africanism, 47, 281, 283; political activism of, 46–47, 194–95, 246; and political issues, 310–11; segregation of, 46, 246; and self-determination, 194–95; and suffrage, 37–47, 53, 224, 245, 246; and the UNIA, 162
Coloured Voters Act, 246
Comaroff, Jean, 90–91
Commission on Interracial Co-Operation, 147, 213
Committee Against Jim Crow in Military Service and Training, 235
Committee on Racial Equality (CORE), 232, 234–35, 236, 261, 265, 292, 295
Communism: and African independence, 192, 195, 196, 198, 200, 208–9, 273; and African socialism, 56; and alliances with other organizations, 179, 187, 196–97, 198–99, 202, 203–5, 207, 210; and Black Consciousness, 309, 311–12; black suspicions of, 189; and capitalism, 186, 188, 191–92, 194, 195, 198, 202, 211, 222, 223, 224; and class issues, 184–85, 189–90, 203, 215; and colonialism, 184, 186–87, 188–89, 195–96, 203, 208, 217; and Coloreds, 217; and consensus, 179–80; contributions of, 204, 210; and democratic centralism, 179; and economic issues, 200–201, 203, 204; and ideology, 185–89; and imperialism, 186, 187, 188, 189, 191–92, 194, 195, 198, 203, 208, 209, 214, 217; and Indians, 217, 283; interracial aspects of, 180–81, 210–11, 217; and labor issues, 180, 185, 186, 191, 196, 204, 209–10, 215, 216, 221; and land issues, 196, 201, 205; and leadership of local parties, 189, 191, 192, 195, 200, 212–13, 222; and liberalism, 200, 203, 205, 213, 214, 216, 217, 222, 223, 224; and localism vs. internationalism, 181–82, 193, 196–97, 199–200, 219–20; and manipulation of people/organizations, 179, 180, 181–82, 186–87, 266–67; and the March on Washington Movement (Randolph), 233; and militancy, 218–19; and nationalism, 194, 221, 266; and the New Deal, 202–3, 204, 205, 214; and nonviolent resistance, 219, 231, 245; and Pan-Africanism, 190, 192, 198, 279, 281–82, 286; party membership, 179, 180, 191, 201, 204, 211, 212, 215–16, 218–19; and political issues, 206, 207, 212–13, 220–21; and populism, 171, 179, 190; public agenda of, 180; purges in, 200; and racism, 179, 181, 183, 184, 185, 196, 213; and radicals, 180, 192, 200, 204, 215, 224; and religion, 201, 204, 211, 215, 220; in rural areas, 196, 201, 205, 209–10; and the Scottsboro boys, 201–2, 203; and segregation, 180, 185, 191, 193, 199, 203, 205, 206, 207, 220–21; and self-determination, 189–202, 203, 216; and separatism, 210, 224; and shifts in party strategy, 181–82, 192–93, 197–98, 199–200, 202, 203, 214–15, 221; and suffrage, 205–6, 218, 220, 221; suppression of, 183, 200, 221–22, 245, 246, 279, 281; and the United Front, 202–14; and urbanization, 220–21; U.S. fears of, 272–73; and violence/intimidation, 211, 252; and whites, 181, 189–90, 191, 192, 195, 196, 212–13, 215, 217, 219, 224; and World War II, 214–24. *See also* Marxism; *specific person or organization*
Communist Internationals, 187, 190, 194, 195, 202
Community organization, 270, 271, 313–14
Competition, and labor issues, 96, 102, 124, 185, 191

The Condition, Elevation, Emigration, and Destiny of the Colored People of the United States (Delany), 66
Cone, James, 298, 304–6
Congo, 74
Congregationalist Church, 75, 80, 81
Congress Alliance, 251–52, 268, 282, 283–84, 285
Congress of Democrats, 282, 283, 298
Congress of Industrial Organizations (CIO), 180, 204, 215, 220
Congress of the People, 283
Congress of South African Trade Unions (COSATU), 311
Congress Youth League. *See* Youth League
Congressional Black Caucus, 316
Consciousness movement: among Coloreds, 46–47. *See also* Black Consciousness
"The Conservation of Races" (Du Bois), 73, 117
Constitution League, 108
Constitution, South African: and the ANC, 312; and the SANNC, 121; and segregation, 101, 114–15, 116, 123, 125, 136; and the South Africa Act, 114–15; and suffrage, 116, 205, 245
Constitution, U.S., 125–26, 136, 272. *See also* Fifteenth Amendment; Fourteenth Amendment
Constitutions, state, 32
Consumer boycotts, 270
Cooperative commonwealths, 159, 160, 170, 171, 173
CORE. *See* Committee on Racial Equality
Cornish, Samuel, 24
Corruption, 21, 30, 33, 37, 167
COSATU (Congress of South African Trade Unions), 311
Council on African Affairs, 252–53, 278–79
Crime, 235
The Crisis (NAACP magazine), 113–14, 119, 133–34, 143, 148, 169, 253
Crummell, Alexander, 67–72, 73, 74, 75, 79, 301
The Crusader (ABB journal), 192
Cultural assimilation, 45–46
Cultural diversity, 71–72, 91
Cultural nationalism: and Black Consciousness, 308–9; and Black Power, 296, 297, 314; and missions/missionaries, 143; and Pan-Africanism, 113, 280; and religion, 61, 76; and segregation, 124, 132; and separatist churches, 76; and white supremacy, 143. *See also specific person or organization*
Cultural orientation, and nonviolent resistance, 230, 243–44

Dadoo, Yusuf M., 217, 238–40, 246
Darkwater (Du Bois), 151
Davis, Angela, 224
Davis, Jefferson, 72
Davis, John P., 203–4
Davis, John W., 215
Deacons for Defense, 295
Debs, Eugene, 184
"Declaration of Conscience" (King), 274
Declaration of Independence, 15, 32, 60
Defiance Campaign (1952), 241, 245–48, 249, 250, 252, 265, 270, 271, 272, 282
Delany, Martin R., 25–26, 28, 66–67, 75, 78, 290
Democratic centralism, 179
Democratic party: as anti-black, 30; and Black Power, 297; black support of the, 220, 223; and civil rights reform, 222; constituency of the, 98; failure of the, 292; and liberalism, 223; and segregation, 98, 127, 143; and suffrage, 17, 31, 32
Democratic socialism, 175, 212, 214, 215, 265, 268, 283, 292
Dennis, Eugene, 200
Depression, 200–201
Diagne, Blaise, 151
Dininzulu, Chief, 118
Direct action. *See* Nonviolent resistance
"Double consciousness" of blacks, 107
Douglass, Frederick: as an abolitionist, 24, 29, 48, 60, 66; as an assimilationist, 24, 54, 55–56, 71, 73, 76, 107; and black solidarity/pride, 26, 29, 55–56; and the Civil War, 28; death of, 34; and emigration, 26, 65; and equal rights, 19; and Ethiopianism/religion, 60, 76; influence of, 30–31, 34, 144; and integration, 24, 26, 71, 76; and patronage, 30–31; and Reconstruction, 29; and the Republican party, 29, 30–31, 54; and self-help, 29; and separatism, 24, 76; and suffrage, 18, 19, 24, 29
Draft resistance, 234, 239, 288
Drake, St. Clair, 61, 74–75, 92, 93
Du Bois, W.E.B.: and African independence, 151, 152, 154, 166; and black leadership, 106–11; and black solidarity/pride, 53; and capitalism, 150, 152; and class issues, 150, 152; and communism/Marxism, 184, 201, 211–12; and cultural nationalism, 113–14, 132,

Du Bois, W.E.B. (*continued*)
143; and the "double consciousness" of blacks, 107; and economic issues, 120, 143, 212; and emigration, 73–74, 151; Garvey's attacks on, 157; influence of, 47, 55, 113–14, 117, 119, 144, 148, 152, 169, 172, 265; influences on, 68, 72, 73; and integration, 107, 212; and labor issues, 201; and the NAACP, 112–14, 118, 119, 124–25, 143, 152, 158; and nationalism, 117, 287; and the Niagara Movement, 106–11, 124; and nonviolence, 150–51; and Pan-Africanism, 73, 74, 107, 113, 117, 132, 144–45, 149–52, 158, 172, 278–79, 284, 287; and philosophy of race, 72–74, 172; and racism, 150, 212; and romantic racialism, 150; and the SANNC, 119–20; and segregation, 105–14, 124, 127, 129, 132, 133–34, 143; and self-determination, 152; and separatism, 73, 107, 132, 133, 150, 212; and socialism, 152; and suffrage, 35–36, 48, 127, 143; Washington vs., 35, 36, 47, 51, 53, 105–11, 119, 129, 132, 146, 148; and World War I, 133–34; writings of, 105, 151, 169. See also *The Crisis*; Talented tenth

Dube, John R., 87–88, 115, 119, 120, 122, 130–32, 152

Due process clause, 129, 135

Duncan, Patrick, 299

Dutch Reformed Churches, 95

Dwane, Rev./Bishop James, 83, 85, 87

Dwanya, Daniel, 115

Economic issues, 16, 29, 110, 139, 165, 212, 264, 303; and class issues, 25, 32–33; and communism, 200–201, 203, 204; and the Du Bois-Washington controversy, 110; and the future of South Africa, 321, 322; and nonviolent resistance, 231, 236–37, 238, 261, 265, 268; and Pan-Africanism, 280, 283; and political issues, 32–33, 35–36; and racism, 32–33, 35; and the redistribution of wealth, 321; and the sanctions movement, 275; and segregation, 105, 106, 128, 129, 147, 148; and suffrage, 15, 16, 19. *See also* Capitalism; Entrepreneurs; Labor issues; Poverty; *specific person or organization*

Edgar, Robert, 167–68

Education: and black leadership, 40, 45, 109, 110, 122–23; and the Civil Rights Movement, 255, 264; and civilization, 39–41, 49; and class issues, 25; court decisions concerning, 222; and the Du Bois-Washington controversy, 110; and entrepreneurs, 120; and the future of South Africa, 321; and missions/missionaries, 40–41; and the NAACP, 133, 213; and nonviolent resistance, 231, 235, 236, 249, 255; and Pan-Africanism, 146; and SANNC, 119, 120; and self-help, 29; and separatism, 148; and sources of authority, 46; and student activism, 299; and suffrage, 15, 17, 18, 35, 38–49, 52. *See also* Industrial education; *specific person for views*

Eisenhower, Dwight D., 267

Elections: of 1856, 24; of 1948, 223–24, 237, 240, 244; of 1968, 297; of 1994, 319

Emigration: and African independence, 287; and black redemption, 61, 66; and the Civil War, 28; and civilizationism, 78; and nationalism, 25; and Pan-Africanism, 24, 151; and religion/Ethiopianism, 61, 65–66, 67, 74, 77, 78; and the Southern United States, 77–78; and suffrage, 26; and the UNIA, 154, 161, 175, 177; and violence/intimidation, 77–78. *See also* Colonization; *specific person*

Engels, Friedrich, 183

Entrepreneurs, 120, 270, 296; Fortune's views of, 32, 33; and the future of South Africa, 321; and segregation, 106, 128, 129; and self-help, 106, 109; Washington's promotion of, 15, 33, 35, 36, 106, 109, 128, 129, 160, 296

Episcopalian Church, 68, 75

Equal protection clause, 97

Equal rights, 22, 57. *See also* Suffrage

Equality, 14, 15–16, 18, 66, 71–72

Essien-Udom, E. U., 288

Ethiopia, 90, 206

Ethiopian Churches, 81–82, 83, 85, 86, 87, 165, 171

The Ethiopian Manifesto, Issued in Defense of the Black Man's Rights in the Scale of Universal Freedom (Young), 63–64

Ethiopian Orthodox Church, 90

Ethiopianism: and abolitionism, 62–63, 64; and "Africa for the Africans," 161; and African redemption, 68–74, 75–80, 84, 89; and American redemption, 74, 93; and the Bible, 61, 63, 75, 89, 92; and black advancement in the U.S., 75; and black redemption, 61, 73, 74, 76, 90; and black solidarity/pride, 68, 84, 89; and capitalism, 67; characteristics/meaning

of, 61–62; and Christian redemption, 62, 71–72; and civil rights, 71–72; and civilization, 67, 68; and civilizationism, 69–74, 79–80; and class issues, 91, 92; and the clergy, 74–80, 87, 92–93; and colonialism, 69, 86–87; and colonization, 62, 64–65, 66–67; contributions of, 88–89, 91–92; decline of, 88, 91; and democratic liberalism, 92; and emigration, 61, 65–66, 67, 74, 77, 78; and Hegelian idealism, 72; and human solidarity, 72; impact of, 114; and imperialism, 69, 79, 80; and inferiority of blacks, 63, 78; and integration, 65, 73, 93; and Marxism, 72–73, 92; and materialism, 71, 72–73, 74; and millennialism, 168; and missions/missionaries, 62, 67, 74–81, 83, 85, 88, 90; and nationalism, 63, 66–67, 83–84, 93; in the nineteenth century, 61–74; and Pan-Africanism, 63, 67, 68, 71, 73, 74, 82, 83, 84–85, 88, 92, 147; and paternalism, 80, 85; and political activism, 84–88, 90–93; popular, 74–80; and progress, 71–72, 73; and protest movements, 63–64, 86, 87, 89, 91–92; Providential Design theory of, 75; and racial purity, 71–72; and radical Africanism, 89; and religion, 59; revision of, 73; and secularization, 91, 92; and self-help, 64, 66–67, 76, 84, 86; and separatism, 61, 65, 73, 76–80, 86–87, 88, 89–90; and slavery, 63, 76, 79; sources for, 61; and the Swedenborgians, 62–63; and violence/intimidation, 77–78, 86; and whites, 74, 84–85, 86, 87, 90–91. *See also* Romantic racialism; *specific organization*
Ethnicism: and Black Power, 297, 308, 314; and class issues, 184; and democracy, 72; and equality, 15–16, 71–72; and nationalism, 183–84; and populism, 139; and republicanism, 66; and socialism, 183–84; and suffrage, 15–16, 22, 23
Europeanism: and Pan-Africanism, 280, 281
Europeanization, 41, 47, 56
Europeans: and the ANC, 321; innate characteristics of, 70
Evangelicalism, 61

Fair employment practices, 234, 235, 236–37
Fairclough, Adam, 264
Fanon, Frantz, 252, 296
Fard, W. D., 287–88
Farmer, James, 232
Farrakhan, Louis, 322
Federation of South African Women (FSAW), 250
Fellowship of Reconciliation, 232, 233, 256
Ferris, William H., 156, 173
Fifteenth Amendment: enforcement of the, 36, 97; and the "grandfather clauses," 135; mentioned, 143; and the NAACP, 112; and the Niagara Movement, 108; passage of the, 14–15, 19; and segregation, 97, 99, 108, 135, 220; and the South Africa Act, 115; and suffrage, 14–15, 18, 19, 29, 36, 55
First, Ruth, 317
Fischer, Bram, 317
Fisk Jubilee Singers, 82
"Force Bill" (1890), 32, 34, 36
Ford, James, 218
Forman, James, 297, 304
Fort Hare Native College, 281
Forten, James, 24
Fortune, T. Thomas: and the Afro-American League, 51; as assimilationist, 71; and civilization, 55–56; as editor, 31, 50, 51, 56; on education, 33–34, 51; influence of, 144; and integration, 71; and political independence, 31, 32–33, 34, 35, 54, 56; and suffrage, 31, 33, 51, 52–53; writings of, 32–33
Fourteenth Amendment, 29, 112, 143; due process clause of the, 129, 135; and the Niagara Movement, 108; and segregation, 97, 99, 108, 127, 128, 129, 135; and the South Africa Act, 115
Franchise Action Council of the Cape, 245
Franchise and Ballot Act (1892), 43, 46
Free blacks, 23–36
Freedom Charter (Congress Alliance, 1955), 251, 268, 282, 283, 300–301, 311
Freedom Ride (1961), 261
Freedom Summer (1964), 295
FRELIMO, 303, 307
"Friends of the Natives," 20, 43
"Friends of the Negro," 294
Friends of the Soviet Union, 219
Fugitive Slave Act (1850), 64, 66
Fuze, Magema M., 144–45

Gandhi, Mohandas K.: and direct action, 227–28, 229, 234, 258; influence of, 226, 229, 233, 234, 235, 238, 239, 246, 253, 256; and King, 255–63; as leader, 228, 230–31, 251, 271; and Niebuhr, 256–59; and passive resistance, 125, 225–41, 255–

Gandhi, Mohandas K. (*continued*) 59, 271; and *saryagraha*, 228–29, 232, 243; and South Africa, 229, 238–41, 243, 246, 265
Garnet, Henry Highland, 24, 65–66, 67–68, 75
Garrison, William Lloyd, 24, 61, 107–8, 111, 317
Garrow, David, 258
Garvey, Marcus: and the African Blood Brotherhood, 192; on black leadership, 157, 158; as a charismatic figure, 154, 160, 175, 194; deportation of, 160; heroes of, 161; and honest prejudice, 157; imprisonment of, 160, 175; influence of, 138, 162–64, 172, 176, 265; influences on, 153; and the NAACP, 153, 157, 160; personal/professional background of, 153; political naiveté of, 177; release from prison of, 164; Washington compared with, 35; on wealth, 159–60. *See also* "Africa for the Africans"; Universal Negro Improvement Association
Garvey movement. *See* Universal Negro Improvement Association
George, Henry, 32
Gerhart, Gail, 307
Ghana, 278
Glen Gray Act (1894), 43
Goduka, James, 87
Gomas, John, 206
Goodwyn, Lawrence, 138
Gow, F. M., 87
Grandfather clauses, 135
Great Britain, 82, 183, 214; abolition of slavery in, 16, 79; black disillusionment with, 45; black support for, 42, 43, 44; and Cape suffrage, 36–47; and class issues, 15; as a colonial power, 79–80; Colored disenchantment with, 46–47; and Ethiopianism, 79–80; and India, 184, 226, 228, 231; liberalism in, 115, 130–31, 174, 223; and racism, 15, 153; and segregation, 100, 115, 125, 130–31, 134; and suffrage issues, 16–17, 44–45; U.S. compared with, 16–17, 53, 54–55
Group Areas Act, 240, 245, 246
Grumbs, J. H., 169
Gumede, J. T., 164, 195, 199

Hacker, Andrew, 319
Hall, Haywood. *See* Haywood, Harry
Hall, Otto, 193
Hamilton, Charles V., 295, 306–8
Harlan, Louis, 118, 120
Harris, Abram, 212
Harris, Frederick John, 298–99
Harris, John, 131
Hartal, 242, 245
Harvard University, 66, 72, 109
Haywood, Harry, 192–98, 212, 221
Hegel, Georg Wilhelm Friedrich, 72, 74, 258
Herndon, Angelo, 211
Hertzog, J.B.M., 148, 167, 191, 205, 214, 223
Hill, Robert A., 163
History of the Negro Race in America (G. W. Williams), 34, 49
Hitler, Adolf, 202, 214–15, 217, 218, 219, 239
Hofmeyr, Jan H., 37, 38, 41, 43
Hofmeyr, Jan H., Jr., 223
Holiness movement, 90, 93
Howard University, 53, 292, 295
Hudson, Hosea, 211
Huggins, Nathan, 260
Hughes, Langston, 199
Human redemption, 93
Human solidarity, 57–58, 72, 284

ICU. *See* Industrial and Commercial Workers' Union of Africa
"If We Must Die" (McKay), 315
ILD (International Labor Defense), 201–2, 203
Imbumba Yama Nyama organization, 42
Imperialism: and Black Power, 295, 314; and capitalism, 186; and communism, 186, 187, 188, 189, 191–92, 194, 195, 198, 203, 208, 209, 214, 217; and Ethiopianism, 69, 79, 80; and nonviolent resistance, 267; and Pan-Africanism, 150, 152
Imvo Zabatsundu ("Native Opinion"), 42–43, 44, 49
Independent black churches. *See* Separatist churches
Independent Industrial and Commercial Workers' Union of Africa (IICU), 172
India, 184, 187, 189, 226, 227, 228, 231, 249, 258, 275. *See also* Gandhi, Mohandas K.
Indian National Congress, 121, 230
Indian Representation Act, 240
Indians: and the ANC, 312, 322; and Black Consciousness, 310–11; and communism, 217, 283; and the definition of black, 162, 301–2; and nonviolent resistance, 226–27, 229, 230–31, 237–41, 245–46,

248, 253; and Pan-Africanism, 277, 280, 281, 282, 283; and political issues, 310–11; and World War II, 217
Industrial and Commercial Workers' Union of Africa (ICU): accomplishments of the, 177; aims of the, 175–76; and the ANC, 173; and black solidarity/pride, 162; branches of the, 195; and the British Labor party, 171–72; and capitalism, 172; and class issues, 169–70, 174, 175–76; and Coloreds, 194–95; and communism, 169, 170, 171, 175, 191, 195, 199; corruption in the, 171; decline/demise of the, 171–72, 174–75; and economic issues, 173, 174; and Ethiopianism, 176; and labor issues, 168–72, 175–76; leadership of the, 162, 175; and liberalism, 171; and localism, 172; membership of the, 171; and millennialism, 171; and nationalism, 138, 172, 175; and Pan-Africanism, 138, 149, 169, 172; as a populist movement, 138; and race issues, 169–70; repression of the, 174–75, 177; and segregation, 171, 172, 174, 177; and self-determination, 172; and self-help, 172; and separatism, 172; and socialism, 168, 170, 171; spread of the, 171; and the UNIA, 168–72, 173–78; and whites, 171–72, 175, 176
Industrial education, 35, 36, 51, 105, 109, 120, 145, 146
Inferiority of blacks, 18, 27, 290, 313; black belief in the, 302; and the definition of black, 301–2; missionaries' views of, 39–40; and religion/Ethiopianism, 57–58, 63, 78, 80–81, 305. *See also* Racism
Integration: and abolitionism, 26–27; and black solidarity/pride, 25–27; and liberalism, 23; and nationalism, 25–27, 55–56; and the Niagara Movement, 106–7; and religion/Ethiopianism, 58, 59, 60, 65, 73, 93; and South Africa, 58; and suffrage, 22. *See also* Segregation; *specific person*
Intermarriage, 135, 157
International Conference on the Negro (1912), 145
International Labor Defense (ILD), 201–2, 203
Ireland/Irish, 27, 183–84, 231
Islam, 68, 71. *See also* Nation of Islam
Israelite sect, 166, 167
Isserman, Maurice, 218
Izwi Labantu ("Voice of the People"), 43, 49, 50

Jabavu, Davidson Don Tengo, 147–49, 150, 206
Jabavu, John Tengo: and African protest politics, 42–44, 45, 49, 149; background of, 42; and civilizationism, 48, 49, 55–56; and segregation, 121; and self-help, 49; and suffrage, 43–44, 48; and unity, 54, 115
Jefferson, Thomas, 17, 64
Jesus Christ, 79, 232, 304, 305, 306
Jews, 35, 304
Jim Crow: and the Civil Rights Movement, 255, 264; and communism, 193, 205, 221; geopolitical context of, 272–73; legislation banning, 260; NAACP campaign against, 203; and nonviolent resistance, 231, 255, 265; and Pan-Africanism, 140, 148; and public consensus, 260; and scientific racism, 213; and World War II, 218. *See also* Segregation; Suffrage
Johnson, James Weldon, 125, 135, 140–41, 232
Johnson, Lyndon B., 273, 292
Joint Councils of Europeans and Natives, 147, 148, 149, 163, 173, 213
Journey of Reconciliation (1947), 235

Kadalie, Clements, 138, 161, 168–72, 175, 176, 177–78, 191, 195
Kapur, Sudarshan, 232
Karenga, Ron, 296
Katz, Bob (aka N. Nasanov), 193–94
Kelley, Florence, 111
Kelley, Robin, 201, 211
Kennedy, John F., 253, 273, 290
Khaile, Eddie, 191, 199
Khoisan people, 19
King, Kenneth, 146
King, Martin Luther, Jr.: and the "beloved community," 258, 264, 293, 312, 317, 322–23; and the Birmingham campaign, 258; and Black Power, 291, 292, 293, 297; and Christianity, 59, 93, 262–63, 271–72, 304; on civil disobedience, 272; contributions of, 272; and democratic socialism, 265, 268, 292; and "Free at last!," 320; and Gandhi, 255–63; importance to Civil Rights Movement of, 253, 260, 271; influence in South Africa of, 265, 274–75, 276; as a leader, 228, 251, 253, 254, 255, 262, 263, 273; and the Montgomery bus boycott, 253, 254–55, 260; as a Negro Gandhi, 256–57, 260; and Niebuhr, 257–59; Nobel Prize for,

King, Martin Luther, Jr. (*continued*) 274; and nonviolent resistance, 228, 252–65, 271–75, 291, 293, 304; on poverty, 292; and racial reconciliation, 295; and the SCLC, 258, 259–60, 261, 262; on self-defense, 263; and separatism/universalism, 59; and suffrage, 268; writings of, 256, 258
Kotane, Moses, 208–9, 212–13, 218, 222, 245
Kruger, Paul, 83
Ku Klux Klan, 30, 154, 160, 174, 175

La Guma, James, 194–95, 197
Labor issues, 56, 159, 177, 309; and the black middle class, 141; and black migration, 142; and communism/socialism, 180, 185, 186, 191, 196, 204, 209–10, 215, 216, 221; and competition, 96, 102, 124, 185, 191; and an interracial labor movement, 141; and militancy, 140–41; and nonviolent resistance, 233, 234, 236, 270; and the pass laws, 141, 142; and populism, 140–41; and poverty, 292; and racism, 140, 141–42, 169–70, 176, 184, 185, 196, 234, 236; and repression of labor, 141, 142, 185; and segregation, 100, 101, 102, 103, 116, 120, 125, 128, 141–42, 169–70, 185, 191; and separatism, 170; and strikes, 102, 116, 140–41, 142, 168, 175, 185, 195, 218, 221; and urbanization, 138, 140, 142; and violence/intimidation, 142; and World War I, 140. *See also* Rural areas; *specific person or organization*
Labor tenancy, 102
Land issues, 29, 39, 40, 165; and communism, 196, 201, 205; and the future of South Africa, 321; and segregation, 99, 100, 101–2, 104, 120, 126, 127–28, 130–32, 134, 136. *See also* Property issues; *specific legislation*
Lawson, James, 261
Leadership: and black solidarity/pride, 45–46; charismatic, 231; of the Civil Rights Movement, 268, 269–70, 273; and the Du Bois-Washington controversy, 106–11; and education, 40, 45, 109, 110, 122–23; and elevation of the masses, 25; emergence of African, 116; need for African, 124; and nonviolent resistance, 230–31, 259; and Pan-Africanism, 149; and racial purity, 157, 158; secularization of, 92; and the "slave culture," 144; and sources of authority, 46; and the training of leaders, 120. *See also* Niagara Movement; Talented tenth; *specific person or organization*
League of African Rights, 199
League Against Imperialism, 195
League of Struggle for Negro Rights, 199
Legal system, 50, 255
Legassick, Martin, 199
Lembede, Anton, 243, 280–81, 284
Lenin, V. I., 183, 185–89, 193, 198, 208, 211
Lester, Julius, 291, 294
Letter from Birmingham Jail (King), 258
Levison, Stanley, 266
Lewis, David Levering, 72–73, 107
Lewis, John, 268
Lewis, John L., 180, 215
Lewis, William H., 110
Liberal party, 298–99
Liberalism, 30, 56, 92, 96, 121, 139, 165, 237; assumptions of, 15; and Black Consciousness, 301, 302, 303; and Black Power, 292, 293, 294, 296, 297; and the Civil Rights Movement, 273; and communism, 200, 203, 205, 213, 214, 216, 217, 222, 223, 224; and cultural assimilation, 46; and the Democratic party, 223; in Great Britain, 115, 130–31, 174, 223; and human solidarity, 284; and integration, 23; and the New Deal, 213, 214; and Pan-Africanism, 137–38, 145, 147, 152, 281, 284, 299; and poverty, 292; and racial/ethnic pluralism, 147; and racism, 14, 15, 18, 22–23; and segregation, 96, 130–31, 134, 174, 223; and separatism, 174; and suffrage, 14–23, 38, 43, 49, 223, 224; view of history of, 72; and violence/intimidation, 298–99. *See also* Cape liberalism
Liberia, 65, 67, 68, 78
Liberian Exodus Joint Stock Company, 78
Lincoln, Abraham, 42, 111
Lincoln, C. Eric, 59
Literacy, 33, 48, 99
Little, Malcolm. *See* Malcolm X
Lodge, Henry Cabot, 31–32, 34
Lodge, Tom, 220, 247, 274
Loyalty, 18, 91
Lutheran Church, 81
Lutuli, Chief Albert, 247, 251, 268, 274, 283
Lynch, Hollis, 70
Lynching, 51, 142, 154, 161, 203, 205, 213, 221. *See also* Violence/intimidation

M Plan (ANC), 250–51
McAdoo, Orpheus M., 82

McCarthyism, 183, 236–37, 245, 279
McKay, Claude, 315
Mahabane, Z. R., 162–63, 164
Makgatho, S. M., 132, 163
Malcolm X (aka M. Little), 259, 289–91, 293–94, 298, 301, 302, 304
Mamiya, Lawrence H., 59
Mandela, Nelson: and ANC, 245, 250, 274, 275, 282, 319, 320; and Black Consciousness, 281–82, 310, 312, 317; and Christianity, 243–44; and the Defiance Campaign, 246; on democracy, 273; imprisonment of, 322; M Plan of, 250; and nonviolent resistance, 248, 273, 274, 275–76; release from prison of, 275, 276, 319
Mangena, Alfred, 116–17
Manifesto of 1948 (Youth League), 281
Manye, Charlotte, 82
Mapikela, Thomas, 115, 130–31
Marable, Manning, 120, 296, 322
March in Mississippi (1966), 295
March to Montgomery, 263
March on Washington (1963), 263, 268, 289–90
March on Washington (1968), 292
March on Washington (Scottsboro Boys), 231
March on Washington Movement (Randolph), 216–17, 218, 221, 222, 232, 233–34, 235, 242, 254, 265
Marks, J. B., 206, 218, 221, 222, 245, 246
Marks, Shula, 104
Marx, Karl, 183–84, 185–86, 189, 211, 239
Marxism: and Black Consciousness, 312; and black liberation leadership, 137–38, 178, 296, 312; and Black Power, 296; and capitalism, 32–33, 296; and class issues, 284; and colonialism, 184; and Ethiopianism, 72–73, 92; and nationalism, 137; and Pan-Africanism, 280; and populism, 139; and racism, 183; revision of, 185–87; and revolutionary nationalism, 296; and segregation, 97; as a Western materialist philosophy, 280; and whites, 318. *See also* Communism; Marx, Karl; Socialism; *specific person or organization*
Mass Democratic Movement, 276
Massachusetts Antislavery Society, 18
Master-Servants Law (1856), 38
Materialism, 71, 72–73, 74, 93
Matthews, E. K., 267
Matthews, Joe, 267
Matthews, Z. K., 282
Mbeki, Gavin, 310
Mda, A. P., 243
Media: and Black Power, 315; and the Civil Rights Movement, 272
Meier, August, 231
Meredith, James, 293
The Messenger (magazine), 169–70
Methodist Churches, 58, 80, 81, 93
Mfengu people, 39
Mgijima, Enoch, 86, 166, 167
Middle class: and Black Consciousness, 309; and the Civil Rights Movement, 269–70; and class issues, 125; and the clergy, 75; and communism, 203, 215; comparison of African and African American, 122; and corruption, 30; and the Du Bois-Washington controversy, 110; and the ICU, 174; and labor issues, 141; and the NAACP, 112, 140, 141, 152; and nonviolent resistance, 237; and Pan-Africanism, 152; and populism, 139; and Reconstruction, 29; and segregation, 97, 99, 105, 106, 110, 122–23, 127, 129; and the UNIA, 173. *See also* Class issues
Milholland, John, 108
Militancy: and communism, 218–19; and the Du Bois-Washington controversy, 106–14; and labor issues, 140–41; and segregation, 99–100. *See also specific person or organization*
Military service: and black troops in the Civil War, 28; and draft resistance, 234, 239, 288; and segregation, 126, 133, 134, 135–36, 217, 235–36; and suffrage, 52
Mill, John Stuart, 15
Millennialism, 28, 162, 164–66, 171, 262. *See also* Nation of Islam
Miller, Kelly, 53, 110, 134–35
Millner, Steven M., 269
Miscegenation, 70, 135, 157
Missions/missionaries: abolitionists compared with, 39; and African independence, 44; African withdrawal from, 80–81, 164; black ambivalence toward, 67; and Black Consciousness, 305–6; and black redemption, 313; as a cause of dissension in Africa, 83; and civilization, 40–41, 49; and colonialism, 305; and cultural nationalism, 143; and education, 40–41; and Ethiopianism, 62, 67, 74–81, 83, 85, 88, 90; and the International Conference on the Negro (1912), 145; and millennialism, 165, 168; and paternalism, 80, 85; and political issues, 41; and racism, 41, 305; and

Missions/missionaries (*continued*)
segregation, 100; and sources of authority, 46; in the Southern United States, 76; and suffrage, 38–40
Mississippi, March in (1966), 295
Mnika, Alfred, 172
Mofutsanyana, Edwin, 206, 209, 213
Mokgatle, Naboth, 248
Mokone, Mangena, 81–83
Molema, Silas M., 123–24, 125
Montgomery, Alabama: bus boycott in, 237, 253–56, 257, 260, 261, 262, 265, 267, 269; March to, 263; nonviolent resistance in, 270
Montsioa, George, 116–17
Moore, Basil, 303
Moral Man and Immoral Society (Niebuhr), 257
Moroka, J. S., 244, 246
Morris, Aldon, 261
Moses, Wilson Jeremiah, 69
Moskowitz, Henry, 111
Moton, Robert R., 147–48
Mozambique, 303, 307
Msane, Saul, 130–31, 132
Msimang, Richard, 116–17, 230
Mtimkulu, Abner, 87, 88
Muhammed, Elijah (aka E. Poole), 288–90, 304
Mulattoes, 70, 157, 301
Muste, A. J., 259
Mvabaza, Levi, 140–41, 143
My Life and the ICU (Kadalie), 169
"My View of the Segregation Laws" (Washington), 129
Mzimba, P. J., 49, 80–81, 86, 87

NAACP. *See* National Association for the Advancement of Colored People
Naicker, G. M. "Monty," 239–40
Naidoo, H. A., 217
Nasanov, N. (aka Bob Katz), 193–94
Natal: British administration in, 42; Ethiopianism in, 81, 84, 85, 86, 87; and the ICU, 171; Indians in, 239, 241; nonviolent resistance in, 226, 240; and segregation, 95, 104; traditional authority in, 104; Zulu rebellion in, 86
Natal Indian Congress, 229, 239–40
Nation: definition of, 187–89
Nation of Islam, 287–92, 322
National Association for the Advancement of Colored People (NAACP): accomplishments of the, 134–35, 142; as an all-black organization, 210; ANC compared with the, 266; and Black Power, 294; and black solidarity/pride, 322; and the Civil Rights Movement, 263–64; and class issues, 112, 139–40, 141, 142, 143, 152; and communism, 180, 198, 199, 201, 203, 204, 210, 221, 266; conventions of the, 119; and cultural nationalism, 113–14; and economic issues, 112, 142, 204; and education, 213; founding of the, 112; and Garvey, 153, 157, 160; influence of the, 174, 253; and integration, 210, 212; and King, 274; and labor issues, 112, 140, 141, 142; leadership of the, 112–14, 122, 124–25, 138–39, 143, 152, 173, 262, 266; legalistic reformism of the, 125–26, 135–36, 142, 237, 263–64; and liberalism, 108, 112, 115, 137–38, 143, 266; as lobbyists, 316; and local movements, 262, 271; and the March on Washington, 216; membership of the, 112, 124–25, 140, 159, 219; and nonviolent resistance, 260; and Pan-Africanism, 151, 152; Plaatje's address to the, 163; platform/programs of the, 112, 124–25, 141, 180; and political parties, 143; and populism, 138–39, 140, 141, 142, 143; protest orientation of the, 121; and racial purity, 157, 158; and radicals, 113; repression of the, 262, 271; SANNC compared with the, 121–23, 124–26, 136; and the Scottsboro boys, 201, 203; and segregation, 112–14, 125–26, 127, 129, 134–35, 213, 237, 263–64; and self-determination, 210; and suffrage, 112, 135, 213, 264; vindication of the, 222–23; and violence/intimidation, 112; and whites, 108, 112, 113, 143; and World War I, 133–34; and World War II, 218. See also *The Crisis*; *specific person*
National Equal Rights League, 133
National Forum, 311
National Independent Political League, 127
National Negro Academy, 73
National Negro Committee, 112, 115
National Negro Congress, 203–4, 214, 215, 216, 218, 233
National Union of South African Students (NUSAS), 299, 300, 311
National Urban League, 112, 159, 216, 294
Nationalism: and abolitionism, 26–27; African-American, 287; and African redemption, 84; antecedents of, 24; and Black Consciousness, 302, 304; and Black Power, 278, 296, 297, 314; and black solidarity/pride, 144; and

communism/Marxism, 137, 194, 221, 266; and emigration, 25; ethnic, 183–84; and integration, 25–27, 55–56; and millennialism, 165; and nonviolent resistance, 227, 251; and Pan-Africanism, 280, 285, 286, 312; and populism, 161; and protest movements, 25–26; and religion/Ethiopianism, 59, 63, 66–67, 83–84, 93, 243–44, 304; revival of, 287, 304; revolutionary, 296, 314; and segregation, 117, 118; and self-help, 66–67; and separatism, 56, 288, 296; and suffrage, 22; weakness in U.S. of black, 279; and Zionism, 137. *See also* Cultural nationalism; Pan-Africanism; Self-determination; Separatism; *specific person or organization*
Native Administration Act (1917/1927), 126, 132, 136, 175
Native Life in South Africa (Plaatje), 131
"Native republic" thesis, 192–93, 194, 200, 208
Native Segregation, 97, 103–4, 114
Natives' Land Act (South Africa, 1913): and land distribution, 321; and populism, 161, 165; and segregation, 101–2, 104, 116, 120, 126, 128, 130–32, 134, 136, 185
Natives' Representative Council, 205, 207, 220, 241, 242
Naude, Beyers, 303
The Negro (Du Bois), 150
Negro Business League, 106, 154
Negro Commission, 194, 195, 197
Negro Factories Corporation, 159
Negro Sanhedrin, 190
Negro World (UNIA newspaper), 56, 154, 156, 162
Negro-folk theologians, 75
New Deal, 136, 202–3, 204, 205, 213, 214
New Kleinfontein, 89
New York Age (newspaper), 50, 51, 52
Newton, T. Huey, 296, 297, 304, 315
Ngcayiya, Abner, 87–88
Ngubane, Jordan, 299
Niagara Movement, 21, 36, 53, 106–11, 112, 121, 122, 124, 318
Niebuhr, Reinhold, 226, 256–59
Niobe, Bransby, 209–10
Nixon, E. D., 254, 262
Nixon, Richard M., 292, 296
Nkrumah, Kwame, 278, 279, 284, 286
Non-European United Front, 217
Non-European Unity Movement, 248
Nonviolent resistance: and abolitionism, 231; and African rights, 237–52; and *ahimsa*, 227; and Black Power, 295; and black solidarity/pride, 233–34, 259; and capitalism, 266, 267; and the Civil Rights Movement, 252–65; and class issues, 237, 265; as coercion, 226, 228–29, 243, 257, 258, 259; and colonialism, 267; and communism, 231, 245; comparison of African and African American, 265–76; and conversion, 226, 243, 258; and cultural orientation, 230, 243–44; decline of interest in, 265; definition of, 225; as a desperate measure, 238; as direct action, 229; and draft resistance, 239; and economic issues, 231, 236–37, 238, 261, 265, 268; examples of, 225, 228, 231, 234, 235, 242–43; function of, 225–26, 227; and the Gandhian tradition, 125, 164, 225–37; and imperialism, 267; and labor issues, 233, 234, 236, 270; and leaders vs. followers, 229; and leadership, 230–31, 239; and liberalism, 237; and local mobilization, 270, 271; and nationalism, 227, 251; and Pan-Africanism, 150–51; and poverty, 292; and power, 258; pragmatic aspects of, 227–28, 229; and property issues, 236, 239; and radicals, 238, 239; and religion, 227, 228, 229, 231, 243–44, 247, 251, 256, 260, 261–63, 270; repression of, 230, 247, 249, 265, 271, 275; and *satyagraha*, 227, 228, 229, 232, 235, 236, 243, 247, 250, 251, 257, 258, 261; and segregation, 231, 235–36, 237, 238, 239, 240, 242, 244–45; and self-defense, 263; and self-determination, 242, 261; and self-help, 233; as a success, 276; and suffrage, 268; tactics of, 228; as a total way of life, 227; and white fears, 260; and women, 250; and World War II, 235, 239. *See also* Civil disobedience; *specific person, organization, or event*
Notes on Virginia (Jefferson), 64
NUSAS (National Union of South African Students), 299, 300, 311
Nyasaland, 86

Odendaal, Andre, 45, 114
Ohlange Institute, 119, 120
Orange Free State, 21, 46, 50, 85, 116, 130, 141, 171
Orange River Colony Native Congress, 114
"Order of Ethiopia," 87
Organization of Afro-American Unity, 290
Organized labor. *See* Labor issues
Ovington, Mary White, 111, 113
Ownership, of South Africa, 300–301

Padmore, George, 279, 283, 284, 286
Pan-African Association, 149–50
Pan-African Congress, 47, 74, 145, 150–51, 152
Pan-Africanism: and African independence, 151, 152, 174, 206–7, 277, 280, 281, 282; antecedents of, 41; and apartheid, 277; and Black Consciousness, 301, 302; and black redemption, 146–47; and black solidarity/pride, 47, 144, 145–46, 174; and capitalism, 150, 152; and the Civil Rights Movement, 277; and civilization, 149, 150–51; and class issues, 144, 145, 150, 152; and colonialism, 151; and Coloreds, 47; and communism/Marxism, 190, 192, 198, 279, 280, 281–82, 286; conservative, 145–46, 148, 149; and cultural nationalism, 113; decline of, 278; definition of, 150; and economic issues, 280, 283; and education, 146; elite, 143–52; and emigration, 24, 151; and Europeanism, 280, 281; and imperialism, 150, 152; international, 149–52; and leadership, 149; liberal reformist, 145, 149–52, 158; and liberalism, 137–38, 145, 147, 152, 281, 284, 299; and materialism, 71; and multiracialism, 277–78, 282; mystical, 280; and nationalism, 280, 285, 286, 312; and the Niagara Movement, 107; and political issues, 114, 145, 280; and populism, 145, 152–61; and racism, 147, 150; and radicals, 281, 287; and religion/Ethiopianism, 59, 63, 67, 68, 71, 73, 74, 82, 83, 84–85, 88, 92, 147; and romantic racialism, 147, 150; and secularization, 92; and segregation, 132, 147, 148, 149; and self-determination, 152, 192; and separatism, 148–49, 150; and the slave culture, 144–45; and socialism, 152; in South Africa, 132–33, 277–86; and suffrage, 50, 148–49; and violence/intimidation, 150, 285; and whites, 151–52, 281, 282, 283, 284, 285, 301; and World War II, 277. *See also* Pan-Africanist Congress (PAC); *specific person or organization*
Pan-Africanist Congress (PAC): and the ANC, 284–86, 299, 311, 316, 318, 322; banning of the, 286, 298; and Black Consciousness, 298, 300, 301, 310, 311, 316; and Cape liberalism, 298; and Coloreds, 283; and communism, 282–83; conferences of the, 251, 271–72, 278, 283; decline of the, 311; as defunct, 310; and Indians, 283; leadership of the, 285–86; and the meaning of race, 284–85; and the meaning of South Africa, 285; militancy of the, 284–85, 300; and multiracialism, 312; and self-determination, 283; and terrorism, 286; and whites, 285, 299. *See also* Poqo
Pan-Negroism. *See* Ethiopianism; Pan-Africanism
Parks, Rosa, 254
Parliamentary Voters Registration Act (1887), 43
Pass laws, 141, 142, 230, 242, 246, 250
Passive Resistance Campaign (1946), 240, 241, 243, 246
Passive resistance. *See* Nonviolent resistance
The Past and Present Condition, and the Destiny of the Colored Race (Garnet), 65
Paternalism: acceptance of, 41; of African Americans toward Africans, 85; and Black Consciousness, 311; black fears of, 316; and communism, 210, 213; and the inferiority of blacks, 302; and liberalism, 54, 273; and missions/missionaries, 80, 85; and religion/Ethiopianism, 57–58, 80, 85; and segregation, 96, 131
Paton, Alan, 298
Patronage, 30–31
Peasantry, African, 40, 102, 198
Peck, Jim, 293
Pentecostal Churches, 58, 88, 90, 93
People's Front, 207
Peregrino, F.Z.S., 47, 52
"Petition to God," 89
Phelps-Stokes Fund, 146, 147
Phillips, Wendell, 107
Philosophy and Opinions (Garvey), 156, 157
Pickens, William, 128
Pinchback, P.B.S., 50
Pirio, Gregory A., 163
Pityana, Barney, 304
Plaatje, Sol, 119, 122, 130–32, 152, 163
Plessy v. Ferguson (1896), 97, 127
Poe, Clarence, 127–28, 130
Political issues: and African independence, 49; and Black Consciousness, 303, 309–10; and civilization, 48–49; and class issues, 32–33; and Coloreds, 46–47, 194–95, 246, 310–11; and communism, 203–4, 206, 207, 212–13; and economic issues, 32–33, 35–36; and Europeanization, 41; and Indians, 310–11; and missions/missionaries, 41; and Pan-Africanism,

114, 145, 280; and religion/Ethiopianism, 59, 84–88, 90–93, 244; and segregation, 101, 136, 206, 207; and self help, 29, 49; and separatism, 101; and strikes, 274, 286; and suffrage as a means of structuring/organization, 36–37. *See also* Native Administration Act; Political parties; *specific person, organization, or issue*
Political parties, 43, 97–98, 143, 220–21, 296–97. *See also specific party*
Poll taxes, 205
Polygamy, 79, 90
Poole, Elijah (aka E. Muhammed), 288–90, 304
Popular Front. *See* United Front
Populism: anti-white, 149; and the black middle class, 139; and capitalism, 139; and communism/Marxism, 139, 179, 190; and economic issues, 139; failure of, 175; interracial, 176–77; and liberalism, 139; meaning/characteristics of, 138–39; and nationalism, 161; and Pan-Africanism, 152–61; and property issues, 139; and race/ethnicity, 139; and segregation, 98; and self-determination, 139; white, 159, 176. *See also specific organization, e.g.* Industrial and Commercial Workers' Union of Africa (ICU) or Universal Negro Improvement Association (UNIA)
Poqo, 286, 301
Porter, William, 19
Poverty, 291–92, 322
Powell, Adam Clayton Jr., 203
Prayer Pilgrimage (1957), 268
Presbyterian Church of Africa, 80
Presbyterian Churches, 75, 80, 81
Prester John, 62
Prisons, 235
Programme of Action (ANC), 242–43, 245, 250
Progress, 71–72, 73
Progressive party (South African), 43, 44
Progressive party (U.S.), 221, 223
Progressivism, 56, 111–12
Proletariat, 186, 187, 189–90, 196, 200, 208, 211–12
Property issues: and civilization, 48; and class issues, 25; and equality, 16; and the future of South Africa, 321; and liberalism, 16; and nonviolent resistance, 236, 239; and populism, 139; and Reconstruction, 29; and segregation, 128; and sources of authority, 46; and suffrage, 15, 16, 17, 18, 19–21, 23, 43, 46, 48, 52, 116. *See also* Land issues
Protest movements: and African pioneer efforts, 42; leadership of, 23–26, 27; literature of, 63–64; and nationalism, 25–26; as radical reformist, 22; and religion/Ethiopianism, 57–61, 63–64, 87, 89, 91–92; and self-help, 27; in the United States, 23–36. *See also specific movement, organization, person, or topic*
Protestant work ethic, 76
Public facilities/transportation: and the Civil Rights Movement, 263; legislation concerning, 31; and the NAACP, 112; and nonviolent resistance, 231, 235, 263; and segregation, 46, 94, 97, 99, 102, 105, 110, 112, 128, 129, 132, 135. *See also* bus boycotts
Purified National party, 214
Purvis, Robert, 24

Raboteau, Albert, 75
Race: and class issues, 27, 150, 169–70, 176–77, 184–85; and equality, 71–72; and the hierarchy of races, 70; and Leninist doctrine, 186–89; meaning of, 284–85; and populism, 139; and socialism, 183–85
"The Race Problem in America" (Crummell), 71
Race riots, 154, 174, 216, 235
Racial capitalism, 310
Racial cooperation movements, 147–48
Racial exclusivity, 293, 311. *See also specific organization*
Racial purity, 70, 71–72, 157–58, 162
Racism: and abolitionism, 27; and American redemption, 93; anti-racist, 278; and black solidarity/pride, 278; black version of, 283, 289; and blaming the victims, 27; and the Civil Rights Movement, 289; and civilization, 15; and communism/socialism, 179, 181, 183, 184, 185, 196, 213; death of, 319, 320; and economic issues, 32–33, 35; in Great Britain, 15, 153; and inferiority of blacks, 27; and labor issues, 140, 141–42, 169–70, 176, 184, 185, 196, 234, 236; and liberalism, 14, 15, 18, 22–23; and the meaning of race, 285; and missions/missionaries, 41, 305; and Pan-Africanism, 147, 150; and romantic racialism, 70; scientific, 70, 213; and segregation, 123; and self-help, 27; and separatism, 287; and slavery, 27–28; and

Racism (*continued*)
suffrage, 14, 15, 18, 19–20, 48–49, 224; Washington on, 35; and white populism, 176. *See also* Inferiority of blacks; Protest movements
Radical Africanism, 89
Radical Reconstruction, 28, 34, 48
Radical Republicans, 54, 95, 99, 112
Radicals: and African socialism, 56; and Black Consciousness, 301, 302, 303, 309; and communism, 180, 186, 192, 200, 204, 215, 224; and Ethiopianism, 89; as heroes, 318; and nonviolent resistance, 238, 239; and Pan-Africanism, 281, 287; suppression of, 245, 298. *See also specific person or organization*
Railroads. *See* Public facilities/transportation
Rampersad, Arnold, 109
Rand Rebellion, 185, 191
Randolph, A. Philip: and Black Power, 292, 297; and communism, 215, 236; and democratic socialism, 268; and the March on Washington, 216–17, 218, 232, 233–35, 237, 254; and the military, 234, 235–36; and the National Negro Congress, 203, 215, 216, 233; and nonviolence, 231, 232–37, 243, 254, 256; and the sleeping-car porters, 141, 177, 203, 233, 254
Reconstruction, 33, 34, 201, 231, 273; overview of, 28–30; Radical, 28, 34, 48; and segregation, 95, 96, 98; and suffrage, 36, 48; Third, 321–22
Reconstruction Acts (U.S., 1867, 1868), 17–18, 29, 51
Reddy, E. S., 252
Redeemers, 19, 31, 33
Redemption: of African Americans, 288; American, 74, 93; Christian, 62, 64, 71–72; human, 93; of whites, 306. *See also* African redemption; Black redemption
Reform Act (Great Britain, 1932), 17
Religion: and abolitionism, 57, 60; and African independence, 243–44; ambivalences in, 59, 60; and Black Consciousness, 303, 305, 314–15; and Black Power, 306; and Black Theology, 304–6; and the Civil Rights Movement, 261–62, 271–72; and civilization, 48, 56; and communism, 201, 204, 211, 215, 220; and cultural assimilation, 46; and cultural nationalism, 61, 76; and Ethiopianism, 59; and inferiority of blacks, 57–58; and integration, 58, 59, 60; and millennialism, 165–66; and nationalism, 59, 243–44, 304; and nonviolent resistance, 227, 228, 229, 231, 243–44, 247, 251, 256, 260, 261–63, 270; as an obstacle or stimulus for black liberation, 59–60; and Pan-Africanism, 59, 82, 147; and paternalism, 57–58; and political activism, 59, 84–88, 90–93; and power, 60; and protest movements, 57–61; and the role of the state, 60; and segregation, 95; and separatism, 60, 304; and slavery, 57, 58, 60–61; and suffrage, 38–40, 42. *See also* Christianity; Clergy; Ethiopianism; Islam; Missions/missionaries; Nation of Islam; Separatist churches; *specific sect or denomination*
Remond, Charles, 24
Republican party: and the black middle class, 29; and Black Power, 296–97; black support for the, 24, 30–31, 42; clergy in the, 77; constituency of the, 98; and corruption, 30; Fortune's views of the, 31, 33; and the NAACP, 143; and patronage, 30–31; and the post-Reconstruction period, 30–31, 33; and Reconstruction, 28, 29, 30; and segregation, 97, 98, 127; and suffrage, 18–19, 24, 31–32, 37–38; Washington's influence in the, 106. *See also* Radical Republicans
Republicanism, 17, 66
Residential/territorial segregation, 128–30, 132, 135, 136, 147, 222, 240, 264, 265, 291
"The Resurrection of the Negro" (Garvey), 157
Revolution: and Black Consciousness, 308; and suffrage, 22
Revolutionary nationalism, 296, 314
Rhodes, Cecil, 43, 44
Rhodes University, 299
Rich, Paul B., 147
Riots. *See specific event*
Robeson, Paul, 222, 279, 287
Robinson, Joanne, 254, 262
Romantic racialism: and civilizationism, 69–74; and moral superiority, 64; and Pan-Africanism, 147, 150; and progress, 73; and racism, 70; and religion/Ethiopianism, 62, 67, 69–74, 79, 93; and the UNIA, 158. *See also* Ethiopianism
Romanticism: and Black Power, 279
Roosevelt, Eleanor, 213
Roosevelt, Franklin D., 202, 213, 216, 234
Roosevelt, Theodore, 50, 106
Roux, Edward, 197

Roy, M. N., 187
Rubusana, Walter, 43, 45, 87, 115, 130–31
Rude, George, 156
Rudwick, Elliott, 231
Rural areas: and communism, 196, 201, 205, 209–10; and labor issues, 171, 196, 201, 205, 209–10; and segregation, 95, 130
Russian Revolution, 185–89
Rustin, Bayard, 232, 235, 236, 256, 257, 268, 292, 297

SACTU (South African Congress of Trade Unions), 274, 285
St. Louis, Missouri, race riots in, 154
Sanctions movement, 275, 314
SANNC. *See* South African Native National Congress
Sartre, Jean-Paul, 278
SASO. *See* South African Student Organization
Satyagraha, and nonviolent resistance, 227, 228, 229, 232, 235, 236, 243, 247, 250, 251, 257, 258, 261
Schreiner, William P., 115, 121
Scientific racism, 213
SCLC. *See* Southern Christian Leadership Conference
Scottsboro boys, 201–2, 203, 231
Secret ballots, 17
Secularization, 91, 92
Segregation: benefits of, 101, 103; and the black middle class, 97, 99, 105, 106, 110, 122–23, 127, 129; and black migration, 103; black mobilization against, 104–14; and black solidarity/pride, 110; and the Civil Rights Movement, 270; and the civil service, 126–27, 134–35; and class issues, 96–97, 98, 123, 124; and the clergy, 75; and communism/socialism, 96, 97, 180, 185, 191, 199, 203, 206, 207, 220–21; comparison of African and African American, 123–24; and competition, 96, 102, 124; court decisions concerning, 97, 127, 237, 255; and cultural nationalism, 124, 132; de facto, 264; and the Du Bois-Washington controversy, 106–11; and economic issues, 96–97, 105, 106, 128, 129, 147, 148; and entrepreneurs, 106, 128, 129; and ethnic status, 95–96; failure of original, 104; and the Fifteenth Amendment, 97, 99, 108, 135, 220; and the Fourteenth Amendment, 97, 99, 108, 127, 128, 129, 135; and Great Britain, 100, 115, 125, 130–31, 134; implementation of, 101; and labor issues, 100, 101, 102, 103, 116, 120, 125, 128, 141–42, 169–70, 185, 191; and land issues, 99, 100, 101–2, 104, 120, 126, 127–28, 130–32, 134, 136; and leadership, 122–23, 124; legalization of, 94, 95, 97, 98–100, 101, 104, 105, 106, 108, 123, 134, 147; and liberalism, 96, 121, 122–23, 130–31, 134, 174, 223; macro-, 104; making of, 94–104; meaning of, 96; micro-, 104; and militancy, 99–100; and the military, 126, 133, 134, 135–36, 217; and missions/missionaries, 100; and nationalism, 117, 118; in the 1930s, 213; and nonviolent resistance, 231, 235–36, 237, 238, 239, 240, 242, 244–45; and Pan-Africanism, 132, 147, 148, 149; and paternalism, 96, 131; persistence of, 319; and political issues, 101, 136, 206, 207; and political parties, 97–98, 127, 143; and populism, 98; in prisons, 235; and Progressivism, 111–12; and property issues, 128; purpose of/reasons for, 103, 123–24; and racism, 123; and Reconstruction, 95, 96, 98; and religion, 95; resisting the high tide of, 126–36; in rural areas, 95, 130; and self-help, 99, 106, 128, 132, 148; and separatism, 95, 101, 105, 110, 129, 132–33; and the South African constitution, 101, 114–15, 116, 123, 125, 136; and suffrage, 101, 104–5, 127, 135, 148; and urbanization, 95–96, 103, 128–30; and the U.S. constitution, 136; and violence/intimidation, 99; whites defying, 246; and World War I, 126, 133–34, 135–36. *See also* Apartheid; Bus boycotts; Jim Crow; Separate but equal; Separate-and-unequal; *specific form of segregation, organization or person*
The Segregation Fallacy (Jabavu), 148
Self-determination: and Coloreds, 194–95; and communism, 189–202, 203, 216; and cultural assimilation, 45; deemphasis on, 203; and definition of nation, 187–89; Haywood's views of, 193–94; and the Leninist doctrine, 186; and nonviolent resistance, 242, 261; and Pan-Africanism, 152, 192, 283; and populism, 139; and progressivism, 56; Stalin's views about, 187–89; victory for South African, 319. *See also* Black Consciousness; Black Power; Nationalism; "Native republic" thesis; "Self-determination for the black belt"; *specific organization*

"Self-determination for the black belt," 192–93, 194, 196, 197, 199, 216, 221
Self-help: and Black Consciousness, 300, 308–9, 313–14; and Black Power, 295, 296, 313–14; and black redemption, 66; and civilization, 49; and the Du Bois-Washington controversy, 110; and economic issues, 29; and education, 29; and entrepreneurs, 106, 109; and the future of South Africa, 321, 322; and land issues, 29; and leadership, 110; and nationalism, 66–67; and nonviolent resistance, 233; and political issues, 29, 49; and racism, 27; and Reconstruction, 29; and religion/Ethiopianism, 64, 66–67, 76, 84, 86; and segregation, 99, 106, 128, 132, 148; and suffrage, 15, 47. *See also specific person or organization*
Self-made men, 25
Selma, Alabama, 263, 271
Seme, Pixley ka Izaka, 116–18, 119, 121, 122, 132
Separate but equal, 103, 105, 128, 147, 255
Separate and unequal, 94–95, 133, 220, 241
Separatism: and African independence, 287; and African redemption, 287; and Black Power, 296, 314; and communism, 210, 224; and disillusionment with the U.S., 287; and economic issues, 212; and education, 148; and integrationist nationalism, 55–56; and labor issues, 170; and liberalism, 174; and millennialism, 165, 168; and nationalism, 56, 288, 296; and Pan-Africanism, 148–49, 150; and political issues, 101; and racism, 287; and religion/Ethiopianism, 60, 61, 65, 73, 76–80, 86–87, 88, 89–90, 304, 305; and segregation, 95, 101, 105, 110, 129, 132–33; and suffrage, 54, 148–49; voluntary, 302. *See also* Separatist churches; *specific person or organization*
Separatist churches: and African independence, 44; and Black Consciousness, 305–6; and cultural nationalism, 76; and Ethiopianism, 76–77, 86–87, 88, 89–90; and political activism, 86–87, 244; types of, 90; in the U.S., 58–59, 75
Sermon on the Mount, 227, 256, 262
Sharecroppers: and communism, 201, 205, 210, 211; in South Africa, 102, 130; in the U.S., 140, 142, 201, 210, 211
Sharecropper's Union, 201
Sharpeville massacre, 274, 275, 286, 298, 302
Shaw, Robert Gould, 107
Sierra Leone, 68, 79–80
Sisulu, Walter, 243, 245, 246, 281–82, 310
Skota, T. D. Mweli, 163–64
"Slave culture," 144–45
Slave rebellions, 63
Slavery: abolition of, in the U.S., 16; and Christianity, 68; and inferiority of blacks, 27; and racism, 27–28; and religion/Ethiopianism, 57, 58, 60–61, 63, 76, 79; in South Africa, 57. *See also* Abolitionism
Slovo, Joe, 317
Smiley, Glenn E., 256, 257, 259
Smith Act, 183, 222
Smuts, Jan Christiaan, 167, 205, 221, 223, 240
SNCC (Student Non-violent Coordinating Committee), 260–61, 266, 292, 295, 314
Sobukwe, Robert, 283–86, 312
Social Darwinism, 156–57, 158
Socialism: African, 56; and communism, 185–87; emergence of, 183; and the ICU, 168, 171; and labor issues, 185; and Pan-Africanism, 152; and populism, 139; and race/ethnic issues, 183–85; and racism, 184; and segregation, 96, 185; and the UNIA, 175. *See also* Democratic socialism; Marxism
Soga, Alan Kirkland, 41, 43, 45, 48, 49–56, 87, 115
Soga, Tiyo, 40–41, 49, 80
SOS RACISME (France), 316
Sotho, 116, 200, 230
Souls of Black Folks (Du Bois), 105–6, 124
South Africa: future of, 320–23; as a multi-racial and multi-cultural society, 91, 285, 322–23; ownership of, 300–301; protest movements in, 21–22. *See also specific geographical section or topic*
South Africa Act, 114–15, 121
South African Bill of Rights, 323
South African Coloured People's Organization, 282
South African Congress of Trade Unions (SACTU), 274, 285
South African Indian Congress, 241, 245–46, 248, 282
South African Labor party, 185, 203
South African National Congress, 162–63
South African Native Affairs Commission, 48, 84, 85, 87, 100–101, 102
South African Native Congress (SANC), 44–45, 50, 53, 54, 87
South African Native Convention (1909), 114–16

South African Native National Congress (SANNC): African-American influence on the, 118, 119–20; and black solidarity/pride, 118; and class issues, 139–40; constitution of the, 121, 229; decline of the, 136; and the Du Bois-Washington controversy, 119–20; and economic issues, 120; and education, 119, 120; and Ethiopianism, 121; forerunners of the, 114, 115; formation of the, 87, 116–18, 121; and labor issues, 140, 141; lack of accomplishments of the, 134, 136; leadership of the, 116–18, 122, 124–25, 138–39; and liberalism, 121, 137–38, 174; membership of the, 121; NAACP compared with the, 121–23, 124–26, 136; and nationalism, 118, 120; and nonviolent resistance, 229–30; and Pan-Africanism, 121; program of the, 118–19; protest orientation of the, 120–21; purpose of the, 117–18; reformist assumptions of the, 125; and segregation, 122–23, 125, 130–33, 134, 136; and self-help, 118, 119; and separatism, 120, 132–33; tensions within the, 121; and the tribal chiefs, 121–22; and tribalism, 118; and whites, 120, 121; and World War I, 133–34. *See also* African National Congress; *specific person*

South African Party (SAP), 43, 116, 205

South African Races Congress, 121

South African Spectator (newspaper), 47

South African Student Organization (SASO), 300, 303, 306–7, 311

South African Students' Movement, 309

South African War (1899–1902), 21, 42, 44, 46, 100, 189

South Negro Youth Congress, 216

Southern Christian Leadership Conference (SCLC), 258, 259–60, 261, 262, 271, 295

Southern Conference for Human Welfare, 205, 214

Soviet Union, 272. *See also* Communism

Soweto uprising, 309–10, 311, 314

Spingarn, Joel, 113, 133

Spoils system, 30

Springfield, Illinois, race riot in (1908), 111

Stalin, Joseph, 187–89, 194, 195, 198, 202; pact with Hitler of, 214–15, 217, 219, 239

Stein, Judith, 159

Stowe, Harriet Beecher, 62

Stride Toward Freedom (King), 256, 274

Strikes: and Aggett's death, 318; and Black Consciousness, 309, 311; and draft resistance, 235; and labor issues, 102, 116, 140–41, 142, 168, 175, 185, 195, 218, 221; political motivation for, 274, 286; student, 309–10; and World War II, 218

Stubbs, Aelred, 303

Stuckey, Sterling, 144

Student Non-violent Coordinating Committee (SNCC), 260–61, 266, 292, 295, 314

Students: and the Civil Rights Movement, 269; in South Africa, 295, 299–300, 304, 309–10, 311. *See also* Biko, Steve; *specific organization*

Suffrage: and abolitionism, 48; and African-American protest politics, 23–36; and the African constitution, 205; and Afrikaners, 20, 37–47; assurance of, for African Americans, 320; and black solidarity/pride, 55; and black voting participation, 291; in the Cape Colony/Province, 36–56; and the Civil Rights Movement, 263, 264, 268, 269; and the Civil War, 28; and civilization, 20, 39–41, 44–45, 48–49; and class issues, 15, 19–20; and Coloreds, 37–47, 53, 224, 245, 246; and communism, 205–6, 218, 220, 221; comparison of African and U.S., 20–21, 48–56; comparison of British and U.S., 16–17; and corruption, 21, 37; early African organizations concerning, 42; and economic issues, 15, 16, 19; and education, 15, 17, 18, 35, 38–49, 52; and emigration, 26; and enforcement of voting rights, 29, 55; and equality, 14, 15–16, 18; and ethnicity, 15–16, 22, 23; for ex-slaves, 14–15; and fear of blacks, 20; and free blacks, 23–26; and the "grandfather clauses," 135; and illegality of white primaries, 222; importance of the, 22, 55; and inferiority of blacks, 18; influences on South African, 34, 37; and integration, 22; and liberalism, 14–23, 38, 43, 49, 223, 224, 298; and loyalty, 18; and military service, 52; and nationalism, 22; in the nineteenth century, 14–23; and nonviolent resistance, 263, 268; origins of American struggle for, 14, 16–19; and the origins of black protest politics, 36–47; origins of South African struggle for, 14–15, 19–23; and Pan-Africanism, 50, 148–49; and political parties, 31–32; and property issues, 15, 16, 17, 18, 19–21, 23, 43, 46, 48, 52, 116; and protest politics in South Africa, 36–47; and protest thought, 23–36; qualifications for, 15, 16,

Suffrage (*continued*)
17–18, 20, 35, 37, 38, 42, 43, 45, 46, 48, 51, 52–53, 116; and racism, 14, 15, 18, 19–20, 48–49, 224; and Reconstruction, 28–29, 36, 48; and religion, 38–40, 42; and the Republican party, 37–38; and republicanism, 17; as a reward, 34–35; as a right vs. privilege, 16, 17, 19, 20; and secret ballots, 17; and segregation, 101, 104–5, 127, 135, 148; and self-help, 15, 47; and separatism, 54, 148–49; and social revolution, 22; and the South African constitution, 116, 245; universal manhood, 16, 17, 18, 19, 36, 37, 45; and violence/intimidation, 19, 21, 30, 31–32, 50; who should control the, 53; and women, 16, 18. *See also* Cape liberalism; Fifteenth Amendment; Jim Crow; *specific legislation, person, or organization*
Sumner, Charles, 107
Sundkler, Bengt, 90, 91
Suppression of Communism Act (South Africa, 1950), 245, 246
Suppression of Communism Act (South Africa, 1962), 298
Swazi, 200
Swedenborgians, 62–63
Sweet, Leonard, 26

Taft, William Howard, 127
"Talented tenth," 109, 110, 114, 120, 122, 124, 138, 139, 151–52, 173
Tambo, Oliver, 243, 245, 274, 282
Tema, S. S., 230
Tembu National Church, 81
Territorial segregation. *See* Residential/territorial segregation
Thaele, James, 162, 164, 169, 173
Thema, Richard Selope, 131–33
Third Reconstruction, 321–22
Tile, Nehemiah, 81
Tocqueville, Alexis de, 94
Toll, William, 150
Tolstoy, Leo, 227
Tonjeni, Elliot, 209–10
Tourgée, Albion W., 107, 317
Transkei: and African independence, 166; Ethiopianism/religion in, 41, 81, 83, 86–87; land issues in, 43; millennialist movements in, 162; and segregation, 100; and suffrage, 20, 38; and the Wellington movement, 166–67
Transvaal: British administration in the, 45; Coloreds in the, 46; Ethiopianism/religion in the, 81, 85, 87, 89; and the ICU, 171; and leadership for African rights, 116; nonviolent resistance in, 226, 229–30; protest movements in the, 87, 89; and segregation, 100; and suffrage, 21, 43, 50
Transvaal Indian Congress, 239–40
Transvaal Native Congress, 140–41
Trapido, Stanley, 40
Treaty of Vereeniging (1903), 44, 50, 54, 116
Tribal chiefs, 121–22
Tribalism, 46, 118, 165, 279
Trotter, William Monroe: as abolitionist, 48, 53, 108; and NAACP, 112–13; and Niagara Movement, 108, 109, 112; and segregation, 127, 133; and suffrage, 36, 55; and unity, 53, 54
Truman, Harry S, 222, 223, 235, 267
Tswana, 116, 200, 230
Tule, John, 83
Turner, Bishop Henry M., 59, 76–80, 83–84, 85, 93, 290
Turner, Nat, 107
Tuskegee Institute, 35, 119, 120, 145, 146, 148, 153, 163
Tutu, Desmond, 276

UCM. *See* University Christian Movement
UDF. *See* United Democratic Front
"The Ultimate Effects of Segregation" (Pickens), 128
Umkonto We Sizwe organization, 252, 286, 298, 312
Uncle Tom's Cabin (Stowe), 62
UNIA. *See* Universal Negro Improvement Association
United Auto Workers, 233
United Democratic Front (UDF), 276, 310–11, 317, 322
United Front, 202–14, 216
United Nations, 240, 253, 267
United party, 205, 214
United States: disillusionment with the, 287; foreign policy of the, 267; Great Britain compared with the, 53, 54–55; as multicultural, 316; protest movements in the, 21–36; utopian views of the, 297
Universal Negro Improvement Association (UNIA): and African independence, 153, 154, 158–59, 161, 166, 287; and African/black redemption, 155, 168, 287; aims/purposes of the, 153, 175–76; and the ANC, 162, 163–64, 176; and black solidarity/pride, 146, 153, 158, 160, 161;

and capitalism, 159; and class issues, 159, 173; and the clergy, 92–93; and communism/socialism, 164, 175, 190, 192, 198; conventions of the, 163; and cultural nationalism, 158–59; decline of the, 160, 174–75; and economic issues, 154, 159–60; effects of the, 161; and emigration, 154, 155–56, 161, 175, 177; and Ethiopianism, 92–93, 158–59, 161; failure of the, 178; Fortune and the, 56; founding of the, 146, 153; and the ICU, 168–72, 173–78; and ideology, 156–59, 175–76; and the KKK, 154, 160, 174; and labor issues, 159, 168–72, 175, 177; leadership of the, 159, 173–74, 175; and localism, 156, 160; membership of the, 152–53, 159, 160, 162; motto of the, 163; and nationalism, 138, 175; and Pan-Africanism, 138, 145, 146, 149, 152–61, 174, 175, 280, 281, 284, 290, 301; as a populist movement, 138, 145, 152; and racial purity, 70, 157–58, 162; and racism, 157, 161; and religion, 165–66; repression of the, 174–75; and segregation, 154, 160, 161; and self-determination, 154,163; and self-help, 154, 160, 161, 175; and separatism, 56, 93, 154, 161, 175, 194, 287; and social Darwinism, 156–57, 158; spread of the, 138, 154, 160; and suffrage, 35, 154; and urbanization, 138; and violence/intimidation, 154; and whites, 154, 156, 157, 161, 164, 176; and World War I, 154

University Christian Movement (UCM), 299–300, 303

University of Natal, 299

Up from Slavery (Washington), 153

Urban Areas Act (South Africa, 1923), 142

Urbanization: and communism, 220–21; and labor issues, 138, 140, 142; and segregation, 95–96, 103, 128–30; and the UNIA, 138

Utopianism, 297

Victoria (queen of Great Britain), 42, 82, 149

Vietnam War, 267, 292

Villard, Oswald Garrison, 108, 111–12, 113, 127

Violence/intimidation: and Black Consciousness, 298; and Black Power, 293, 314; against blacks, 21, 23, 30, 50, 291; as a catharsis, 252; and the Civil Rights Movement, 264, 295; and communism, 211, 252; and emigration, 77–78; and Ethiopianism, 77–78, 86; and farm workers, 201; increase in, 77; and labor issues, 142; and liberalism, 298–99; and Pan-Africanism, 150, 285; and segregation, 99; and stereotypes of blacks, 260; and suffrage, 19, 21, 30, 31–32, 50; in the U.S., 19, 21, 30, 31–32; against whites, 252, 285, 286

Virginia Jubilee Singers, 82

Voting Rights Act (South Africa, 1994), 14

Voting Rights Act (U.S., 1965), 291

Wald, Lillian, 111

Waldron, J. Milton, 113

Walker, David, 63–64, 65, 66, 72, 78

Wallace, Ernest, 167

Wallace, George, 292

Walling, William English, 111

Walshe, Peter, 164

Walters, Bishop Alexander, 53, 111, 113, 149

Washington, Booker T.: as an accommodationist, 33, 35–36, 48–49, 50–52, 105–6, 130, 149, 161, 232, 295; "Atlanta Compromise" speech of, 105; and black solidarity/pride, 145–46; and Christianity, 93; and civilization, 34–35, 55–56; criticisms of, 35, 36, 51, 52; death of, 153; Du Bois vs., 35, 36, 47, 51, 53, 105–7, 109, 110–11, 119, 129, 132, 146, 148; and economic issues, 35, 36, 105, 160; and education, 15, 33, 36, 51, 105, 109, 120, 146; and entrepreneurs, 15, 33, 35, 36, 106, 109, 128, 129, 160, 296; Garvey compared with, 35; influence of, 34, 35, 47, 51–52, 53, 106, 109, 110, 111, 119–20, 144, 148–49, 153, 161, 175, 265; legacy of, 35; and the NAACP, 112; and the Niagara Movement, 106–11; and nonviolent resistance, 231; and Pan-Africanism, 145–46, 148, 175; on racism, 35; and the SANNC, 118, 119–20; and segregation, 105–6, 107, 108, 128–29, 130, 132, 149; and self-help, 15, 29–30, 33, 34, 35, 36, 49, 86, 109–10, 118–19, 128, 148, 160, 175, 295; and separatism, 35, 105, 148; and suffrage, 15, 34–36, 48–49, 50–53; support for, 51, 53. *See also* Industrial education; Tuskegee Institute

Watts riots (1965), 291

Weber, Max, 91

Wellington, Dr. (aka E. Butelezi), 166–67, 168, 171

Wellington movement, 166–68, 171

Wells-Barnett, Ida, 113
Whipper, William, 24, 25
White, Walter, 142, 266
Whites: and African independence, 281; attitudes toward Zionist Churches of, 91; backlash of, 291; and Black Consciousness, 301, 302, 303, 309, 310–11, 312, 317–18; and Black Power, 292–97, 301; blacks as role models for, 93; blacks as superior to, 64, 65, 260; and the Civil Rights Movement, 273, 289, 292–93, 295; and communism/Marxism, 189–90, 191, 192, 195, 196, 212–13, 215, 217, 219, 224, 318; defying segregation laws, 246; and Ethiopianism, 74, 84–85, 86; failures of, 151; Fortune's views of, 33; innate characteristics of, 72; and the International Conference on the Negro (1912), 145; and the March on Washington Movement (Randolph), 233; moral depravity of, 64, 65; and the Nation of Islam, 288–89, 290; and the Niagara Movement, 107, 108, 111; and Pan-Africanism, 151–52, 281, 282, 283, 284, 285, 299, 301; and the People's Front, 207; redemption of, 306; and the standards of civilization, 151; violence against, 252, 286. *See also specific organization*
Wilberforce University, 82, 86
Williams, George Washington, 34, 36, 49
Williams, Henry Sylvester, 149
Wilson, Woodrow, 126, 127, 134, 135–36
Winthrop, John, 60
Wofford, Harris, 253
Women, 16, 18, 250, 254
Women's League, 250
Women's Political Council, 254
Woodson, Lewis, 24
Woodward, C. Vann, 97, 98
The Workers' Herald (ICU newspaper), 169, 170
Workers' Party, 192
Working-class blacks, 25
World War I, 126, 133–34, 135–36, 137, 140, 154
World War II, 214–24, 235, 239, 277
Wright, Nathan Jr., 294–95
Wright, Richard, 279

Xhosa, 37, 38, 39, 43, 116, 168, 200, 230
Xuma, Albert B., 126, 219, 240–42, 266

Yergan, Max, 278–79
Young, Robert Alexander, 63
Youth League: and the ANC, 241–43, 245; and capitalism, 266; and communism, 221, 281, 299; founding of the, 221, 241; leadership of the, 243–44; and liberalism, 266; and the Manifesto of 1948, 281; and nationalism, 221; and Pan-Africanism, 241, 280–81; and the Programme of Action, 242–43; and self-determination, 221

Zellner, Bob, 293
Zion, Illinois, 90
Zionism, 137, 154
Zionist Churches, 88, 90–91, 244, 271
Zulu Congregational Church, 81
Zulus, 39, 81, 86, 100, 116, 200, 230, 241. *See also* Natal